■ Collected Works, Volume I

Collected Works

VOLUME I

*Scientific Rationality,
the Human Condition, and
20th Century Cosmologies*

Adolf Grünbaum

EDITED BY

Thomas Kupka

OXFORD
UNIVERSITY PRESS

OXFORD
UNIVERSITY PRESS

Oxford University Press is a department of the University of Oxford.
It furthers the University's objective of excellence in research, scholarship,
and education by publishing worldwide.

Oxford   New York
Auckland   Cape Town   Dar es Salaam   Hong Kong   Karachi
Kuala Lumpur   Madrid   Melbourne   Mexico City   Nairobi
New Delhi   Shanghai   Taipei   Toronto

With offices in
Argentina   Austria   Brazil   Chile   Czech Republic   France   Greece
Guatemala   Hungary   Italy   Japan   Poland   Portugal   Singapore
South Korea   Switzerland   Thailand   Turkey   Ukraine   Vietnam

Oxford is a registered trademark of Oxford University Press in the UK and certain other
countries.

Published in the United States of America by
Oxford University Press
198 Madison Avenue, New York, NY 10016

© Oxford University Press 2013

Library of Congress Cataloging-in-Publication Data
Grünbaum, Adolf.
[Works. Selections]
Collected works / Adolf Grünbaum ; edited by Thomas Kupka.
3 volumes ; cm
ISBN 978–0–19–998992–8 (hardcover : alk. paper)   1. Science–Philosophy.   I. Kupka,
Thomas.   II. Title.
Q175.3.G78 2013
501–dc23
2012040625

ISBN 978–0–19–998992–8

1 3 5 7 9 8 6 4 2
Printed in the United States of America
on acid-free paper

I dedicate this trilogy of my writings on the philosophy of science to my cherished, long-term friend and colleague, Gerald J. Massey, Distinguished Service Professor of Philosophy at the University of Pittsburgh, who very ably leavened my academic endeavors there and ran interference for me through thick and thin.

Adolf Grünbaum

■ CONTENTS

■ PROVENANCE OF CHAPTERS

1. "Is Falsifiability the Touchstone of Scientific Rationality? Karl Popper Versus Inductivism." Reprinted from R. S. Cohen et al. (eds.), *Essays in Memory of Imre Lakatos, Boston Studies in the Philosophy of Science* 39, Dordrecht, The Netherlands: D. Reidel Publishing Co., 1976, pp. 213–252. This essay is based on research supported by the National Science Foundation.

2. "The Degeneration of Popper's Theory of Demarcation." Reprinted from *Epistemologia* 12 (1989), pp. 235–260. Prior publication: F. D'Agostino and I. C. Jarvie (eds.), *Freedom and Rationality: Essays in Honour of J. Watkins*, Dordrecht, The Netherlands: D. Reidel Publishing Co., 1989, pp. 141–161. This paper draws on a section of the sixth of my eleven *Gifford Lectures*, delivered at the University of Saint Andrews, Scotland, in early 1985 on the topic "Psychoanalysis and Science." The *Gifford Lectures* as a whole will be published for the first time in Vol. III of the present collection.

3. "The Falsifiability of Theories: Total or Partial? A Contemporary Evaluation of the Duhem–Quine Thesis." Reprinted from *Synthèse* 14 (1962), pp. 17–34. Portions of this essay are drawn from earlier publications by the author as follows: "The Duhemian Argument," *Philosophy of Science* 27 (1960), pp. 75–87; and "Geometry, Chronometry and Empiricism," in H. Feigl and G. Maxwell (eds.), *Scientific Explanation, Space, and Time, Minnesota Studies in the Philosophy of Science* 3, Minneapolis: University of Minnesota Press, 1962, Section 7.

4. "Free Will and Laws of Human Behavior." Reprinted from H. Feigl et al. (eds.), *New Readings in Philosophical Analysis*, New York: Appleton-Century-Crofts, 1972, pp. 605–627. This is a revised version of the article in *American Philosophical Quarterly* 8 (1971), pp. 299–317.

5. "Historical Determinism, Social Activism, and Predictions in the Social Sciences." Reprinted from *British Journal for the Philosophy of Science* 7 (1956), pp. 236–240.

6. "In Defense of Secular Humanism." Reprinted from "The Poverty of Theistic Morality," in K. Gavroglu et al. (eds.), *Science, Mind and Art: Essays on Science and the Humanistic Understanding in Art, Epistemology, Religion and Ethics, in Honor of Robert S. Cohen. Boston Studies in the Philosophy of Science* 165, Dordrecht, The Netherlands: Kluwer Academic Publishers, 1995, pp. 203–242.

■ PREFACE AND ACKNOWLEDGMENTS

The present book is the first volume of a three-volume collection of essays and lectures spanning more than five decades of my academic authorship. Their selection, made in close collaboration with Thomas Kupka of Bremen, Germany, was guided by two principal considerations.

First, we selected those essays that were most cited in the literature. Second, we wished to make these books useful to several sorts of readers—notably philosophers of science, philosophers generally, and interested laymen—as well as to natural scientists who value the philosophy of science.

Hence, the three volumes are grouped under the major topics on which I have worked: Volume I deals with scientific rationality, determinism and the human condition, and theological interpretations of twentieth-century physical cosmologies. Volume II pertains to my work in the philosophy of physics and, more specifically, in the philosophy of space and time. Last, Volume III features my lectures on the philosophy of psychology and psychoanalysis, including my 1985 Gifford lectures, *Psychoanalytic Theory and Science*, which were delivered at the University of St. Andrews in Scotland. They are published here for the first time.

I am most grateful to Thomas Kupka for reviving this long-term publication project and for his initiative in bringing it to fruition. For each volume, he also provides an introduction. Furthermore, I deeply and cordially thank Leanne Longwill, my long-time assistant, for her highly scrupulous and knowledgeable production of these books. Without her punctilious care, they could not have been produced or published. A generous grant from the Harvey and Leslie Wagner Foundation facilitated the editorial work immensely. And John Earman and Peter Achinstein gave helpful advice on chapter selection.

For their conscientious transcription and proofreading of chapters originally written in the precomputer era, thanks are due to Brianna McDonough, Kathryn Tabb, Tomas Bednar, and Megan Kenna.

Last, but not least, my very warm thanks go to Peter Ohlin of Oxford University Press for his patience and cooperation during the long haul on this project.

Adolf Grünbaum

Introduction

Thomas Kupka

Adolf Grünbaum's philosophical work covers an unusually broad scope of scholarly topics, ranging from space–time and quantum physics, through the analysis of physical cosmologies and their theistic interpretation, to a thorough critique of psychoanalysis. Thus, he was very pleased when the *New York Times* journalist Jim Holt paid tribute to him, writing:[1]

> In the philosophical world, Adolf Grünbaum is a man of immense stature. He is arguably the greatest living philosopher of science. In the 1950s, Grünbaum became famous as the foremost thinker about the subtleties of space and time. Three decades later, he achieved a wider degree of fame—and some notoriety—by launching a sustained and powerful attack on Freudian psychoanalysis. This brought down on him the wrath of much of the psychoanalytic world and landed him on the front page of the science section of the *New York Times*.

Hence, we are happy to present here to the interested public a selection of his most influential essays and lectures in a three-volume collection.

The essays of the present volume are grouped into three subdivisions. The papers in Part I, *Scientific Rationality*, address the requirements to be met by a theory to qualify it as scientific; they ask whether one can distinguish clearly between science and nonscience, and they examine the well-known Duhem-Quine thesis regarding conceptual holism in scientific theory formation.

Part II, *Determinism and the Human Condition*, is primarily concerned with free will and its compatibility with the observable regularities in human behavior as described in terms of natural or statistical laws. Furthermore, it develops that *historical* determinism does not preclude social planning and activism, and an important conclusion is then drawn in favor of a secular humanism.

Last, Part III concerns theistic interpretations of twentieth-century physical cosmologies.

In the following paragraphs, brief summaries are given for each of the essays.

Essay 1, "Is Falsifiability the Touchstone of Scientific Rationality? Karl Popper Versus Inductivism," challenges Popper's claim that "testability is falsifiability." In particular, Grünbaum contests Popper's historiography of inductivism and shows in detail that his portrayal of psychoanalytic theory as a paradigm of nonfalsifiability is ill conceived. Contrary to Popper, Grünbaum argues that a careful inductivist investigation of psychoanalysis reveals that some of its hypotheses are indeed falsifiable, although a cogent appeal to inductivism seriously impugns the evidential credentials of Freudian theory.

Essay 2, "The Degeneration of Popper's Theory of Demarcation," then takes closer issue with Popper's well-known claim that we can draw a clear line of demarcation between science and nonscience. Here again, à propos of Popper's contention that psychoanalysis has no potential empirical falsifiers, Grünbaum now examines more specifically the relation between the causal and the so-called meaning connections of mental states.[2]

Essay 3, "The Falsifiability of Theories: Total or Partial. A Contemporary Evaluation of the Duhem-Quine Thesis," grapples with the claim, promulgated by F. S. C. Northrop in the 1940s, that there is an important asymmetry between the verification and the refutation of empirical hypotheses, in the sense that refutation is presumably categorical or decisive, while verification is deemed to be irremediably inconclusive. Duhem, however, and later Quine contested this claim, arguing that a given observational consequence is deduced not from an empirical hypothesis alone, but rather from the conjunction of this hypothesis with the relevant set of *auxiliary* assumptions. Hence, the failure of the observational consequence does not deductively refute the hypothesis by *modus tollens,* when taken by itself, but discredits only its conjunction with the pertinent auxiliaries. However, Grünbaum then shows that *holism* in empirical theory formation is untenable when taken nontrivially.[3] In a letter to Grünbaum of June 1, 1962, Quine admits that the so-called Duhem-Quine thesis *is* indeed trivial.

So here, we find an interesting agreement between Grünbaum's scrutiny of falsifiability and Quine's well-known evolving thoughts, which then led Quine to refine the holism he had espoused at the end of "Two Dogmas." At the request of Adolf Grünbaum, and with Quine's kind permission, we include Quine's 1962 letter as an appendix to Essay 3.[4]

Essay 4, "Free Will and Laws of Human Behavior," deals with the issue of determinism versus indeterminism, which has a long history in both the natural sciences and humanities. Its discussion was especially lively in the 1960s–'70s, and has acquired even more vigor in response to the advances in the neurosciences during the last two decades. What is at issue is the philosophical conception of free will in the conflict between *compatibilism*, that is, the view that free will is compatible with determinism, on the one hand, and *incompatibilism*, which denies it, on the other. Essay 4 develops and defends the specific view of a *humane determinist.*[5]

Grünbaum offers a conclusion of the debate between determinists and indeterminists, which often culminates in the pros and cons of punishment in criminal law. A *humane determinist*, he says, would argue that his opponent is:

> being gratuitously vengeful, on the grounds that the indeterminist is committed by his own theory to a retaliatory theory of punishment. The indeterminist cannot consistently expect to achieve anything better than retaliation by inflicting punishment; for were he to admit that punishment will causally influence all or some of the criminals, then he would be abandoning the basis for his entire argument against the determinist. As Leszek Kolakowski has rightly remarked, "it is an obvious truth ... that if one believes punishment ... can be effective, then one posits by that very fact some kind of determinism of human behavior."[6]

This statement may be seen as a crisp formulation of a major problem in moral theory: the problem of social planning and prediction.[7]

Essay 5, "Historical Determinism, Social Activism, and Predictions in the Social Sciences," expands on this issue and explores whether a deterministic socio-political theory can consistently advocate a social activism with the aim of thereby bringing about a future state whose eventuation the pertinent theory regards as determined by historical causation.

Related to the determinism debate is also the question of what constitutes the so-called human condition. Here Essay 6, "In Defense of Secular Humanism," argues that theism is neither indispensable for the well-founded articulation of moral imperatives, nor motivationally necessary to assure adherence to moral standards. On the contrary, as Grünbaum shows in great detail, it is what he calls the "moral sterility" of theism that has both theoretically and historically led to very unfortunate consequences. Just think of Martin Buber's hypothesis, Grünbaum notes, that there is a caring God who, however, "goes into eclipse from time to time"—as he purportedly did, for instance, during the Nazi Holocaust.

After the very questionable consequences of a theistic morality have been laid out, Part III is devoted to the related dubieties of the theistic interpretations of modern physical cosmologies. *Essay 7*, "The Poverty of Theistic Cosmology," concentrates on the famous Leibnizian question, "Why is there something rather than nothing?" which inspired the theistic doctrine of *creatio ex nihilo*. Grünbaum responds to the Leibnizian question with the counter-question: "But why *should* there be nothing rather than something?" and unmasks it with great philosophical clarity as a pseudo-problem (*Scheinproblem*).

This essay originated in the first two of Grünbaum's *Leibniz Lectures* (Hanover, Germany 2003), the third of which, titled "Does Psychoanalysis Deserve a Second Century of Influence?" we include as *Lecture 6* of Volume III of the present collection.

Essay 8, "Theological Misinterpretations of Current Physical Cosmology," is the prequel to Essay 9, "A New Critique of Theological Interpretations of Physical Cosmology." Both essays show how Grünbaum developed his later views on the subject from a history and philosophy of science perspective. The main conclusion here is that the Big Bang models of (classical) general relativity, as well as the original 1948 versions of the steady state cosmology, are each logically incompatible with the theological doctrine that perpetual divine creation (*creatio continuans*) is required in each of these theorized worlds.

Furthermore, Grünbaum again challenges the received theological doctrine that there must be, as a *ratio essendi*, a divine creative cause (as distinct from a transformative one) for the very existence of the physical world. And he takes issue with the views of theists like Richard Swinburne and Philip L. Quinn, the latter a former doctoral student of Grünbaum's, who hold that the specific content of the scientifically most fundamental laws of nature requires supra-scientific explanation and that a satisfactory explanation of this sort is provided by the hypothesis that God *willed them* to be exactly what they are. Generally, we can see in Adolf Grünbaum's work on these issues not only an example of thorough philosophical writing but also a lighthouse model for great scholarly pedagogy.

Last, Essay 10, "Pseudo-Creation of the Big Bang," considers the scientific credentials of the Big Bang model compared with competing cosmological theories.

* * * * *

Let me make some editorial remarks: In this collection we primarily *document* the included essays. They give evidence of the development of Adolf Grünbaum's thought during the period of their original publication. Hence, we confined ourselves to just correcting mistakes and misprints; rarely did we add, delete, or change words or phrases for purposes of clarification. Otherwise, the essays have been left as they were originally written.

These particular versions elicited critiques. Hence, it would be unfair to their authors, if we were now to present these essays with substantial changes. Cases in which Grünbaum has changed his mind over the decades are documented sufficiently by the essays themselves. For example, in presenting his papers on religion and cosmology in Part III, we begin with the most recent essay of 2004 (Essay 7), which states Grünbaum's fully developed view on the subject. But then we revert to his earlier essays, because they reveal the route by which he reached his conclusions. Yet since these papers have been widely discussed in the literature, we are making them now available to the reader in a single volume.

We occasionally took the opportunity to condense references, when we changed footnotes or bracketed textual references into endnotes for the purpose of a unified style of referencing. This procedure applies especially to *Essay 8*, which originally appeared in the 1998 volume of the journal *Philo*.

■ NOTES

1. Jim Holt, *Why Does the World Exist?* New York: W.W. Norton & Company 2012, p. 63. There have been three substantial *Festschriften* in honor of Adolf Grünbaum's work to which I shall confine myself in this introduction: Robert S. Cohen and Larry Laudan (eds.), *Physics, Philosophy, and Psychoanalysis*, Dordrecht, The Netherlands: Kluwer Academic Publishers, 1983 [hereafter *Festschrift I*]; John Earman, Allen I. Janis, Gerald J. Massey, and Nicholas Rescher (eds.), *Philosophical Problems of the Internal and External Worlds*, Pittsburgh, PA: University of Pittsburgh Press, 1993 [*Festschrift II*]; Aleksander Jokić (ed.), *Philosophy of Religion, Physics, and Psychology*, Amherst, NY: Prometheus Books, 2009 [*Festschrift III*]. The latter *Festschrift* also contains Grünbaum's "Autobiographical-Philosophical Narrative" (pp. 11–155), which gives an incisive digest of his academic career. An admirably concise yet telling biographical sketch of Grünbaum is given by Richard T. Hull in the volume of *Presidential Addresses of the American Philosophical Association, 1900–2000* (forthcoming) à propos of the reprinting of Grünbaum's December 1982 Presidential Address to the American Philosophical Association, Eastern Division, delivered at its meeting in Baltimore, Maryland.

2. For careful assessments of Popper's demarcation thesis, its legacy, and Grünbaum's critique of it, see Larry Laudan, "The Demise of the Demarcation Problem," in *Festschrift I*, pp. 111–127; and Robert E. Butts, "Sciences and Pseudosciences: An Attempt at a New Form of Demarcation," in *Festschrift II*, pp. 163–185.

3. For more on Grünbaum's scrutiny of scientific holism, see John Worrall, "Falsification, Rationality, and the Duhem Problem: Grünbaum Versus Bayes," in *Festschrift II*, pp. 329–370.

4. In a letter to Grünbaum dated March 24, 1970, Quine requested that, in case of its publication, this letter be quoted in full, "since it will help to clear up misunderstandings of my position." Of course, we gladly comply with his request.

5. Grünbaum has defended his compatibilism as of 1952 in "Causality and the Science of Human Behavior," *American Scientist* 40 (1952), pp. 665–676; and in his "Science and Man," *Perspectives in Biology and Medicine* 5 (1962), pp. 483–502. The essay included in this volume has precursors in his articles "Free Will and the Laws of Human Behavior," *L'Age de la Science* 2 (1969), pp. 105–127; and "Free Will and the Laws of Human Behavior," *American Philosophical Quarterly* 8 (1971), pp. 299–313.

6. The reference is to Leszek Kolakowski, "Determinism and Responsibility," published in his *Toward a Marxist Humanism*, New York: Grove Press 1968, p. 188, an article that was much discussed at the time.

7. Thoughtful commentary on Grünbaum's views regarding the determinism versus indeterminism debate is offered by Philip Quinn, "Grünbaum on Determinism and the Moral Life," in *Festschrift I*, pp. 129–151; Arthur Fine, "Indeterminism and the Freedom of the Will," in *Festschrift II*, pp. 551–572; and John Watkins, "Adolf Grünbaum and the Compatibilist Thesis," in *Festschrift II*, pp. 573–588.

PART ONE

Scientific Rationality

1

Is Falsifiability the Touchstone of Scientific Rationality? Karl Popper Versus Inductivism

There is already a sizeable literature in which the question posed in the title of this essay is answered in the negative. My justification for now undertaking to add to that literature is as follows: Popper's very influential version of falsificationism asserts the primacy of falsifiability in at least *four* distinct major epistemological or methodological theses. I claim that each of these four central theses is beset by fundamental and seemingly irremediable difficulties of its own. But so far as I know, most of my cardinal objections to these four Popperian doctrines have not been raised by others. Since present-day exponents of these doctrines have found the prior critiques by others unconvincing, I hope that such exponents will be challenged by my critical scrutiny of Popper's views in the present essay.

To avoid misunderstanding of my intent, let me say that the philosophical moral I draw from my critique of Popper's falsificationism is *not* the *wholesale* agnosticism espoused by, for instance, Imre Lakatos, who wrote: "Can we ... learn from experiment that some theories are false? ... We cannot learn from experience the falsehood of any theory."[1] A statement of my reasons for not embracing this universal *falsificational* agnosticism must await another occasion. On the other hand, Lakatos did espouse some of the major anti-inductivist tenets of Popper's which I shall contest below.

Let me outline the four major groups of Popperian falsificationist theses which I shall challenge:

(1) In his quest for a criterion of demarcation between science and nonscience, Popper invokes his own historiography of inductivism and concludes the following: In 1919 the account of testability furnished by all forms of inductivism was such that "there clearly was a need for a different criterion of demarcation."[2] Popper then enunciates falsifiability as the linchpin of the *scientific entertainability* of a hypothesis. Thus, falsifiability by potential negative instances is claimed to play a distinguished role to the exclusion of inductive supportability or probabilistic confirmability by positive instances. Says he in his *Conjectures and Refutations*: "Testability is falsifiability" (C & R, p. 36). I shall speak of Popper's "demarcation asymmetry" in order to allude to the asymmetry in the roles which he assigns to the falsifiability and the supportability of a hypothesis vis-à-vis its scientific status.

According to Popper, Freudian psychoanalytic theory does not meet his standards of falsifiability and should therefore be regarded as unscientific. Moreover, this unfavorable appraisal of psychoanalysis is claimed to illustrate the superiority of Popper's falsificationist criterion of demarcation over the inductivist requirements for scientific status.

(2) In his theory of *quantitative* verisimilitude and quantitative content of a hypothesis, Popper invokes *a priori* or "logical" probabilities of universal hypotheses which are *non*zero. But he rejects *a posteriori* probabilities of specifically *such* hypotheses as absurd. Hence he rejects Bayesian probabilistic inductivism. And he offers an account of corroboration in which the corroboration of a theory T by an observable fact F which was successfully predicted by T is *parasitic* on the logical relations of F to T's *known* rivals R_1, R_2..., R_n as follows: For each known rival R, the accepted fact F *either* deductively falsifies R *or* F is *novel* with respect to R in the sense of being logically independent of R. In other words, a successfully predicted positive instance F of T corroborates T, if for each R, the fact F is either contrary to R or not predicted by R but without being contrary to R. Such knowledge as we have is therefore held to rest probatively solely either on refutations of rival theories by accepted negative instances or on novel facts. This is the cornerstone of Popper's proposed solution of the problem of induction. As against Pierre Duhem, Popper asserts an asymmetry of fallibility between the falsification and the corroboration or confirmation of a hypothesis.

(3) According to Popper, the comparative degrees of falsifiability of a theory and of a modified form of it containing a new auxiliary hypothesis are the yardstick for the admissibility of the auxiliary hypothesis and for whether it is *ad hoc*.

(4) In *Objective Knowledge*, falsifiability is claimed to play a fundamental role in the quest for theories of greater verisimilitude: The method of bold conjectures and *refutations* is held to be *the* method of science, because of its purported conduciveness to theories of greater verisimilitude.

Due to limitations of space, I shall now deal with only the first two of these four distinct assertions, that is, with the epistemological and methodological primacy of falsifiability. My critique of the remaining two is given elsewhere.[3]

■ 2. POPPER'S HISTORIOGRAPHY OF INDUCTIVISM AND THE TEST CASE OF FREUDIAN PSYCHOANALYTIC THEORY

Philosophers whom Popper disapprovingly calls "inductivists" try to use supportive instances of hypotheses or theories to make either absolute or relative *credibility assessments* of them. An example of an absolute credibility judgment would be to say that a given hypothesis is more likely to be true than false. A relative credibility judgment might take the form of saying that a theory is more credible than a certain rival. *Some* inductivists propose to "probabilify" hypotheses on given evidence by holding that, in principle, hypotheses can be assigned numerical degrees of confirmation which are mathematical probabilities in the sense of satisfying the mathematical calculus of probability. But other inductivists deny that either absolute or relative degrees of credibility of hypotheses *must* be construed as mathematical probabilities.[4]

Popper's historiography of inductivism is unfortunately oblivious of a paramount fact. It is the fact that there are very important differences among inductivist epistemologies concerning the requirements which must be met by an observational finding, if this finding is to count as a *bona fide supportive* instance of a hypothesis. And, as we shall see, just these crucial differences render quite untenable Popper's declaration that "the fundamental doctrine which underlies all theories of induction is *the doctrine of the primacy of repetitions....* According to this doctrine, repeated instances furnish a kind of *justification* for the acceptance of a universal law."[5] Indeed, the test case of psychoanalysis will illustrate the following conclusion, which will emerge from my impending critique of Popper's historiography of inductivism: The inductive use of supportive instances to "credibilify" hypotheses in one way or another does *not* automatically commit inductivism as such to grant *credible scientific status* or even to lend at least *some* credence to a hypothesis H *merely* because there are *numerous* observationally true consequences of H and no known negative instances. For, as will become clear, the mere quest for credibilifying a hypothesis by supporting evidence does *not* require at all that *every* positive instance of a hypothesis be automatically held to be supportive of the hypothesis to some *nonzero* degree or other! *A fortiori*, the program of inductive credibilification *as such* in no way requires that all positive instances count as *equally* supportive to some *nonzero* degree. Furthermore, I shall argue that whereas Popper's own criterion of demarcation does not entitle him to indict psychoanalysis as unscientific, certain classical inductivist canons do seriously impugn the scientific credentials of Freud's therapeutic theory and also have the capability of similarly discrediting Freud's psychodynamics.

In the first chapter of Popper's C & R, he adduces astrology, Freudian psychoanalysis, the Marxist theory of history, and Adlerian psychology in an endeavor to show the following: (1) All forms of inductivism are logically committed to grant scientific status to these four theories in the face of the numerous positive instances marshaled by them; and (2) whereas the requirements of inductive supportability are hopelessly incompetent to derogate the theories in question as pseudo-scientific, an alternative demarcation criterion based on falsifiability succeeds in excluding them from the body of science.[6]

Let us begin our appraisal of Popper's historiography of inductivism by considering the views which the arch-inductivist Francis Bacon (1561–1626) expressed on astrology three centuries before Popper.

Popper (C & R, pp. 112, 255–256) illuminatingly calls attention to major weaknesses in Bacon's methodological prescriptions, qua purported methods for *discovering* theories and for certifying reliably that theories thus discovered must be *true*. Nonetheless, Popper credits Bacon to the extent of noting that "the problem of *drawing a line of demarcation*" between science and pseudo-science "has agitated many philosophers since the time of Bacon" (C & R, p. 255). But then Popper tries to discredit Bacon's inductivism as an *answer* to the problem of demarcation by writing as follows:

> Many superstitious beliefs ... have no doubt often been based on something like induction. Astrologers, more especially, have always claimed that their "science" was based upon a great wealth of inductive material. This claim is, perhaps, unfounded; but I have never heard of any attempt to discredit astrology by a critical investigation of its alleged inductive material. (C & R, p. 256)

Yet when we turn to Bacon's writings, we find that he explicitly refers to astrology when discussing the "idols of the tribe" in his *Novum Organum* (Book I, Section 46). Speaking of the relative importance of attention to negative and positive instances of a theory, Bacon condemns astrology as superstitious because, among other things, it neglects evidence *unfavorable* to it. The Baconian aphorism dealing with astrology merits quotation in its entirety because it is delicious. Bacon wrote:

> The human understanding when it has once adopted an opinion (either as being the received opinion or as being agreeable to itself) draws all things else to support and agree with it. And though there be a greater number and weight of instances to be found on the other side, yet these it either neglects and despises, or else by some distinction sets aside and rejects; in order that by this great and pernicious predetermination the authority of its former conclusions may remain inviolate. And therefore it was a good answer that was made by one who when they showed him hanging in a temple a picture of those who had paid their vows as having escaped shipwreck, and would have him say whether he did not now acknowledge the power of the gods,—"Aye," asked he again, "but where are they painted that were drowned after their vows?" And such is the way of all superstition, whether in astrology, dreams, omens, divine judgments, or the like; wherein men, having a delight in such vanities, mark the events where they are fulfilled, but where they fail, though this happen much oftener, neglect and pass them by. But with far more subtlety does this mischief insinuate itself into philosophy and the sciences; in which the first conclusion colors and brings into conformity with itself all that come after, though far sounder and better. Besides, independently of that delight and vanity which I have described, it is the peculiar and perpetual error of the human intellect to be more moved and excited by affirmatives than negatives; whereas it ought properly to hold itself indifferently disposed towards both alike. Indeed in the establishment of any true axiom, the negative instance is the more forcible of the two. (*Novum Organum* I, 46)[7]

Long before Popper's injunction that "confirmations should count only if they are the result of *risky predictions*" (C & R, p. 36), Francis Bacon made a vital contribution toward distinguishing *merely positive* from *supportive* instances of a theory by emphasizing that some kinds of positive instances can differ radically from others in *evidential value* and treating the evidentially significant ones under the heading of "Prerogative Instances" (*Novum Organum* II, Section 21ff.). I shall say in more modern parlance that an instance is a "positive" one with respect to a nonstatistical theory *T*, if its occurrence or being the case can be deduced from *T* in conjunction with suitable initial conditions. But an instance is supportive of *T*, if it is positive *and* has the probative significance of conferring a stronger truth presumption on *T* than *T* has without that instance. As will be shown in our impending discussion of the very important case of causal hypotheses, post-Baconian inductivists appealed to the distinction between merely positive and supportive instances of a theory: they did so in order to guard against unsound causal inferences such as *post hoc ergo propter hoc* and against other ravages of the unbridled use of the hypothetico-deductive method in inductive inference. The very odd one-sidedness of Popper's account of the Baconian inductive method of comparative instances becomes further evident from the exposition given, for example, in the 1916 standard book on *Logic* by the English philosopher Horace William Brindley Joseph.[8]

As Joseph stresses, Bacon was grievously mistaken in supposing that scientists could devise an *exhaustive* finite list of all logically possible alternative theories relevant to a given phenomenon. Having made that unsound assumption, Bacon felt justified in maintaining that scientists could then irrevocably establish the truth of *one* of these specified theories by refuting all of its rivals in the purportedly exhaustive disjunctive class of relevant theories. But while rejecting this Baconian conclusion as ill-founded, Joseph also points out that, according to Bacon, the confident affirmation of the truth of the one remaining theory "would rest *not* on the positive testimony" of its positive instances "but upon the fact that we had disproved all possible rival theories."[9] In short, for Bacon these positive instances were prerogative or *supportive* of the one theory, because these same instances were *also* negative ones for its rival theories. Says Joseph:

> We must proceed then by *exclusions*. Where a hundred instances will not *prove* an universal connexion, one will *disprove* it. This is the corner-stone of his [Bacon's] method: *maior est vis instantiae negativae* [negative instances have more force].[10]

Not only Bacon's tables of comparative instances but also J. S. Mill's "inductive methods" for appraising *causal* hypotheses had long ago led inductivists to demand *controlled experiments* or so-called *controls* as an indispensable check on whether positive instances do have the probative significance of being supportive instances! But in fairness to Popper, it should be pointed out that there has been one important school of inductivists among both philosophers and eminent scientists who championed the doctrine that *any* positive instance of a hypothesis also necessarily qualifies as a *supportive* instance of the hypothesis. This doctrine that any positive case of a hypothesis is automatically also supportive to *some* degree is sometimes called "the instantiation condition" (Nicod's criterion).[11] In an *unbridled* use of the instantiation condition, any positive instance of a hypothesis (e.g., one resulting from the hundredth repetition of the same experiment) will *increase* the credibility of the hypothesis *as much as any other*. And hence in *such* an invocation of the instantiation condition, one considers the mere *number* rather than the relative weight of positive instances when assessing the credibility of a hypothesis. On the latter *stronger* version of the instantiation condition, degrees of credibility could not be construed as mathematical probabilities, since a sufficiently large finite number of positive instances would then yield a "probability" *greater* than unity.

But it was none other than the confirmed inductivist J. S. Mill who emphatically rejected the instantiation condition after Bacon had rejected induction by simple enumeration of positive instances as puerile. Thus, when discussing the case of a *spurious* plurality of causes in his *Logic*, Mill *denies* the probative value of *mere* repetitions of positive instances just as much as Popper does (LSD, p. 269), and indeed for much the same reasons.[12] Mill stresses the need to find positive instances of a specified causal hypothesis such that these instances *also* refute one or more *rival* hypotheses as to the cause. And he writes:

> The tendency of unscientific inquirers is to rely too much on number, without analyz-
> ing the instances; without looking closely enough into their nature to ascertain what
> circumstances are or are not eliminated by means of them. Most people hold their
> conclusions with a degree of assurance proportioned to the mere *mass* of the experi-
> ence on which they appear to rest; not considering that by the addition of instances to
> instances, all of the same kind, that is, differing from one another only in points already
> recognized as immaterial, nothing whatever is added to the evidence of the conclusion.
> A single instance eliminating some antecedent which existed in all the other cases [i.e.,
> an instance which refutes a *rival* hypothesis as to the cause] is of more value than the
> greatest multitude of instances which are reckoned by their number alone.[13]

Clearly, there is at least one historically influential inductivist conception of testability which makes three assertions as follows: (i) it calls attention to the difference between the genus of positive instances of a hypothesis *H* and the species *S* of those special positive instances which serve to eliminate one or more *specified* rivals of *H*; (ii) it rejects the instantiation condition by denying that every member of the genus of positive instances automatically qualifies as supportive; and (iii) it regards the species *S* of positive instances as supportive in the sense of conferring some degree of credibility on *H*, even if the amount of available supportive evidence is *insufficient* to make it more likely that *H* is true rather than false. Note that, among these three assertions, only the third endorses inductive inference. Yet Popper is so preoccupied with dissociating himself from the third of these assertions (LSD, p. 419) that he ignores the first two. Thus, he is prompted to misportray *all* post-Baconian theories of induction as assigning *supportive* probative significance to all positive instances *alike*. In this way, he is *misled* into charging that all such versions of inductivism are unable to indict psychoanalysis, for example, as unscientific in the face of its many positive instances. He gives a concise statement of this misportrayal by making the following previously cited statement (LSD, p. 420, italics in original): "The fundamental doctrine which underlies all theories of induction is *the doctrine of the primacy of repetitions* .... According to this doctrine, repeated instances furnish a kind of *justification* for the acceptance of a universal law." But this claim is untenable in the face of the espousal of assertions (i) and (ii) by the eliminative inductivism of Bacon and Mill. And, as Section 3 will show, Popper's portrayal of all forms of inductivism is likewise refuted by the fact that Bayesian probabilistic inductivism disavows the instantiation condition. Popper's preoccupation with opposing assertion (iii), to the exclusion of appreciating assertions (i) and (ii), is made evident by his appraisal of the eliminative inductivism of Bacon, Whewell and Mill. Speaking of the latter on the immediately preceding page of LSD (p. 419, italics in original), he says:

> The sole purpose of the elimination advocated by all these inductivists was to *establish as firmly as possible the surviving theory* which, they thought, must be the *true* one (or perhaps only a *highly probable* one, in so far as we may not have fully succeeded in eliminating every theory except the true one).
>
> As against this, I do not think that we can ever seriously reduce, by elimination, the number of the competing theories, since this number remains always infinite.

When we discuss Popper's theory of corroboration in Section 3, the last sentence of this quotation from him will return to haunt his own theory!

For brevity, I shall use the term "instantionist inductivism" to refer to the very special version of inductivism which does espouse the instantiation condition. Then I can say that Popper's characterization of all theories of induction as being instantionist also runs counter to the fact that Mill demanded a check by the "Joint Method of Agreement and Difference" on the *probative* significance of the positive instances which had been accumulated by the merely *heuristic* "Method of Agreement." Hence, we reach the following important conclusion: At least some post-Baconian inductivists *deny* that every positive instance of a hypothesis H is automatically supportive of H to some nonzero degree or other. And *a fortiori* these inductivists—as distinct from the exponents of *unbridled instantionist* inductivism!—deny that all positive instances count as *equally* supportive. In view of the importance of this conclusion, let us restate it by saying that all *anti*-instantionist versions of inductivism espouse the following cardinal epistemic principle: *The ability of a theory T to "explain" and/or predict certain phenomena deductively* (with the aid of suitable initial conditions) *is generally only a necessary and NOT a sufficient condition for qualifying these phenomena to count as T-supporting instances over and above being merely positive instances of T!*

Note here that, within the genus of positive instances, the term "confirming instance" is disastrously ambiguous as between a supportive and a nonsupportive species of positive instances. By the same token, logical mischief has been wrought by the weasel word "verification," and by the equivocal verb "verify." Popper's misportrayal of inductivism as universally instantionist is abetted by the ambiguity of the terms "confirmation" and "verification," when he says misleadingly: "It is easy to obtain confirmations, or verifications, for nearly every theory—if we look for conformations" (C & R, p. 36). Contrary to Popper, we shall see very soon that whereas it is easy to find positive instances for the therapeutic claims made by psychoanalysts, it is at best moot whether there exists any *bona fide* or *significant amount* of genuine inductive support for these claims, let alone for the theory undergirding psychoanalytic treatment.

So much for my examination of Popper's historiography of inductivism, which prompted his claim that "there clearly was a need for a different criterion of demarcation" (C & R, p. 256). Imre Lakatos echoes Popper's grossly one-sided concentration on the least sophisticated parts of Bacon's inductivism. Bacon was surely oblivious of the fact, later noted by Kant, that "percepts without concepts are blind," yet he stressed rightly that theoretically aseptic fact collecting is either a delusion or sterile (*Novum Organum*, Book I, Aphorism 95). But this patent deficiency does not justify the imbalance in Lakatos's account of Bacon's conception of scientific inference. When speaking of Popper's role as "the scourge of induction" (PDI, p. 258), Lakatos distinguishes three strands in Popper's anti-inductivist campaign and says of the first, which is of interest here:* "there is the campaign against the *inductivist logic of discovery*. This is the Baconian doctrine according to which a discovery is scientific only if it is *guided* by facts and not *misguided* by theory" (ibid.). Having thus one-sidedly dealt with Baconian doctrine, Lakatos can then declare (PDI, p. 259) that "at least among philosophers of science Baconian method is now only taken seriously by the most provincial and illiterate."

The Illustrative Test Case of Psychoanalysis

Turning now to Freudian psychoanalysis, we note that psychoanalysis offers not only a psychodynamic theory to explain behavior but also a type of treatment—prejudicially dubbed "therapy"—to be practiced for the alleviation of psychological disturbances. I shall focus on the *therapeutic* theses of psychoanalysis as a test case for those of Popper's allegations against inductivism on which he bases his demarcation asymmetry. In so doing, I make due allowance for the following: If it were shown that psychoanalytic therapeutics is not entailed by Freud's psychodynamics, then Freud's theory *may* conceivably have a correct account of the etiology and pathogenesis of certain disorders while failing to provide successful therapeutic prescriptions for them. Generally speaking, it is not incumbent upon

---

* *Editor's Note*: The two other strands that Lakatos mentions are: (ii) "Popper's attack … against the program of an *a priori* probabilistic inductive logic or confirmation theory" (ibid., p. 259), and (iii): "[A] tacit but stubborn refusal to accept any *synthetic* inductive principle connecting Popperian *analytic* theory-appraisals (like content and corroboration) with verisimilitude" (ibid., p. 260).

a psychodynamic theory to produce effective recommendations for psychotherapy in order to qualify as etiologically explanatory, any more than a theory which *explains* carcinogenesis can be expected to *guarantee* that cancer is curable.

Using his falsificationist criterion of demarcation, Popper felt entitled to indict psychoanalysis as unscientific after having addressed the following challenge to Freudians:

> But what kind of clinical responses would refute to the satisfaction of the analyst not merely a particular analytic diagnosis but psycho-analysis itself? And have such criteria ever been discussed or agreed upon by analysts (C & R, p. 38)?

Whereas Popper is concerned with the scientific entertainability of a hypothesis, the inductivist is concerned with its scientific *credibility*. Hence, let us now see quite specifically how the Bacon-Mill inductivist can impugn the scientific credentials of psychoanalytic therapeutics by his particular requirements for *supporting* evidence. It will then be clear how, *mutatis mutandis*, inductivist checks on the probative significance of positive instances can prevent Freudian *psychodynamics* from adducing such instances uncritically with impunity in favor of its etiological claims concerning, say, hysteria or fetishistic sexual behavior. Thus, it will emerge, for example, how the inductivist can challenge Freudian claims about the homosexual etiology of paranoia. In this way, the inductivist will be seen to match Popper's ability to disparage the scientific status of, say, the Freudian Oedipus hypothesis, which Popper does by pointing to the failure of psychoanalysis to specify what kind of behavior (of sons toward their fathers) would warrant its abandonment.

Just how do the time-honored inductivist canons of Bacon and Mill enjoin us to employ *experimental controls* as a curb on giving *undeserved* causal credit to psychiatric treatment for such improvement as is shown by patients during or after treatment? As is well-known, though apparently overlooked by Popper, their inductivist canons discount instances of subsequently improved patients as nonsupportive, *unless* the incidence of improvement among treated patients *exceeds the spontaneous remission rate!* Existing sets of cases of *improved* psychoanalytic patients may indeed qualify as *positive* instances of the hypothesis that Freudian treatment is effective in a specified diagnostic category. But this improvement does *not* inductively redound to the credit of this treatment, unless its incidence exceeds the spontaneous remission rate, which is practically always positive. In particular, the *mere* fact that *at least* some psychoanalytically treated neurotics in a certain diagnostic category get better *after* treatment does *not* show inductively that this treatment is effective, unless *other evidence* shows the spontaneous remission rate for that diagnostic category to be *zero*. Thus, in the absence of further information, positive instances of improvement can fail to be supportive. Indeed, it is conceivable that patients who undergo treatment have a remission rate which is *lower* than the spontaneous one, thereby giving rise to the possibility that the treatment made at least some of them worse! In psychiatry no less than in somatic medicine, there can be iatrogenic or doctor-induced disease.

Furthermore, by Popper's standards, the claim that Freudian treatment is effective is falsifiable. For Popper could hardly deny that the failure to surpass the spontaneous remission rate would serve to refute the hypothesis of therapeutic effectiveness under the assumption of suitable auxiliary claims. And according to Popper's demarcation criterion, falsifiability is *sufficient* for scientific status though *not*, of course, for being a *corroborated* scientific hypothesis. Therefore, by *Popper's* standards the thesis that psychoanalytic treatment is therapeutic does qualify for scientific status without prejudice to whether this therapeutic claim will turn out to be a *corroborated* scientific claim by Popper's requirements.

As for the inductivist, I have been concerned to note so far that the question as to the scientific *credentials* of psychoanalytic therapeutics is tantamount to the following question: What are the actual empirical findings as to the success of psychoanalytic treatment in *surpassing* the spontaneous remission rate in any given diagnostic category? But I must now emphasize that there is a host of difficult subsidiary questions here which must be answered before this question becomes susceptible of a *meaningful* and empirically supported reply. For example, there is much imprecision in the diagnostic categories of neuroses and in certifying membership of patients in one of these. Also, it is unclear what criteria are to be employed in assessing *improvement* in a given diagnostic category such as depressives. Is there improvement in a previously depressed man who is full of joy after treatment, if he also now beats his wife and children when they cannot keep up with his energetic demands?

Whatever these various subsidiary questions, which are well-known to those who are familiar with treatment outcome research, the inductivist places the logical burden of formulating a *meaningful* claim of therapeutic success for Freudian treatment on the *advocates* and *dispensers* of psychoanalytic or psychiatric treatment. And I would add that the *moral* burden of doing so likewise belongs squarely on the shoulders of these advocates and dispensers. In sum, if we are inductivist sceptics, it is emphatically *not* incumbent on the rest of us to show that psychoanalytic treatment is ineffective! In this vein, psychologists such as Hans Jürgen Eysenck,[14] Stanley Rachman,[15] and others have argued that there is no satisfactory evidence showing that psychoanalytic treatment succeeds in surpassing the spontaneous remission rate, although others have sharply challenged their analysis and consider the question moot.[16]

In any case, one can only be baffled and distressed by the immense epistemological crudity encountered in *some* of the literature of psychoanalysis in this connection. Thus, in his 1970 book *The Crisis in Psychoanalysis,* Erich Fromm acknowledges the justification for a crisis of confidence in psychoanalysis. But when Fromm turns to the issue of the causal efficacy of psychoanalytic treatment, he gives not even a hint of being aware of the need for a control group in assessing that efficacy. This, even though anyone who ever studied medicine could hardly be unaware of the need for a control group if the therapeutic efficacy of a drug or chemical agent were at issue, for example. Indeed, Fromm reasons in accord with *post hoc ergo propter hoc* that Freudian psychotherapy is causally efficacious, because psychoanalysts have "observed" improvement in their patients. One might as well argue that coffee consumption cures colds, because we can all observe *with our own eyes* that when we have a cold we get better every time after we drink coffee for a sufficient number of days! Fromm writes:

> Indeed, the facile denial of the therapeutic success of psychoanalysis says more about the difficulty of some fashionable authors to grasp the complex data with which psychoanalysis deals than about psychoanalysis itself. Criticism by people with little or no experience in this field cannot stand up against the testimony of analysts who have observed a considerable number of people relieved of troubles they complained about. Many patients have experienced a new sense of vitality and capacity for joy, and no other method than psychoanalysis could have produced these changes.[17]

Extra ecclesiam nulla salus!

The exponent of anti-instantionist inductivism says to the psychoanalyst: Suppose that Eysenck is right and psychoanalytic treatment does not improve on the spontaneous remission rate. Then how, if at all, can you nonetheless hope to adduce such improvement as is shown by your patients in support of your claim that Freudian psychotherapy *is* effective? Allen Bergin has argued that the studies used by Eysenck and others in reaching their conclusion dealt with *average* effects of treatment and thus overlooked evidence indicating the following: After treatment, the improvement of *some* of the treated patients does exceed the gains exhibited by the untreated patients in the control group, but other treated patients get *worse* after treatment than the untreated controls. Says Bergin:

> When these contrary phenomena are lumped together in an experimental group, they cancel each other out to some extent, and the overall yield in terms of improvement (in these particular studies) is no greater than the change occurring in a control group via "spontaneous remission factors."[18]

Significantly, however, Bergin goes on to concede that, on his view, it is not the treatment *as prescribed by the therapeutic theory* which is effective in a certain subclass of cases but only certain *dispensers* of the treatment having as yet unknown special characteristics or knacks whose efficacy is not understood *and cannot be credited to the therapeutic theory!*

Let me suggest a different putative reply which might be offered by the psychoanalyst and then appraise it according to "neo-Baconian" inductivist standards. Very loosely put, he might conjecture that, in at least some diagnostic categories, patients who seek and receive Freudian treatment are, in some unspecified way, *sicker* than those neurotics who do not seek such treatment and remain untreated by a psychoanalyst. Thus, he might go on to venture the counterfactual claim that if the presumably sicker subclass of treated neurotics had been left *untreated*, then they would have exhibited a remission rate which is LOWER than the spontaneous one among the actually untreated neurotics. In short, the putative conjecture is that the treated and untreated neurotics do *not* have the same capacity for spontaneous remission, the treated neurotics having been initially sicker.

But to confer credibility on this counterfactual claim by the stated inductivist standards, the psychoanalyst would *at least* have to produce evidence showing the following: Neurotics in a given diagnostic category who seek but are deliberately denied psychoanalytic treatment—say by being put on a waiting list—do exhibit the allegedly lower remission rate. Until and unless this is done, the claim of therapeutic effectiveness stands indicted, according to the specified inductivist canons, as gratuitous and hence as devoid of scientific credibility or credentials. Moreover, even if the putative psychoanalyst did furnish the required evidence of *lower spontaneous* remission in the treated subclass, the Baconian inductivist would go on to issue the following challenge to him: In the absence of further information, the positive instances provided by this evidence cannot *redound to the credit of the theory* undergirding psychoanalytic therapeutics. Before such credit can be given, at least the *rival hypothesis* of autosuggestive "faith healing" or *attention placebo effect* would need to be refuted. Note that it is *not* incumbent on the inductivist critic or sceptic to *establish* the rival hypothesis of placebo effect. But it *is* incumbent upon the proponent of psychoanalytic theory to *refute* it! The latter refutation would presumably require showing that in the allegedly sicker subclass, various quite different kinds of psychotherapy *fail* to produce as high a remission rate as psychoanalysis does. To my knowledge, there is no such evidence of any superiority of psychoanalytic treatment. On the contrary, a careful 1970 book by Meltzoff and Kornreich[19] concludes that there is no difference among a whole gamut of different schools of traditional psychotherapy in regard to outcome success. And this important finding refutes the aforecited claim by Erich Fromm that "no other method than psychoanalysis could have produced these changes" [remissions].[20]

Clearly, contrary to Popper, post-Baconian inductivism has at least the capability of challenging the scientific credentials of psychoanalytic therapeutics by issuing the stated twofold challenge to the aforementioned putative defender of its therapeutics. Thus, post-Baconian inductivism does command the logical resources to derogate the scientific credentials of psychoanalysis in the face of the positive instances adduced by Freudians. Inductivism has that capacity in regard to psychoanalysis just as it has the capacity to question, for example, whether the relatively low incidence of coronary heart disease among *athletic* adult males shows that physical exercise contributes to cardiovascular health, unless one can rule out the possibility that precisely those who are otherwise healthy cardiovascularly are the ones who engage in physical exercise to begin with.

Let me merely mention, but not rehearse, the criticism in the literature that Popper's deductive falsifiability of a hypothesis is much too strong a requirement for scientific entertainability, as illustrated by the *mere example* of the hypothesis, "All men are mortal." For the latter says that, for each man, *there is some time or other* at which he dies, that is, that no man lives forever. To *falsify* this deductively by a test statement, we must produce at least one man who *never* dies. For no matter how long any Methusalah lives, he may still die later on. Hence, no basic observation statement can deductively falsify the old saw that we are all mortal. Other examples are furnished by assertions of nonexistence concerning *perpetual* motion machines of the first and second kinds, as in the first two laws of thermodynamics.

Surely I am not being captious if I conclude from my analysis that Popper's demarcation asymmetry is either unsound or too strong: It is unsound if it is claimed to have demarcational capabilities with respect to psychoanalysis and astrology, for example, which are not also possessed by the requirements of anti-instantionist inductivism, or it is too strong as just explained. As shown by my examples, the mere fact that anti-instantionist inductivists try to use supportive instances to "probabilify" or credibilify hypotheses does *not* commit them to granting credible scientific status to a hypothesis *solely* on the strength of existing positive instances, however numerous. Indeed, as shown by thousands of cases of people who are improved *after* psychiatric treatment, Mill's inductivism can discount even such a multitude of positive instances as *non*supportive of the hypothesis of therapeutic effectiveness. But I now need to forestall a possible misunderstanding of the moral I draw from my comparison of Popper's demarcation asymmetry with the conception of scientificality advocated by an inductivist who rejects the instantiation condition.

As already noted, Popper is primarily concerned with scientific *entertainability*, whereas the inductivist is intent upon the scientific *credibility* of a hypothesis. For Popper, the mere falsifiability of a hypothesis suffices for according scientific status to it, but the inductivist may be prepared to grant no more than *potential* scientific status to it in virtue of its inductive support*ability* in principle. And Popper's *corroborated* scientific hypothesis is the *counterpart* of the inductivist's *actually* scientific hypothesis. Hence, if one wished, one could treat Popper's falsifiability and the inductivist's genuine support*ability* as counterpart criteria in their respective endeavors to effect a demarcation between nonscience, on one hand, and those theories which are at least scientifically *entertainable*, on the other.

Can it be held that the inductivist's criterion avoids any defect corresponding to the weakness of Popper's falsifiability requirement, which fails by being too strong? I think not, because the inductivist's genuine supportability likewise fails to the extent that there is no satisfactory general theory of evidential support which states unambiguous necessary and sufficient conditions that are successfully applicable to *every* concrete case. We must be mindful in this connection of the fact that science covers a whole gamut of kinds of claims which pose problems of empirical validation: Comprehensive theories like neo-Darwinism or general relativity, causal hypotheses like "Asbestos is carcinogenic," statistical hypotheses, and even simple generalizations like "All ravens are black." Consider just the last of these examples, which lends itself to the statement of Carl Hempel's paradox of confirmation. If the inductivist avoids that paradox by rejecting the instantiation condition, then it is still incumbent upon him to tell us under what conditions a positive instance of "All ravens are black" *does* count as supportive. Yet no satisfactory general answer is available, as far as I know. Nor is it clear that inductivism can furnish a generally satisfactory method for assigning *nonzero* posterior mathematical probabilities to even the empirically most successful *universal* law hypotheses (I shall turn to that in Sections 3.1 and 3.2). Hence, I think that the inductivists have no more succeeded than Popper in stating *general* criteria for effecting a *neat* demarcation of science from nonscience.

Thus, the upshot of my comparison of inductivist conceptions of scientifical-ity with Popper's is *not* the claim that there is a viable inductivist counterpart to Popper's defective demarcation criterion for scientific entertainability. Instead the moral I draw is the following: Popper was seriously mistaken in claiming that, *in the absence of negative instances*, all forms of inductivism are necessarily commit-ted to the (probabilified) scientific credibility of a theory, merely because that the-ory can adduce numerous positive instances.

As for psychoanalysis in particular, contrary to Popper, Freudian therapeutic theory is no less falsifiable by Popper's own standards than it is in principle genu-inely supportable inductivistically. But in regard to actual scientific credibility, the inductivist can indict that theory is at best gratuitous, because its positive instances do not have the probative force of being sufficiently supportive and perhaps are not even supportive at all. Thus, the inductivist's willingness to either probabilify or somehow credibilify theories which *can* marshal genuinely supportive positive instances does *not* render the inductivist helpless to dismiss the positive instances adduced by psychoanalysts as nonprobative. And, as we shall soon see, the induc-tivist can justly complain that Popper's pure deductivism has no nontrivial answer to the question: "What does it mean to say that a successful risky prediction *counts* in favor of the theory that made it?"

■ 3. POPPER'S THEORY OF CORROBORATION VERSUS BAYESIAN PROBABILISTIC INDUCTIVISM AND DUHEMIAN HOLISM

3.1 Popper on the Probability of Universal Laws

One of Popper's major reasons for alleging the "impossibility of an inductive probability" is as follows: In a universe containing infinitely many distinguishable things or spatio-temporal regions, "*the probability of any (nontautological) univer-sal law will be zero*" (LSD, Appendix vii, p. 363, italics in original). Commenting on the meaning of the term "probability" in this sentence, he says (LSD, p. 364):

> By "probability," I mean here either the *absolute* logical probability of the universal law, or its probability *relative to some evidence*; that is to say, relative to a singular statement,

or to a finite conjunction of singular statements. Thus if a is our law, and b any empirical evidence, I assert that

(1) $p(a)=0$

and also that

(2) $p(a,b)=0.$

These formulae will be discussed in the present appendix. The two formulae, (1) and (2), are equivalent.

And for the case in which the universal statement a is interpreted as entailing an infinite conjunction of singular statements, he then goes on to give several arguments for concluding that $p(a) = 0$. He regards one of these arguments to be "incontestable" (LSD, p. 366). But Colin Howson has offered various rebuttals to such a strategy.[21]

The same conclusion is reiterated and amplified in C & R, where Popper writes:

> In view of the high content of universal laws, it is neither surprising to find that their probability is zero, nor that those philosophers who believe that science must aim at high probabilities cannot do justice to facts such as these: that the formulation (and testing) of *universal laws* is considered their most important aim by most scientists. (C & R, p. 286)[22]

Furthermore, he contrasts the inevitably vanishing inductive probabilities of universal law statements with the nonvanishing degrees of corroboration of which such statements are capable on appropriate evidence in his theory of corroboration, saying:

> But it can be shown by purely mathematical means that *degree of corroboration can never be equated with mathematical probability.* It can even be shown that all theories, including the best, have the same probability, namely zero. But the degree to which they are corroborated (which, in theory at least, can be found out with the help of the calculus of probability) may approach very closely to unity, i.e. its maximum, though the probability of the theory is zero. (C & R, pp. 192–193)

But Popper's conclusion that, for any universal statement a, $p(a) = 0$ can now be shown to be incompatible with his *quantitative* theory of the content and verisimilitude of a hypothesis. For when we turn to the latter theory (C & R, pp. 218, 234, 390–397), we find the following:

(1) Both in C & R and in his *Objective Knowledge*,[23] Popper defines the *measure* ct(a) of the content of a theory a as ct(a)=1−p(a) where 'p(a)' is "the logical probability" of a. He further characterizes the latter probability very meagerly (OK, p. 51) as "the logical probability that it [a] is true (accidentally, as it were)."

(2) When speaking of the logically *incompatible* theories of Einstein and Newton—denoted by E and N, respectively—Popper tells us without any explanation that "the content measures $ct(N)$ and $ct(E)$" bear out the intuition that, as between these two theories, "Einstein's has the greater content" (OK, p. 53). But if $ct(E)> ct(N)$, then $p(N)>p(E)$, and hence $p(N)>0$ even though N is a conjunction containing universal statements and should therefore have *zero* probability according to Popper's cited claim. Let us now see how this inconsistency carries over into his theory of quantitative verisimilitude.

(3) Popper states his view as to what outcome of attempted falsifications of E would epistemically warrant the conjecture that the quantitative falsity content of E does not exceed the corresponding falsity content of N (OK, p. 53). And he then deduces (OK, p. 53) that E has greater quantitative verisimilitude than N from the conjunction of the following three assertions: (i) the aforementioned claim that $ct(E)>ct(N)$; (ii) the stated conjecture as to the comparative falsity contents of E and N; and (iii) the contention that "the stronger theory, the theory with the greater content, will also be the one with the greater verisimilitude *unless its falsity content is also greater*" (OK, p. 53). I should mention incidentally that, in the case of *logically incompatible* theories such as E and N, Popper's justification for this third assertion breaks down: His 1966 theorem on truth content, which serves as a premise in his derivation of the third assertion (OK, pp. 52–53), is *not* proven for the purported content relation $ct(E)>ct(N)$ of *incompatible* theories such as E and N and but only for the corresponding proper inclusion relation among the *Tarskian* logical contents of a pair of theories, one of which unilaterally entails the other.[24] In any case, his cardinal example of greater verisimilitude makes the claim that $ct(E)>ct(N)$.

But since N and E are each replete with universal statements, consistency with his claims in LSD (Appendix vii) requires Popper to say that both $p(N)=0$ and $p(E)=0$. And since $ct(a)=1-p(a)$, this has the consequence that $ct(E)$ and $ct(N)$ have the *same* value unity. Yet, as we just noted, he told us in OK (p. 53) that $ct(E)>ct(N)$. Furthermore, Popper asserts (C & R, p. 218) that (1) "with increasing content, probability decreases, and vice versa"; and that (2) if one (universal) theory B *unilaterally entails* another (universal) theory A, then $ct(B)>ct(A)$ (OK, pp. 51, 53). These assertions (1) and (2) require that if B unilaterally entails A, then $p(A)>p(B)$, and hence that at least one of the two probabilities $p(A)$ and $p(B)$ be *nonzero*. But for universal nontautological theories A and B, the latter requirement contradicts the cited LSD claim that *both* of them must be zero. Moreover, even for the case in which B *unilaterally* entails A, the probability calculus requires only that $p(A)\geq p(B)$ rather than that $p(A)>p(B)$, as Popper does here.

Thus, Popper hoists himself on his own petard: If he is to furnish an indictment of inductive probabilities by his LSD claim that $p(a)=0$ and $p(a,b)=0$ for any nontautological universal a and any empirical evidence b, then his theory of quantitative content and quantitative verisimilitude is untenable; conversely, if the latter is affirmed, the claim $p(a)=0$ is no longer available as a premise for a *reductio ad absurdum* argument against inductive probabilities.

3.2 Popper on Probability Vis-à-Vis the Aims of Science

One of the citations in the preceding subsection contains Popper's complaint that the aim to achieve high probabilities as in Bayesian inductive inference *fails* to vindicate the quest for *universal* laws: As Imre Lakatos has summarized it (PDI, p. 259): "If inductive logic was possible, then the virtue of a theory was its improbability rather than its probability, given the evidence." Concerning this complaint, Ronald Giere has aptly made a personal remark to me to the following effect: The probabilistic inductivist does *not* say that the aim of science is or ought to be to play it safe. Instead, this inductivist tells us (i) what risks we take, *if* we choose *not* to play it safe, and (ii) that *if* we wish to play it safe, then we should act on the *less* daring of two theories whenever possible. On this understanding of probabilistic inductivism, Popper's complaint no longer applies.

3.3 Comparison Between Popper's Methodology and Bayesian Inductivism

It has been or ought to have been well-known since the nineteenth century that the prior probabilities in Bayesian inference pose formidable if not insoluble difficulties, at least if they are to be construed nonsubjectively. Hence, if I now proceed nonetheless to assess the capabilities of Bayesian inductivism to implement Popper's methodological prescriptions, I do *not* do so in the spirit of Jean-Baptiste d'Alembert, who is said to have declared, "Allez en avant, et la foi vous viendra." Rather I do so, because I hope to show that this assessment is instructive, although it is moot whether the status of Bayes's prior probabilities will be satisfactorily clarified in the future. I shall inquire: Can Popper do any better?

3.3.1 The Epistemic Capabilities of Popper's Methodology

The following two epistemic contentions of Popper's are pertinent here:

(i) "Confirmations should count only if they are the result of *risky predictions*; that is to say, if, unenlightened by the theory in question, we should have expected an event which was incompatible with the theory—an event which would have refuted the theory" (C & R, p. 36).

(ii) "When trying to appraise the degree of corroboration of a theory we may reason somewhat as follows. Its degree of corroboration will increase with the number of its corroborating instances. Here we usually accord to the first corroborating instances far greater importance than to later ones: once a theory is well corroborated, further instances raise its degree of corroboration only very little. This rule however does not hold good if these new instances are very different from the earlier ones, that is, if they corroborate the theory in a *new field of application*. In this case, they may increase the degree of corroboration very considerably" (LSD, p. 269).

Popper clarifies the latter of these two demands by writing elsewhere (C & R, p. 240) as follows:

(iii) "A serious empirical test always consists in the attempt to find a refutation, a counter example. In the search for a counter example, we have to use our background knowledge; for we always try to refute first the *most risky* predictions, the '*most unlikely* ... consequences' (as Peirce already saw [footnote omitted here]); which means that we always look in the *most probable kinds* of places for the *most probable* kinds of counter examples—most probable in the sense that we should expect to find them in the light of our background knowledge. Now if a theory stands up to many such tests, then, owing to the incorporation of the results of our tests into our background knowledge, there may be, after a time, no places left where (in the light of our new background knowledge) counter examples can with a high probability be expected to occur. But this means that the degree of severity of our test declines. This is also the reason why an often repeated test will no longer be considered as significant or as severe: there is something like a law of diminishing returns from repeated tests (as opposed to tests which, in the light of our background knowledge, are of a *new kind*, and which therefore may still be felt to be significant). These are facts which are inherent in the knowledge-situation; and they have often been described—especially by John Maynard Keynes and by Ernest Nagel—as difficult to explain by an inductivist theory of science. But for us it is all very easy. And we can even explain, by a similar analysis of the knowledge-situation, why the empirical character of a very successful theory always grows stale, after a time" (C & R, p. 240).

Before commenting on the capabilities of the Bayesian inductivist to implement these methodological prescriptions of Popper's, let me point out why I think that Popper himself cannot justify these demands within his own deductivistic framework.

(i) As for Popper's demand (i) that positive instances should count only if they are the results of successful *risky* predictions, let us ask: What does the word *count* mean in his deductivist framework, when Popper declares, "Confirmations should count only if they are the result of *risky predictions*"? Count toward or for what? Qua *pure* deductivist, can Popper possibly maintain without serious inconsistency, as he does, that *successful* results of initially risky predictions should "count" in favor of the theory making the risky prediction in the sense that in these "crucial cases ... we should expect the theory to fail if it is not true"

(C & R, p. 112)? The latter statement is part of his characterization of "crucial cases" in which he says:

> A theory is tested not merely by applying it, or by trying it out, but by applying it to very special cases—cases for which it yields results different from those we should have expected without that theory, or in the light of other theories. In other words we try to select for our tests those crucial cases in which we should expect the theory to fail if it is not true. (C & R, p. 112)

Note Popper's claim that—on the basis of background theories—we should expect the [new] theory to fail [a "crucial" test] if it is *not* true. As against this assertion, I say that what is warranted *deductivistically* is the following:

(a) The older theories (*cum* initial conditions) predict an outcome contrary to the result C predicted by the new theory H, or the older theories are at least deductively noncommittal with respect to C, thereby making the prediction of the new theory "risky" in *that special sense.* Popper is here concerned with the more highly risky kind of prediction C, which is *contrary* to each of the older theories rather than logically independent of them. For this reason as well as in order to simplify the discussion, I shall sometimes concern myself with just the more highly risky kind of C.

(b) If the new theory H does *not* fail the crucial test but passes it, clearly nothing follows deductively about *its* truth. What does follow from the observation statement C is the truth of the "infinitely" weaker disjunction D of *all* and only those hypotheses which individually entail C. For C itself is one of the disjuncts in the infinite disjunction D. And thus C entails D in addition to being entailed by D.[25]

Note that for any true disjunct in D, the latter has an infinitude of *false* disjuncts which are pairwise *incompatible* with that true disjunct in D and, of course, are also incompatible with the background knowledge. Yet—contrary to the background knowledge—each of this infinitude of false theories predicts C no less than H does, and indeed H may be false as well. Hence of what avail is it to Popper, *qua deductivist*, that by predicting C, H is one of an infinitude of theories in D incompatible with those *particular* theories with which scientists had been working by way of historical accident and that scientists happened to have thought of the particular disjunct H in D?

According to Popper's definition of the term "severe test," the experiment E which yielded the riskily predicted C does qualify as a "severe test" of H. But surely the fact that H makes a prediction C which is incompatible with the prior theories constituting the so-called background knowledge B does *not* justify the following contention of Popper's: A *deductivist* is entitled to expect the experiment E to yield a result *contrary* to C, *unless H is true.*

Indeed, Popper's reasoning here is of a piece (as I showed in *Brit. J. Phil. Sci.* 27 (1976), 108–110) with the reasoning of the Bayesian proponent of *inductive* probabilities which has been stated by Wesley Salmon as follows:

> A hypothesis risks falsification by yielding a prediction that is very improbable unless that hypothesis is true. It makes a daring prediction, for it is not likely to come out right

unless we have hit upon the correct hypothesis. Confirming instances are not likely to be forthcoming by sheer chance.[26]

It would seem, therefore, that when Popper assessed the epistemic significance of severe tests in the quoted passage, he was unmindful of his own admonition against Baconian eliminative inductivism. As we recall from our citation of LSD in Section 2, this admonition reads as follows:

> The sole purpose of the elimination advocated by all these inductivists was to *establish as firmly as possible the surviving theory* which, they thought, must be the *true* one (or perhaps only a *highly probable* one, in so far as we may not have fully succeeded in eliminating every theory except the true one).
>
> As against this, I do not think that we can ever seriously reduce, by elimination, the number of the competing theories, since this number remains always infinite. (LSD, p. 419, italics in original)

What then would be a *deductivistically* sound construal of Popper's injunction that "confirmations should count only if they are the result of *risky predictions*"? It would seem that deductively speaking, C can be held to *count for* H only in the following Pickwickian sense: C deductively refutes or *counts against* those *particular* rivals of H which belong to B while leaving fully intact an infinitude of other rivals in the disjunction D. Given the latter infinitude of relevant competing theories rightly stressed by Popper, I see no rational basis on which a *pure deductivist* can even hope for a good chance that *severe* tests will weed out *false* theories any better than *non*severe tests. And, hence, *I cannot see any purely deductivistic rationale for advocating severe tests as being conducive to the discovery of true theories.* What then is the purportedly superior deductivistic rationale for Popper's corroborations as contrasted with neo-Baconian eliminative inductions? I shall return to this point in Section 3.4.

So much for the first of Popper's demands, stated under (i).

(ii) Is it "all very easy" for him in keeping with his major anti-inductivist tenets to espouse the *reasons* offered by him for *not* repeating the same corroborating experiment *ad nauseam* and for claiming that "there is something like a law of diminishing returns from repeated tests" (C & R, p. 240)? Alan Musgrave has argued cogently that it is not. And Musgrave makes the following telling points:

(a) Let p qualify as a *very risky* kind of prediction with respect to the initial background knowledge B_0, and let p be tested successfully, say, ten times, thereby yielding ten corroborating instances for the new theory T entailing p. Musgrave says:

> After each performance of the test, the particular results are incorporated into our background knowledge, changing it successively from B_0 to B_1, to B_2, and so on, up to B_{10}. Popper's claim seems to be that the probability of the prediction in the next instance *gradually rises* in the light of our successively augmented background knowledges.

Hence the severity of the successive repetitions of the same test *gradually falls*, and so does the degree of corroboration afforded by them to T.

But this clearly involves a straightforward *inductive argument*. It involves the inductive argument that each positive instance of the universal test-implication *p* increases its probability. (Alternatively, if *p* is construed as a singular prediction, it involves the inductive argument that the accumulation of past instances renders the next instance of the same kind more and more probable.) It would seem, therefore, that if we eschew inductive arguments, then the incorporation of past results of a severe test into background knowledge can do nothing to reduce the severity of future performances of that test. It would seem, in other words, that if Popper's theory of corroboration really is a noninductivist one, then it cannot provide us with diminishing returns from repeated tests.[27]

(b) Now suppose that Popper were to attempt to *escape* the charge of inductivism as follows: On grounds whose specification remains urgent, he takes the aforementioned ten corroborations of *p* to be "sufficiently many" repetitions of the experiment to accept *p*. Commenting on this putative approach, Musgrave writes:

> Suppose that we test *p* ten times, that we always get the same result, and that ten repetitions is "sufficiently many" for us to accept the universal test-implication *p* and reject the falsifying hypothesis. Suppose that we incorporate into background knowledge, not the *particular* results of our ten experiments, but the *universal* statement *p*. Now we need no inductive argument to make the severity of future tests of *p* decline. Indeed, it will decline sharply to *zero*, because *p* follows trivially from our augmented background knowledge as well as from the theory T.
>
> This account is perfectly consistent with Popper's claim to have a noninductivist theory of corroboration. But it can hardly be said to provide us with *diminishing* returns from repeated tests, if this means *steadily diminishing* returns. Instead, we have a "one-step function": before "sufficiently many" repetitions of the same test have been performed, each one has the same severity; after "sufficiently many" repetitions, the severity of all future ones is zero. Perhaps this is what Popper had in mind all along: for he only claimed that he could provide *something like* a law of diminishing returns, and one step down is, I suppose, something like a gradual slide.[28]

(c) But the latter "saltation" version of the law of diminishing returns leaves unsolved for Popper the problem of "*how* often is sufficiently often" when repeated performances of a corroborating test are to be held adequate for incorporating the *universal* hypothesis *p* into the background knowledge.

The arguments given so far against Popper's ability to accommodate his own demands (i) and (ii) might be no more than *tu quoque* arguments, unless the Bayesian inductivist can do better in regard to justifying the gradual diminution of the epistemic returns from repeated trials. I shall now try to show in what sense the latter can indeed do better. We already saw earlier (Section 2) that the kind of inductivism which espouses the *unbridled* invocation of the instantiation condition can indeed *not* do better. For in *this* particular version of inductivism, all positive instances alike, including those resulting from repetitions of the same kind of test, will raise the credibility of the relevant hypothesis by the *same* amount. In this *latter* sense, there is warrant for the concessions by Keynes and Nagel, which are mentioned by Popper.

Hence, let us now look at the capabilities of the Bayesian framework.

3.3.2 Bayesian Inference: The Rejection of the Instantiation Condition and the Law of Diminishing Epistemic Returns from Repeated Tests

In comparing Bayesian inference with Popper's epistemology, it behooves us to see first that the Bayesian schema rejects even the weaker version of the instantiation condition. After furnishing the specifics of that rejection, we shall see how the Bayesian obtains diminishing epistemic returns from repetitions of the same kind of experiment.

We are about to consider a form of inductivism which regards degrees of credibility to be mathematical probabilities and thus holds Bayes's theorem to be applicable to the probability of hypotheses. But I disregard here whether a subjectivistic or an objectivistic construal is given to these probabilities and will use the term "Bayesian inductivism" in a sense neutral to the difference between them.

Let B assert an initial condition relevant to the universal statement or deterministic (causal) law hypothesis H, and let B also assert some other as yet unspecified background knowledge. Let C be the phenomenon which is predicted deductively by B and H. Assume that C occurs.

Then in the Reichenbach notation, where $P(X, Y)$ is the probability *from* X *to* Y, Bayes's theorem (division theorem) can be written for this case as

$$P(B \& C, H) = \frac{P(B,H)}{P(B,C)} \times P(B \& H, C)$$

where $P(B\&H,C)=1$.

Now assume that $P(B,C) \neq 0$: Unless this assumption is made, the right-hand side (RHS) of the equation, and hence the posterior probability would be undefined, and it would be incoherent to have gone ahead with a test for C. And be mindful of the fact that, in our case of a universal (nonstatistical) hypothesis H, $P(B \& H, C)=1$. Then the necessary and sufficient condition that the occurrence of the positive instance C will yield a *posterior* probability which is *equal* to H's *prior* probability is as follows: Either $P(B,H)=0$ or $P(B,C)=1$. Given our assumption that $P(B,C) \neq 0$, the first of these two disjuncts assures that both the prior and the posterior probability of H vanish.

Clearly, *if* one can find a rationale for rejecting the view that universal law statements H necessarily have a vanishing probability, the first disjunct $P(B,H)=0$ is clearly a very extreme and hence uninteresting example of the violation of the instantiation condition in the Bayesian version of inductivism.[29]

Is the second disjunct $P(B,C)=1$ a similarly extreme and dull example? At first glance, it might be thought that it *clearly* is *not*. For suppose that the background knowledge B is now held to comprise an older hypothesis H_0 which is incompatible with H but such that under the initial condition asserted in B, the older H_0 entails C no less than the newly proposed H. In that case, it would be nontrivial that $P(B,C)=1$. Moreover, since B is now presumed to comprise H_0, the conjunction B & H is an inconsistent system, and we also have $P(B,H)=0$, while $P(B \& H,C) = 1$ as before. Thus, in this putative way of assuring $P(B,C)=1$, we would obtain $P(B \& C, H) = 0$. And it would seem that the Bayesian version of inductivism would then implement with a vengeance the Popperian requirement (C & R, p. 36, item (1)) that a positive instance C of H should count in H's favor only if B entails a prediction which is *contrary* to C or if C is novel for B (C & R, p. 390).

But this way of nontrivializing the case of $P(B,C)=1$ (i.e., by the incorporation of H_0 in B) must meet the following challenge: It is at least unclear how the old hypothesis H_0 can ever be supplanted by the new one H within the framework of the Bayesian formalism, once the old hypothesis H_0 has been made an integral part of the background knowledge that is brought to bear on the evaluation of H. William Harper[30] has proposed a way of meeting this challenge of subsequent exportation of H_0 from B. If Harper's proposal is viable, then the case of $P(B,C)=1$ comprises repudiations of the instantiation condition by Bayesian inductivism which are at once nontrivial and splendidly meet Popper's stated requirement. Once H_0 has become dubious, it should be exported from B.

In any event, given $P(B,C)\neq0$, we know that in the case of a hypothesis H for which $P(B \& H, C) = 1$, the positive instance C will always be somewhat supportive of H except when either $P(B,C)=1$ or $P(B,H)=0$. For in all other cases in the stated category, the ratio of the posterior and prior probabilities of H will always exceed 1 even if only slightly, so that the occurrence of C will assure that the posterior probability of H is greater than its prior probability.

When $P(B \& H,C)$ is *less* than 1, we must get clear in what sense C is or is not a *positive* instance of H. We are interested in the case when C does so qualify as positive. Then we can say the following: Given both $P(B,C)\neq0$ and $P(B,H)\neq0$, we have

$$\frac{P(B \& C,H)}{P(B,H)} = \frac{P(B \& H,C)}{P(B,C)}.$$

Thus, in this case a sufficient condition for the equality of the posterior and prior probabilities of H is that the ratio on the RHS has the value 1, that is, that the evidence C be equally probable whether or not H is true.[31]

Let us suppose for argument's sake that the previously discussed way of assuring nontrivially that both $P(B,C)=1$ and $P(B \& C,H)=0$ fails after all, say, because Harper's proposed method of "exporting" H_0 from B is beset by an unexpected difficulty. And return to the case $P(B$ & $H,C)=1$. But now suppose that $P(B$ &$C,H)>P(B,H)$.

Then we are concerned with supportive rather than nonsupportive positive instances, and we ask: To what extent, if any, can the Bayesian inductivist comply with Popper's two stated demands? As we recall, these were that positive instances should count just if they are furnished by successful risky predictions and that repeated trials yield diminishing epistemic returns.

Thus as before, let B assert an initial condition relevant to the law hypothesis H *and* also assert other background knowledge. Let C be the phenomenon which is predicted deductively by B and H. Assume that C occurs.

Then in the Reichenbach notation, Bayes's theorem (division theorem) can be written for this case as before:

$$P(B \& C,H) = \frac{P(B,H)}{P(B,C)} \times P(B \& H,C)$$

where $P(B$ & $H,C)=1$.

Assume that neither of the two prior probabilities in the fraction on the RHS vanishes. Then we can write

$$\frac{P(B \& C,H)}{P(B,H)} = \frac{1}{P(B,C)}.$$

We are now concerned with the case of a successful *risky* prediction in which the probability of C on the background knowledge and given initial condition *without* H is *very* low—say 10^{-6}—so that $P(B,C)$ is *near* zero. Then the RHS is *very* large, say 1,000,000. But in that case the *ratio* of the posterior probability of H to its prior probability is likewise huge, viz. 1,000,000.

But this means that according to this theorem of the probability calculus, the following is the case: If a hypothesis H predicts a phenomenon C whose occurrence is very unlikely without that hypothesis on the basis of the background knowledge alone, then the occurrence of C confers strong support on H in the sense that the *factor* by which C *increases* the probability of H is enormous, say a *factor* of 1,000,000.

It is important to note here that a huge *factor* of probability increase need not be tantamount to a large *amount* of increase in the probability of H. To see this, note first from an expanded equivalent of the term $P(B,C)$ in an alternate form of Bayes's theorem, that in our putative case of $P(B\&H,C)=1$, the value of $P(B,H)$ will typically be even smaller than $P(B,C)$. This is also intuitive because hypotheses *other than* H whose prior probability on B is likewise very low will also entail the prediction C. Hence, for a sufficiently small value of $P(B,H)$—say 10^{-8}—even a *million-fold* increase in the probability of H will yield an *amount* of increase of even less than 1/100 in the probability of H.[32] For the *difference* between the prior and posterior probabilities of H will be only 10^{-2}–10^{-8}.

This *caveat* concerning the distinction between the *factor* by which the probability of H changes and the *amount* of such change should now be borne in mind when considering the Bayesian ability to implement Popper's demand for diminishing epistemic returns from repeated trials. We now ask: What is the Bayesian *supportive* significance of *repeated* occurrences of the positive kind of event instantiated by C (hereafter "C-like" event)? Can the Bayesian succeed where unbridled instantianist inductivism failed and show why the same confirming experiment should *not* be repeated *ad nauseam*?

In the Bayesian inductivist framework, each occurrence of a C-like event can be held to change the *content* of the background knowledge B such that the prior probability $P(B,C)$ of the next C-like event will *increase*, so that its reciprocal will *decrease*. Therefore, the *ratio* of the posterior and prior probabilities of H will *decrease* with such repeated positive occurrences. Hence, the successive *factors* by which repeated occurrences of C-like events will increase the probability of H will get ever smaller. But this decrease in the successive *factors* of probability increase does *not* itself suffice at all to assure a *monotonic* diminution in the *amounts* of probability increase yielded by repeated occurrences after the first occurrence C! For suppose as before that prior to the first such occurrence $P(B_0,H)=10^{-8}$, so that just before the *second* such occurrence the revised prior probability $P(B_1, H)$ becomes 10^{-2} after the aforementioned putative million-fold increase. Suppose that before the second occurrence, $P(B_1,C)$ were, say, 1/50. Then the second occurrence would increase the probability of H by a factor of only fifty as contrasted with a million. But the *amount* of such probability increase effected by the first occurrence would be less than 1/100, whereas the amount of increase yielded by the second occurrence would have the much greater value 49/100. It is true that the posterior probability can become at most 1, so that the successive amounts of probability increase can neither keep growing on and on nor even indefinitely remain above some nonzero lower bound. But the latter restriction by itself is a far cry from *assuring* a *monotonic* diminution in the successive amounts of probability increase. Such a diminution might be assured, however, if the decrease in the successive values of the *ratios* of the posterior and prior probabilities of the hypothesis were sufficiently drastic each time.

Clark Glymour (*Journal of Philosophy*, August 1975) has argued that there are cases in scientific practice in which the successful outcome of a "severe" test nonetheless justifiably does *not* redound to the credibility of the hypothesis in the scientific community. For example, measurements of the gravitational red shift predicted by the general theory of relativity (GTR) would seem to qualify as a "severe" test of the latter's field equations in the sense that such a red shift is not predicted by the earlier Newtonian rival theory of gravitation. But Glymour adduces Eddington's analysis of this case to exhibit a justification for the view of physicists that the gravitational red shift is a very weak test of the GTR, while other phenomena are rather better tests, although the latter do not qualify as equally "severe."

The foregoing comparison between Popper's methodology and Bayesian inductivism in regard to the capability of *implementing* Popper's avowed "law of diminishing returns from repeated tests" shows the following: Vis-à-vis Bayesian inductivism, Popper is hardly entitled to claim that "it is all very easy" for him to show that his methodology has a superior capability of justifying that "law."

One important point remains to be considered in our comparative appraisal of Bayesian inductivism: Does Bayesianism sanction the credibilification of a new hypothesis by a sufficient number of positive instances which are the results of *non*risky predictions? For example, would a sufficient number of cases of people afflicted by colds who drink coffee daily for, say, two weeks and recover not confer *a posterior* probability greater than 1/2 on the new hypothesis that such coffee consumption cures colds? And how, if at all, does the Bayesian conception of inductive support enjoin scientists to employ a *control group* in the case of this causal hypothesis K with a view to testing the rival hypothesis that coffee consumption is causally irrelevant to the remission of colds?

As for the first question, note that in the case of a test of H by a *non*risky prediction, $P(B,C)$ is even initially quite *high*, that is, $P(B,C)>1/2$, as in our example of the recoveries from colds. And as we accumulate positive instances C_i (i = 1, 2, 3 ...), the content of B changes after each new positive instance C_i such that $P(B,C)$ increases further. Hence, the latter's reciprocal will decrease. But that reciprocal is the previously discussed *factor* $1/P(B,C)$ of probability increase of the hypothesis under test. In our example of the hypothesis K concerning the therapeutic benefits of coffee drinking, the *factor* of its probability increase will therefore itself decrease. The *general* question is whether the probability increases in the hypothesis which ensue from the accumulation of indefinitely many positive instances will ultimately raise its probability to nearly 1 or whether these increases will be such as to yield a posterior probability which is *asymptotic to only very much less than 1/2*.

Unfortunately, apart from some rather restricted conditions which Allan Gibbard has investigated (unpublished), it would seem that more typically the Bayesian formalism does indeed permit the following: In the case of those successful but *non*risky predictions which *are* supportive, a sufficient number of them will credibilify a hypothesis to be more likely true than false. And it would seem that one must likewise *supplement* the formalism by the injunction to employ experimental controls in testing causal hypotheses like K: Such an injunction could be based on Risto Hilpinnen's version of the principle of *total* evidence which enjoins us to *seek out* as much potentially relevant evidence as possible.[33]

In any case, the Bayesian probabilistic form of inductivism, no less than the Baconian version, is able to *deny* that a mere mass of positive instances of a hypothesis must automatically be held to confer a substantial measure of credibility on the hypothesis. And in Popper's theory of corroboration, the corroboration of a theory T by an observable fact F which was successfully predicted by T is *parasitic* on the logical relations of F to T's *known* rivals $R_1, R_2 ..., R_n$, as follows: For each known rival R, the accepted fact F *either* deductively falsifies R *or* F is *novel* with respect to R in the sense of being logically independent of R. In other words, a successfully predicted positive instance F of T corroborates T, if for each R, the fact F is either contrary to R or not predicted by R but without being contrary to R (C & R, pp. 36, 390).

On the basis of our previous comparative scrutiny of Popper's methodology vis-à-vis two major versions of inductivism, it will be apparent why I conclude the following: Popper is not justified if he claims to have advanced over all forms of post-Baconian inductivism "in explaining satisfactorily what is ... a 'supporting instance' of a law" (OK, p. 21, see also item (5) on p. 20 there). In saying this, I am also being mindful of my previous arguments that Popper has given us no *purely deductivistic* rationale for advocating *severe* tests as being at all conducive to the discovery of true theories. Indeed, we saw that his critique of Baconians boomerangs massively.

3.4 The Inductivist Capstone of Corroboration in "Objective Knowledge"

I now need to call attention to an important *tacit* inductivist commitment in Popper's assignment of *epistemic* significance to his concept of truth preference for corroborated theories over refuted ones.

He tells us that "since we are searching for a true theory, we shall prefer those whose falsity has not been established" (OK, p. 8). And he articulates his concept of corroboration as well as the latter's relation to his notion of preference with respect to truth in the following passages:

> By the degree of corroboration of a theory I mean a concise report evaluating the state (at a certain time *t*) of the critical discussion of a theory, with respect to the way it solves its problems; its degree of testability; the severity of tests it has undergone; and the way it has stood up to these tests. Corroboration (or degree of corroboration) is thus an evaluating *report of past performance*. Like preference, it is essentially comparative: in general, one can only say that the theory A has a higher (or lower) degree of corroboration than a competing theory B, in the light of the critical discussion, which includes testing, *up to some time t*. Being a report of past performance only, it has to do with a situation which may lead to preferring some theories to others. *But it says nothing whatever about future performance, or about the "reliability" of a theory* (OK, p. 18, italics in original).

It may perhaps be useful to add here a point about the degree of corroboration of a statement *s* which belongs to a theory *T*, or follows logically from it, but is logically much weaker than the theory *T*.

Such a statement *s* will have less informative content than the theory *T*. This means that *s*, and the deductive system *S* of all those statements which follow from *s*, will be less testable and less corroborable than *T*. But if *T* has been well tested, then we can say that its high degree of corroboration applies to all the statements which are entailed by it, and therefore to *s* and *S*, even though *s*, because of its low corroborability, could never on its own, attain as high a degree of corroboration.

This rule may be supported by the simple consideration that the degree of corroboration is a means of stating *preference with respect to truth*. But if we prefer *T* with respect to its claim to truth, then we have to prefer with it all its consequences, since if *T* is true, so must be all its consequences, even though they can be less well tested separately.

Thus I assert that with the corroboration of Newton's theory, and the description of the earth as a rotating planet, the degree of corroboration of the statement *s* "The sun rises in Rome once in every 24 hours" has greatly increased. For on its own, *s* is not very well testable; but Newton's theory, and the theory of the rotation of the earth are well testable. And if these are true, *s* will be true also.

A statement *s* which is derivable from a well tested theory *T* will, *so far as it is regarded as part of T*, have the degree of corroboration of *T* …. (OK, pp. 19–20, italics in original)

Consider the statement *s*, "The sun rises in Rome once in every 24 hours," and its rival statement *u*, "The sun rises in Rome once *biennially*." It would seem that, according to Popper's stated criteria, the augmented Newtonian theory T which entails *s* now has a very much higher degree of corroboration than some equally well-tested rival theory H which entails *u*. Given a particular time *t* at which we assess the past performance of the theories T and H, the statements *s* and *u* pertain to the *future* of *t* no less than to its past. Thus both *s* and *u* have entailments involving times which are in the future of *t*. Therefore, we can say that, if we identify the time *t* as "now," *s* and *u*, and hence T and H, *respectively*, entail the following two rival future-tensed statements *p* and *r*, made at time *t*: "In the future, the sun will rise in Rome once every 24 hours," and "In the future, the sun will rise in Rome once biennially." Since Popper holds that the relative degree of corroboration is preserved under logical entailment, the future-tensed statement *p* "inherits" a very much higher degree of corroboration from T than the likewise future-tensed statement *r* inherits from H. Thus, while alike being about the future, the incompatible statements *p* and *r* differ greatly in degree of corroboration and *therefore* in preferability with respect to truth (empirical verisimilitude). Yet Popper told us previously that relative degree of corroboration and hence difference in empirical verisimilitude *"says nothing whatever about future performance, or about the 'reliability' of a theory"* (OK, p. 18, italics in original).

How then are we to understand as a *compatible* trio of statements the following claims made by Popper if we are mindful of the important fact that many theoretical law statements are *universally quantified with respect to the time?*

(i)   "Obviously, the degree of corroboration at the time *t* (which is a statement about preferability at the time *t*) says nothing about the future .... It is just a report about the state of discussion at the time *t*, concerning the logical and empirical preferability of the competing theories" (OK, p. 19).

(ii) The relative degree of corroboration of a hypothesis is preserved under logical entailment.

(iii) "The degree of corroboration is a means of stating *preference with respect to truth*" (OK, p. 20, italics in original).

It would seem to follow on the strength of (ii) and (iii) that for the case of two rival theories, both of which do have logical entailments with respect to the future but possess very different degrees of corroboration, there can be *future*-tensed entailments, such as our statements *p* and *r*, which differ greatly in corroboration *and* hence in truth preferability or empirical verisimilitude. If that is so, *what does it mean* to say with Popper that, because of their relatively different degrees of corroboration, we prefer the future-tensed statement *p* to the future-tensed statement *r with respect to truth* while *also* saying the following: Despite their differing strengths as candidates for being true, according to (i) there is complete parity of "reliability" between *p* and *r* in the sense that for each one alike, its degree of corroboration at the time *t* "says nothing about the future"? Can Popper tell us how his claims (i), (ii), and (iii) can all be both meaningful and compatible? I cannot see how, in the stated context, they can be meaningful without an incompatibility between (i) on one hand and the conjunction of (ii) and (iii) on the other.

Let us suppose that, for the sake of securing compatibility, Popper were willing to weaken the very strong denial (i) in order to preserve his much more justified claims (ii) and (iii). Vis-à-vis Hume's challenge, Popper's epistemic position would then become very similar to that of the (Bayesian) inductivist. Indeed, in the context of law statements, each of which pertains to all times but differs in corroboration, Popper's criteria (ii) and (iii) do seem to commit him to some form of inductivism in different terminological garb. For in the stated context, these criteria require that some *future-tensed* statements are truth preferable to others or have greater Popperian empirical verisimilitude—"reliability"—than others.

3.5. Duhem Versus Popper: Epistemological Comparison of Refutations and Corroborations

Shortly after the 1959 English publication of LSD, I noted that Popper had misportrayed Duhem's position in that book.[34] There Popper wrote (LSD, p. 78, fn. 1): "Duhem denies (English translation p. 300) the possibility of crucial experiments, because he thinks of them as verifications, while I assert the possibility of crucial *falsifying* experiments." Thus, here Popper erroneously depicts Duhem as *allowing* that, *for the kind of hypothesis typically at issue in science*, there is an *asymmetry of fallibility* between its verification and its falsification such that falsifications have logical credentials which verifications fail to possess. Having made this imputation in LSD, Popper asserted the compatibility of Duhem's position with his own championship of falsifications as preeminently logically reliable vis-à-vis verifications as tests of a hypothesis (theory).

Before the 1962 publication of his C & R, Popper had been made aware of the need to correct his exegesis of Duhem. Thus, in 1962, Popper acknowledged Duhem's espousal of symmetry of fallibility, saying (C & R, p. 112, fn. 26): "Duhem, in his famous criticism of crucial experiments (in his *Aim and Structure of Physical Theory*), succeeds in showing that crucial experiments can never *establish* a theory. He fails to show that they cannot *refute* it."

How are we to construe Popper's concurrence with Duhem in regard to verifiability and his avowal of disagreement with him in regard to falsifiability? Surely Popper is *not* charging Duhem with unawareness that *modus tollens* is a deductively valid form of inference while commending him for appreciating the deductive fallaciousness of affirming the consequent! If someone were to insist on documentation that Duhem is cognizant of the validity of *modus tollens* no less than the rest of us, we would need only to point to his statement that "when the experiment is in disagreement with his [the physicist's] predictions, what he learns is that at least one of the hypotheses constituting this group [i.e., the comprehensive theory being tested experimentally as a whole] is unacceptable and ought to be modified."[35] Thus, it is clearly agreed on all sides that there is an asymmetry of deductive validity between affirming the consequent and *modus tollens*. I shall speak of this asymmetry as "the deductive asymmetry."

How then are we to interpret Popper's 1962 divergence from Duhem? The 1962 statement of disagreement with Duhem which we quoted previously from Popper is appended as a footnote to the following much fuller text in which Popper declares (C & R, p. 112):

> But while Bacon believed that a crucial experiment may establish or verify a theory, we shall have to say that it can at most refute or falsify a theory. [It is here where Popper states his previously quoted disagreement with Duhem in the footnote.] It is an attempt to refute it; and if it does not succeed in refuting the theory in question—if, rather, the theory is successful with its unexpected prediction—then we can say that it is corroborated by the experiment ....
>
> Against the view here developed one might be tempted to object (following Duhem) that in every test it is not only the theory under investigation which is involved, but also the whole system of our theories and assumptions—in fact, more or less the whole of our knowledge—so that we can never be certain which of all these assumptions is refuted. But this criticism overlooks the fact that if we take each of the two theories (between which the crucial experiment is to decide) *together* with all this background knowledge, as indeed we must, then we decide between two systems which differ *only*

over the two theories which are at stake. It further overlooks the fact that we do not assert the refutation of the theory as such, but of the theory *together* with that background knowledge; parts of which, if other crucial experiments can be designed, may indeed one day be rejected as responsible for the failure. (Thus we may even characterize a *theory under investigation* as that part of a vast system for which we have, if vaguely, an alternative in mind, and for which we try to design crucial tests.)

Here Popper alleges that Duhem has overlooked two points. To state and evaluate Popper's rebuttal, suppose that a theory T_1 composed of a major hypothesis H_1 and an auxiliary A_1 entails a consequence C_1, while the conjunction $H_2 A_2$ constituting a *rival* theory T_2 entails a consequence C_2 *incompatible* with C_1. And suppose further that a purportedly crucial experiment yields evidence which is taken to be favorable to the truth of C_1 but adverse to the truth of C_2. Popper has us consider two such comprehensive theoretical systems T_1 and T_2, which, respectively, yield correct and incorrect predictions and whose respective auxiliaries A_1 and A_2 are identical. Then his first point is that the common auxiliary A will not be at issue between T_1 and T_2 and that the latter systems differ only with respect to the disputed component hypotheses H_1 and H_2. I cannot see that Duhem "overlooked" Popper's first point. For the latter is unavailing against Duhem's important claims that (i) the experimental falsification of T_2 via *modus tollens* does *not* have the logical force of a like falsification of H_2 itself, yet it is the pair H_2 and H_1 rather than T_2 and T_1 which is at stake in the given context of scientific controversy; and (ii) there is *symmetry* of fallibility between the falsification of H_2 itself and the verification of H_1.

Plainly, Duhem did not deny that *if* the (common) auxiliary A is *not* at issue *and* is presumed to be true, then the false prediction made by T_2 will permit the valid deduction of the falsity of H_2 via *modus tollens*. Instead, Duhem maintained that precisely because we cannot have any guarantee of the truth of A, we cannot construe the experimental decision against T_2 and in favor of T_1 as logically tantamount to a crucial refutation of H_2 itself. Far from being gainsaid by Popper's first point, the soundness of this Duhemian contention is explicitly conceded and even endorsed in the second point that Popper goes on to make. For there Popper tells us that "parts" of the background knowledge A "may indeed one day be rejected as responsible for the [experimental] failure" of T_2.

As for Popper's second point, Duhem recognized explicitly, as we saw, that under the posited conditions, at least one of the component hypotheses H_2 and A of the comprehensive theory T_2 "is unacceptable and ought to be modified." Hence, Duhem has no quarrel with Popper, if Popper is indeed content to assert the crucial falsifiability or refutability of the total system T_2 vis-à-vis T_1, in contradistinction to claiming this kind of refutability for such component hypotheses as H_2, which are typically the foci of scientific controversy. Oddly enough, Popper does come close to contenting himself with only *global* falsifiability of a "*whole system*" (LSD, p. 76) without mentioning Duhem at this point. Yet only two pages later (LSD, p. 78, fn. 1), he refers to Duhem and says, "I assert the possibility of crucial *falsifying* experiments," where he presumably means that *component* hypotheses are crucially falsifiable.

■ 4. CONCLUSION

In the light of the several groups of considerations presented in this paper, how can Popper be warranted in maintaining that falsifiability is the touchstone of scientific rationality, to the exclusion of inductive supportability? And how can Popperians justify adhering to Popper's indictment of inductivism while maintaining that he has given us a viable epistemological alternative on genuinely deductivist foundations?

■ NOTES

1. Lakatos, "The Role of Crucial Experiments in Science," *Studies in History and Philosophy of Science* 4 (1974), 310; hereafter this paper will be cited within the text as RCES.

2. Popper, K. R., *Conjectures and Refutations*, 1962, p. 256. Hereafter this work will be cited as C & R within the text.

3. Grünbaum, A., "Can a Theory Answer More Questions Than One of Its Rivals?" *British Journal for the Philosophy of Science* 27 (1976), 1–24; and the author's subsequent articles in that volume in the June and December issues.

4. Cf. Russell, B., *Human Knowledge*, 1948, p. 381.

5. Popper, K. R., *The Logic of Scientific Discovery*, 1959, p. 420 (italics in original). Hereafter this work will be cited within the text as LSD.

6. According to Lakatos [RCES, p. 315 and "Popper on Demarcation and Induction," (hereafter PDI), in: P. A. Schilpp (ed.), *The Philosophy of Karl Popper*, 1974, Book I, pp. 245–246], Popper *tailored* his demarcation criterion to the requirement of *not* according scientific status to the aforementioned four theories.

7. Burtt, E. A. (ed.), *The English Philosophers From Bacon to Mill*, 1939, p. 36. Any quotations from Bacon are taken from Burtt's edition. I am grateful to Ernan McMullin for calling my attention to Aphorism 95 in Book I of Bacon's *Novum Organum*, where Bacon seems to stress the sterility of such fact collecting as is *not* guided by theory. We shall recall this view of Bacon's when noting later that it runs counter to Popper's portrayal of Bacon as echoed by Lakatos.

8. Joseph, H. W. B., *An Introduction to Logic*, 2nd rev. ed., 1916.

9. Ibid., p. 393.

10. Ibid; cf. also p. 565, n. 1.

11. Cf. the illuminating discussion in R. Giere, "An Orthodox Statistical Resolution of the Paradox of Confirmation," *Philosophy of Science* 37 (1970), 354–362, of the resolution of Hempel's paradox of confirmation in statistical theory by means of *rejecting* a certain version of the instantiation condition. But note the *caveat* in fn. 31 concerning the difference between my concept of "positive instance" and the corresponding concept relevant to Giere's analysis.

12. In Section 3 we shall consider to what extent, if any, Popper's own methodology entitles him to claim that *mere* repetitions of an initially corroborating type of instance increase the corroboration of a hypothesis only very little, if at all.

13. Mill, *A System of Logic*, 8th edition, 1887, p. 313. Thus, in §2 of chapter X, Mill explains how his inductive methods can invalidate actually false claims of a plurality of causes. But in §§1 and 3 of ch. X, he contends that there are genuine cases of multiple causation. And in §3, he maintains that his inductive methods can handle the latter as well.

14. Eysenck, H. J., *The Effects of Psychotherapy*, 1966; and "The Effects of Psychotherapy," *International Journal of Psychiatry* 1 (1965), 97–178.

15. Rachman, S., *The Effects of Psychotherapy*, 1971.

16. Cf. Bergin, A. E., "The Evaluation of Therapeutic Outcomes," in: A. E. Bergin and S. L. Garfield (eds.), *Handbook of Psychotherapy and Behavior Change*, 1970, pp. 217–270; see also the important additional references given on pp. 217–218. H. J. Eysenck and G. D. Wilson, *The Experimental Study of Freudian Theories*, 1973, pp. 378–379, point out that, whereas Freud's theory emphatically denies the existence of spontaneous remission, the latter is one of the best attested findings of psychiatry.

17. Fromm, E., *The Crisis of Psychoanalysis*, 1970, pp. 3–4. Emanuel Peterfreund's discussion of "The Nature of Therapeutic Changes in Psychoanalysis," *Information Systems, and Psychoanalysis*, 1971, pp. 351–358, does not improve on Fromm, despite Peterfreund's strongly *revisionist*, learning-theoretic stance toward Freud.

18. Bergin, A. E., op. cit., pp. 246–247. For a criticism of this claim, see Rachman, S., op. cit., ch. 3 and p. 16.

19. Meltzoff, J. and Kornreich, M., *Research in Psychotherapy*, 1970, p. 200. See pp. 258, 113–114, 190 for placebo effects.

20. Fromm, E., op. cit., p. 4.

21. Howson, C., "Must the Logical Probability of Laws Be Zero?" *British Journal for the Philosophy of Science* 24 (1973), 153–160. See also Hintikka, J., "Carnap and Essler Versus Inductive Generalization," *Erkenntnis* 9 (1975), 235–244.

22. I am indebted to Noretta Koertge for some of these Popper references.

23. Popper, *Objective Knowledge*, 1973. p. 51. Hereafter this work will be cited as OK.

24. Popper, "A Theorem on Truth-Content," in: P. Feyerabend and G. Maxwell (eds.), *Mind, Matter and Method*, 1966, Theorem 1, p. 350. For a discussion of relevant details, see Section 2 (i) of the first of my articles cited in note 3 and p. 134 of the *second* of these articles.

25. That D is indeed equivalent to C becomes intuitive by reference to Tarskian logical contents as follows: The logical content of a disjunction is the *intersection* of the respective contents of the disjuncts. And the intersection of the contents of the infinitude of disjuncts in D will be just the content of C. To illustrate, take the *entire* infinitude of pairwise different curves, all of which go through or contain each of a finite set S of points in the x–y plane. This totality T of curves or point sets will have exactly the set S as its set-theoretical intersection. For no point outside of S can belong to *all* of the curves in T. The points in S play the role of "data points" in analogy to the observation statement C. And the curves in T play the role of the hypotheses in D, each of which entails C.

26. Salmon, W. C., *The Foundations of Scientific Inference*, 1966, p. 119.

27. Musgrave, A., "Popper and 'Diminishing Returns from Repeated Tests,'" *Australasian Journal of Philosophy* 53 (1975), 250.

28. Ibid., p. 251.

29. In the paper by Howson cited in note 21, he has offered a counterexample on pp. 161–162 to the claim that p (a) = 0 for every universal a. See also Hintikka's paper in note 21.

30. Harper, H. W., "Rational Belief Change, Popper Functions and Counterfactuals," *Synthèse* 30 (1975), 221.

31. Good, I. J., "The White Shoe Is a Red Herring," *British Journal for the Philosophy of Science* 17(4) (1967), 322, has given a perhaps far-fetched example having the following features: C states that a randomly selected bird is a black raven, H states that "all ravens are black," and P(B & H, C)<1 (i.e., $10^2/10^6$). Good takes a black raven to be a "case" (positive instance) of H *in a sense different from our aforementioned sense*, since his B does *not* assume the initial condition that the randomly selected bird is a raven! And despite the universality of H, the special feature of Good's example then is that not only is P(B & H, C)<1 but also P(B & H, C) < P(B, C), so that the ratio on the RHS is *less than* 1. But this means that the "case" C of the hypothesis yields a *posterior* probability of H which is *smaller* than its prior probability: a perhaps somewhat far-fetched but even more resounding repudiation of the instantiation condition than the case of *equal* prior and posterior probabilities of H. I am indebted to Wesley Salmon and William Harper for helpful comments relating to the violation of the instantiation condition in Bayesian inference and I thank Laurens Laudan for stimulating criticisms of Bayesian inference.

32. The need to distinguish the *factor* of probability increase from the *amount* was overlooked in this context in Salmon's *The Foundations of Scientific Inference*, op. cit., pp. 118–120.

33. Hilpinnen, R., "On the Information Provided by Observations," in: J. Hintikka and P. Suppes (eds.), *Information and Inference*, 1970, Section II, esp. pp. 100–101. I am indebted to Teddy Seidenfeld not only for this reference but also for very clarifying comments on Allan Gibbard's results, which I already mentioned.

34. Grünbaum, "The Duhemian Argument," *Philosophy of Science* 27 (1960), 76, fn. 1. Cf. also Grünbaum, *Philosophical Problems of Space and Time*, 1963, p. 109, n. 4. The second enlarged edition of the latter work (Dordrecht: Reidel, 1973) has the same comment in the same place.

35. Duhem, *The Aim and Structure of Physical Theory*, 1954, p. 187.

2 The Degeneration of Popper's Theory of Demarcation

1. INTRODUCTION

Karl Popper, on one hand, and the descriptive or hermeneutic phenomenologists, on the other, have offered radically different, influential diagnoses of the failure of psychoanalytic theory to pass scientific muster. As Popper would have it, "Freud's theory ... simply does not have potential falsifiers"[1] and is therefore nonscientific. But, in his view, the time-honored inductivist conception of scientific rationality was unable to detect this fundamental flaw. And, thereby, it purportedly failed to give a correct diagnosis of the scientific bankruptcy of the psychoanalytic enterprise.[2] Therefore, Popper concluded: "Thus there clearly was a need for a different criterion of demarcation"[3] between science and pseudo-science, other than the inductivist one. In this way, psychoanalysis served as the gravamen and benchmark of his case for the superiority of his own falsifiability criterion of demarcation.

In my *Foundations of Psychoanalysis* (1984) I argued for the following contrary thesis:

> It is ironic that Popper should have pointed to psychoanalytic theory as a prime illustration of his thesis that inductively countenanced confirmations can easily be found for nearly every theory, if we look for them ... *it is precisely Freud's theory that furnishes poignant evidence that Popper has caricatured the inductivist tradition by his thesis of easy inductive confirmability of nearly every theory!*[4]

But, as he tells us in his *Realism and the Aim of Science* (1983), its Chapter 2 contains his first "published ... *detailed* analysis of Freud's method of dealing with falsifying instances and critical suggestions."[5] When I was completing my *Foundations*, Popper's 1983 volume was not yet available to me. And in my later reply to him,[6] I made only a cursory critical remark about his account of Freud's scientific miscarriage in his *Realism and the Aim of Science*. Here, I shall therefore deal with his most recent and fullest treatment of this issue in adequate detail. And in Section 4, I shall deal with two other flaws in Popper's theory of demarcation that are unrelated to psychoanalysis.

According to Karl Jaspers, Paul Ricoeur, and Jürgen Habermas, Freud misunderstood his own theory "scientistically" by giving far too little weight to "meaning connections" between mental states, as distinct from causal connections. In their view, hermeneutic victory can be snatched from the jaws of Freud's scientific defeat, once we appreciate that the hermeneutic discernment of so-called meaning connections is at the heart of the psychoanalytic enterprise.[7] Elsewhere, I have argued that this criticism basically misidentifies both the source and the import of Freud's scientific shortcomings: far from giving too little explanatory weight to "meaning connections" (thematic affinities), Freud endowed them with much too much explanatory significance, drawing fallacious causal inferences from them.[8]

Therefore, I can confine the present paper to Popper's *Realism and the Aim of Science*. Frank Cioffi has published two ill-tempered replies to my earlier criticisms of Popper's charge of pseudo-science against psychoanalysis.[9] It is a mark of Cioffi's mode of disputation that in the latter of 1988, he makes no mention at all of my "Author's Response," of 1986, where I painstakingly dealt with his objections to my *Foundations*. There, I also documented his exegetical fabrications and straw men.[10] Edward Erwin has given a careful defense of my views against Cioffi's.* And I can now refer the reader to my "The Role of the Case Study Method in the Foundations of Psychoanalysis" (1988) for my most detailed treatment of Freud's *Rat Man* case, which undercuts Cioffi's objections to my handling of that case history.[11]

■ 2. POPPER'S 1983 MISDIAGNOSIS OF FREUD'S SCIENTIFIC FIASCO

Let me begin with a retrospect on Popper's treatment of the topic prior to the appearance of the Postscript of his *Realism and the Aim of Science*.

Throughout his career, Popper has repeatedly made two claims: (1) Logically, psychoanalytic theory is irrefutable by any human behavior; and (2) in the face of seemingly adverse evidence, Freud and his followers always dodged refutation by resorting to immunizing maneuvers. According to (1), *none* of the deductive consequences of Freud's hypotheses are refutable by potentially contrary empirical evidence. But clearly, this charge of unfalsifiability against psychoanalytic theory *itself* does not follow from the *sociological* objection that Freudians are not responsive to criticism of their hypotheses. After all, a theory may well be invalidated by known evidence, even as its true believers refuse to acknowledge this refutation. Besides, if Popper were right that "Freud's theory [...] simply does not have potential falsifiers," why would it have been necessary at all for Freudians to *dodge* refutations by means of immunizing gambits? Popper's (1) and (2) seem incoherent.

Ironically, it emerges clearly from some of Popper's other doctrines that the recalcitrance of Freudians in the face of falsifying evidence, however scandalous, is not at all tantamount to the irrefutability of their theory. As he tells us, theories, on one hand, and the intellectual conduct of their protagonists, on the other, "belong to *two entirely different* 'worlds.'"[12] Yet because Popper sometimes discusses them in the same breath, my response to his views on psychoanalysis takes both into account. My principal objection to him pertains to his logical thesis of empirical

---

* *Editor's Note*: The reference originally given hinted at "forthcoming." See now Edward Erwin, "Philosophers on Freudianism: An Examination of Replies to Grünbaum's *Foundations*," in: John Earman et al. (eds.), *Philosophical Problems of the Internal and External Worlds: Essays on the Philosophy of Adolf Grünbaum*, Pittsburgh: University of Pittsburgh Press, 1993, ch. 16.

irrefutability, although I argued in *Foundations* that he also used gross oversimplification to make his case for his sociological objection to Freud's evasive methodological behavior.

Indeed, contrary to Popper, Freud unflinchingly issued some important theoretical retractions in response to acknowledging the emergence of refuting evidence. A telling case in point is furnished by his 1926 landmark revision of his prior views on the causal relation between repression and anxiety. There Freud wrote:

> It was anxiety which produced repression and not, as I formerly believed, repression which produced anxiety.
>
> It is no use denying the fact, though it is not pleasant to recall it, that I have on many occasions asserted that in repression the instinctual representative is distorted, displaced, and so on, while the libido belonging to the instinctual impulse is transformed into anxiety [footnote omitted]. But now an examination of phobias, which should be best able to provide confirmatory evidence, fails to bear out my assertion; it seems, rather, to contradict it directly.... It is always the ego's attitude of anxiety which is the primary thing and which sets repression going. Anxiety never arises from repressed libido.... I believed I had put my finger on a metapsychological process of direct transformation of libido into anxiety. I can no longer maintain this view.[13]

In the Schilpp volume on Popper's philosophy, Popper had claimed—once again—that psychoanalysis is an empirically untestable psychological metaphysics, which does "not exclude any physically possible human behavior."[14] And from this allegation of empirical unfalsifiability, he immediately drew the fallacious inference that psychoanalysis indeed can, in principle, *explain* any actual behavior. Thus, on the heels of saying that Freud's and Adler's theories do not exclude any possible human behavior, Popper tells us that "whatever anybody may do is, in principle, explicable in Freudian or Adlerian terms."[15]

But if a theory, in conjunction with particular initial conditions, *does not exclude* any behavior at all, how can it deductively *explain* any *particular* behavior? To explain deductively is to exclude: as Spinoza emphasized, to assert (entail) p is to deny every proposition incompatible with it. Note that in psychoanalytic theory, just as in Newton's physics, for example, law-like statements cannot explain particular behavior without initial conditions: without *suitable* initial velocity specifications, Newton's laws of motion and gravitation do not yield an elliptical orbit for the earth under the gravitational action of the sun. Hence, analogously, if no potential behavior could falsify psychoanalysis under given initial conditions I, then this theory, cum I, could not explain any actual behavior deductively. *A fortiori*, if the theory T were unfalsifiable, it could not explain *all* such behavior, as Popper believes. Furthermore, if the conjunction T and I fails to explain some particular behavior b deductively, then I and b cannot confirm (support) T hypothetico-deductively. Thus, if psychoanalysis were unfalsifiable, how could any actual behavior—let alone *all* physically possible behavior—be explained by it so as to confirm it inductively, as Popper claims in his *Conjectures and Refutations*?[16] On the contrary, the alleged unfalsifiability would *preclude* such hypothetico-deductive confirmability.

According to psychoanalytic theory, in both genders, repressed homosexual desires are the causal *sine qua non* of paranoia, whatever the variations in the delusional modes of this psychosis. The modes include delusions of persecution, of jealousy, of grandeur, and of heterosexual erotomania.[17] Thus, every sort of paranoiac is an arena for the psychic conflict between homosexual impulses, and the need to banish them from consciousness as objectionable to avoid emotional distress, such as disgust, shame, or guilt.[18] In brief, his or her delusions (of persecution, etc.) are the *unconscious defense* against the conscious emergence of the forbidden homosexual cravings. And this defense is mediated by two defense mechanisms: (1) Reaction-formation, which converts the homosexual love into hatred toward the love-object, i.e., once a dangerous impulse has been largely repressed, it surfaces in the guise of a more acceptable contrary feeling; (2) projection of the ensuing hatred from the lover by imputation to the beloved.

In 1915, Freud published a paper entitled "A Case of Paranoia Running Counter to the Psychoanalytic Theory of the Disease."[19] The pertinent case is that of a woman suffering from persecutory delusions. And, as suggested by the very title of this paper, Freud saw her case history as at least potentially supplying a *refuting instance* or disconfirmation of the particular sexual etiology he had posited for this patient's type of psychosis. In this vein, he notes that the patient's conduct might well supply evidence that she is *not* beset by a conflictual struggle against unconscious homosexual impulses.[20] To take a contemporary putative example, imagine a self-declared lesbian, who is also paranoid and living in San Francisco as the avowed lover of another woman. Suppose also that she is publicly active in the gay rights movement there. Freud emphasizes that if there are indeed empirical indications of such freedom from repressed homosexuality on the part of a paranoiac, then the person in question would *count against* the psychoanalytic etiology of paranoia. The posited case would be a contrary instance, because this etiology declares intense repressed homosexuality to be a sine qua non for paranoia.

As Freud points out furthermore, the hypothesized etiologic role of repressed homosexuality leads "to the necessary conclusion that the persecutor must be of the same sex as the person persecuted."[21] But, if so, it becomes important to determine empirically to what extent the supposed persecutors are actually of the *same* gender as the persecutees. Since the delusions of persecution are themselves conscious, this empirical determination of gender does not require the techniques of psychoanalytic investigation as such, and is therefore extraclinical in that sense. It is, of course, logically possible that any and every paranoiac feels persecuted *only* by members of the *opposite* sex. For this reason alone, Freud's etiology is falsifiable by a finite number of such instances in the face of Popper's own denial of such falsifiability!

What then are the known findings among paranoiacs as to identity or difference of gender? And does Freud conclude that his etiology of paranoia would simply be refuted by fancied persecutors of the *opposite* sex?

His answer to the first question is unequivocal: "In psychiatric literature there is certainly no lack of cases in which the patient imagines himself persecuted by a person of the opposite sex."[22] Indeed, the paranoid young woman of his 1915 case history sought out a lawyer for protection against the imagined persecutions by a young man with whom she had been having an affair. As Freud admits: In light of his homosexual etiology of paranoia, "it … seems strange that a woman should protect herself against having a man by means of a paranoic delusion."[23] How then, if at all, does Freud propose to reconcile such cases with the requirement of gender identity that he himself had deduced from his homosexual etiology of paranoia? His reply takes the form of two claims that are, in principle, empirically testable. First, he reports "as a rule we find [in paranoia] that the victim of persecution remains fixated … to the same sex."[24] Furthermore, he maintains, wherever a supposed persecutor is of the *opposite sex,* then investigation will reveal *this* antagonist to be only a secondary one, whom the paranoiac imagines to be in collusion with an *original primary* persecutor of the *same* sex.

But are there statistics bearing out the first claim that the great majority of paranoiacs believe themselves to be persecuted by members of the *same* sex? Even if there are such statistics, neither Freud nor his followers have provided evidence to support his crucial further thesis that, for any presumed persecutor of the opposite sex, there demonstrably exists a person of the same sex who was the *original* object of the paranoiac's persecutory delusions. Indeed, to the extent that, in the case of oppositely sexed persecutors, psychoanalysts *fail* to find the primary, gender-identical persecutor required by their theory, their etiology of paranoia is *disconfirmed.*

There are further reasons for concluding that the psychoanalytic etiology of paranoia may actually be disconfirmed. As Fisher and Greenberg have pointed out, "The appearance of overt homosexual … acting out in the paranoid would represent a contradiction … of Freud's theory of paranoia."[25] How so? That theory makes the following conditional prediction: If a person does not strongly repress homosexual impulses, then this person will *not* be paranoid. But any individual who is openly and publicly a highly active practicing homosexual surely qualifies as a person who does *not* strongly repress his or her homosexuality. And if that same individual also exhibits strong delusions of persecution toward, say, various members of the opposite sex, then this flamboyant homosexual gives a strong observable indication of being paranoid. Yet, according to the prediction derived from Freud's etiology, just such an openly active homosexual should *not* be paranoid.

It should not be unduly difficult for extraclinical researchers to sample the population of avowedly practicing homosexuals of either sex to determine whether there are any paranoiacs among them. In fact, presumably it is not a very risky bet to suppose that there are. And, if so, their existence casts strong doubt on the psychoanalytic claim that all paranoiacs are *repressed* homosexuals. Yet Harold S. Zamansky believes to have produced experimental support for it,[26] which I find dubious.[27]

In sum, the psychoanalytic etiology of paranoia is empirically falsifiable or disconfirmable, and Freud explicitly said so in his 1915 paper. As he noted, empirical indicators can bespeak the absence of strongly repressed homosexuality as well as the presence of paranoid delusions.[28] Thus, empirical indicators can fallibly discredit the stated etiology.

Hence, this example has an important *general* moral: Whenever empirical indicators can warrant the *absence* of a certain theoretical pathogen *P* as well as a *differential diagnosis* of the *presence* of a certain theoretical disorder *D*, then an etiologic hypothesis of the strong form "*P* is causally necessary for *D*" is *fallibly* refutable. The etiology will be falsified or disconfirmed by any victim of *D* who has *not* been subjected to the avowedly required *P*. For the hypothesis *predicts* that anyone not so subjected will be spared the miseries of *D*, a prediction having significant prophylactic import for child rearing. Equivalently, the etiologic hypothesis *retrodicts* that any instance of *D* was also a case of *P*. Thus, Clark Glymour's account of Freud's famous case history of the "Rat Man" points out that Freud's specific etiology of the Rat Man's obsession with his father's death was fallibly falsified by means of disconfirming the retrodiction that Freud had based on it.[29] (Later, I elaborated on the several auxiliary hypotheses on which this falsification was predicated.[30])

Popper undertook to write on the falsifiability of psychoanalysis as a scientific theory. Yet he was apparently unaware of Freud's 1915 paper on paranoia, whose mere title would have alerted him to the testability of the Freudian theory of psychopathology, or would at least have made him aware of the fact that Freud had entertained its falsifiability. Indeed, it is a measure of the inadequacy of Popper's psychoanalytic scholarship that, by his own account, he was first made cognizant of such testability by William Bartley. As we learn from an insert in Popper's *Realism and the Aim of Science*,[31] in 1980, Bartley pointed out to him that the existence of a homosexually active paranoiac is *ruled out* by Freudian theory, because it demands that the homosexuality of paranoiacs be strongly repressed rather than active. But Fisher and Greenberg had already emphasized this prohibition, as we saw.

It so happens that Bartley gave this example to Popper during the discussion of my own paper at the July 1980 London Symposium on Popper's Philosophy. As I had argued in that paper, Freud's theory implies that the decline in the social taboo on homosexuality in communities like San Francisco should, in due course, issue in a decreased incidence of paranoia in both sexes. This quasi-statistical prediction is quite compatible with Freud's allowance for a "multiplicity of mechanisms of repression proper"[32] which include autochthonous developmental factors that are independent of exogenous social taboos. Hence, I contended that Freud's etiology of paranoia would be disconfirmed epidemiologically by the failure of the quasi-statistical prediction to materialize.

Let us now see how Popper reacted to Bartley's example. In context, the relevant passage in Popper's Postscript to his *Realism and the Aim of Science* reads as follows:

> I cannot think of any conceivable instance of human behaviour which might not be interpreted in terms of either [Freud's or Adler's] theory, and which might not be claimed, by either theory, as a "verification."[33]

The insert I mentioned then reads:

> ** Added 1980 [by Popper]. The last sentence of the preceding paragraph is, I now believe, too strong. As Bartley has pointed out to me, there are certain kinds of possible behaviour which are incompatible with Freudian theory—that is, which are excluded by Freudian theory. Thus Freud's explanation of paranoia in terms of repressed homosexuality would seem to exclude the possibility of active homosexuality in a paranoid individual. But this is not part of the basic theory I was criticizing. Besides, Freud could say of any apparently paranoid active homosexual that he is not really paranoid, or not fully active.*

This retort strikes me as unfortunate for the following reasons:

1. To justify Popper's original thesis of nonfalsifiability, it was necessary to *show*—not just to assert peremptorily—that there simply are no potentially falsifying empirical instances in the case of psychoanalytic theory, as he claimed anew in his 1974 reply to Imre Lakatos.[34] In that reply, Popper asserts that Newtonian physics does have potential empirical falsifiers, which could be neutralized only by using immunizing strategies, such as modifying some auxiliary hypotheses of the Newtonian corpus. Thus, he admits that Newtonian physics can be immunized against falsification. By the same token, his assertion of its falsifiability presupposes "disregarding the possibility of immunizing stratagems." But in 1974 he then claimed a contrast between Newtonian theory and psychoanalysis as follows: "And this is the heart of the matter, for my criticism of Freud's theory was that it simply does not have potential falsifiers."[35] Furthermore, "I cannot describe any state of affairs concerning Mr. Smith—say about his behaviour—which would need immunization [i.e., which would require neutralization by immunization of the theory] in order not to clash with Freudian theory."[36] But despite Popper's own admitted inability to describe such a case, the documented ability of others to do so shows how wrong he was to have inflated his mere suspicion of the empirical irrefutability of psychoanalysis into a thesis of unfalsifiability. Plainly, his own reported inability to think of contrary instances was, at best, grounds for a mere suspicion of nonfalsifiability. Worse, Popper flatly contradicts his thesis of nonfalsifiability: (i) In the Postscript, he tells us in italics that "*anxiety dreams constitute a refutation of the general* [Freudian] *formula of wish-fulfillment*"[37]; and (ii) Popper assumes falsifiability when he complains that Freudians use immunizing gambits to neutralize contrary evidence.

2. It is important to note how Popper tries to practice damage control by qualifying and hedging his more recent admission that the case of a homosexually

active paranoiac does contradict his thesis. Thus, he immediately resorts to *two* immunizing maneuvers, which are disingenuous, I submit, besides being demonstrably unsuccessful, as we shall now see.

He admonishes us that Freud's repression etiology of paranoia "is not part of the basic theory I was criticizing." Yet we are left completely in the dark as to just what does count for him as the "basic" part of psychoanalytic theory, *and why*. In effect, Popper asks us to overlook that, throughout his career, he had leveled the charge of nonfalsifiability *tout court* rather than only against a so-called basic part of the theory, whose identity he strangely fails to specify. Indeed, as we saw, he wrote: "My criticism of Freud's theory was that it simply does not have potential falsifiers."[38]

Vague as it is, even his claim that Freudian etiologic theory is *not* "basic" to psychoanalysis is simply untenable. As any student of the subject knows, both historically and logically, precisely the theory of psychopathology, which features the repression etiology of the neuroses, is the *most foundational* part of Freud's edifice. Thus, as we know, its architect had ample reason to declare in his *History of the Psychoanalytic Movement* (1914): "The theory of repression is the cornerstone on which the whole structure of psycho-analysis rests. It is the most essential part of it."[39]

Not content with being disingenuous, the kettle calls the teapot black when Popper writes against Freud:

> I wish to criticize Freud's way of rejecting criticism. Indeed I am convinced that Freud could have vastly improved his theory, had his attitude towards criticism been different
>
> ....
>
> This self-defensive attitude is of a piece with the attitude of looking for verifications; of finding them everywhere in abundance; of refusing to admit that certain cases do not fit the theory.[40]

In response, I find it hard not to exclaim: "Physician, heal thyself."

Moreover, Popper wants to take out an insurance policy in case his immunizing resort to the alleged non-"basicality" of the theory of paranoia is not seen by others as a vindication. So he comes up with the previously cited remark: "Besides, Freud could say of any apparently paranoid active homosexual that he is not *really* paranoid, or not *fully* active." In this way, Popper wants to claim that this kind of case does not discredit his lifelong thesis of irrefutability about psychoanalysis after all.

To this I say two things: (i) In regard to Freud's scientific honesty, the question is not whether he "*could*" save his etiology of paranoia in this way but only whether he *would* try to do so. And his 1915 paper shows explicitly that, in fact, he did not, although his epistemic reasoning there is problematic in other respects, as I have pointed out; and (ii) Popper's depiction of what Freud *could* say to neutralize the counterexample does not help to sustain Popper's thesis of irrefutability, as I shall now show.

Assume we grant that Freud could evade falsification by the case of the paranoid active homosexual in the far-fetched manner proposed by Popper. Then surely a like gambit is available to a physicist, mutatis mutandis, to neutralize unpalatable contrary evidence, so as to parry the refutation of his theoretical hypothesis, as I shall soon illustrate. Yet Popper claims falsifiability for physics, even though he explicitly *predicates* it on "disregarding the possibility of immunizing stratagems."[41] And what is sauce for the goose is, I submit, also sauce for the gander. Thus if, as Popper claims, the physical scientist does *not* forfeit the falsifiability of his hypotheses in the face of immunizing maneuvers that are surely available to that scientist, why then should Freud have to acknowledge irrefutability merely because he too could adopt such stratagems?

As Popper would have it, Freud "could" evade refutation by pleading that even the most promiscuous homosexual is not "fully active" after all. But Freud's etiology of paranoia had explicitly declared that *strongly repressed*, intense homosexuality is the sine qua non of paranoia.[42] And even an average homosexually active paranoiac violates this declared necessary condition. A fortiori, an erotomanically active paranoid homosexual clearly violates it.

Could Freud still say, as Popper suggests, that the active homosexual is "not *really* paranoid," no matter how delusional his behavior? To deal with this question, let me first recall a relevant moral from a paper by Rudolf Carnap.[43] There he acknowledged over fifty years ago that in physics, no less than in somatic medicine or psychology, empirical data are *not sufficient conditions* for the presence of hypothesized states, which they are taken to manifest. Instead of being sufficient conditions for the theoretical states, empirical indicators *underdetermine* such states and betoken their presence only more or less probabilistically. Thus, it is commonplace, even among laymen, that an x-ray or even the more refined tomogram or CT scan does not guarantee any one somatic diagnosis. Hence, if the repudiation of a psychiatric diagnosis of paranoia is an immunizing option open to Freud, then a like gambit of re-diagnosis is available to a radiologist, when expectations based on the original diagnosis fail to materialize.

By the same token, tracks in Wilson cloud chambers *may* actually bespeak the existence of as yet unknown particles rather than the passage of those postulated entities that physicists now (fallibly!) take them to attest. Carl Hempel has reminded us that even a circular array of iron filings does not guarantee the presence of a magnet deductively. And, as Wesley Salmon noted,[44] the same holds in psychoanalysis. True, even patent delusions of intense persecution are only probabilistic evidence of paranoia. But Popper's appeal to the fallibility of a differential diagnosis and to the possibility of re-diagnosis is misleading as a basis for singling out psychoanalysis as remaining irrefutable, even in the face of the example of a homosexually active paranoiac.

In sum, to the detriment of Popper's vaunted contrast between the testability of physical theory and psychoanalysis, the rescuing option that, according to him, is open to Freud, could likewise be exercised, mutatis mutandis, in physics so as to preclude refutation of *its* hypotheses. For example, if the measured charge of an electron turned out *not* to accord with theoretical expectations or requirements, a physicist *could* say—in Popper's sense of "could"—that the particle in question is "not *really* an electron." Besides, as I have illustrated,[45] when Freud did retract a differential diagnosis to preserve a specific causal hypothesis, he was sensitive to the need for independent evidence in favor of the alternative diagnosis.

Far from supporting his nonfalsifiability thesis, Popper's account of the case of the homosexual paranoiac evidently relies on just the kinds of immunizing maneuvers he is quick to reject when others engage in them. In any case, as we saw, his account seems incoherent: Only a few pages after discussing the paranoia example, he deals with Freud's dream theory and astonishingly declares it to be a "simple fact ... that *anxiety dreams constitute a refutation of the general formula of wish-fulfillment.*"[46] Thus, by Popper's own appraisal, the dream theory must be falsifiable, since it had already *been* falsified by anxiety dreams when Freud first proposed it! And this clear acknowledgment of falsifiability is hardly tempered in the *logical* sense by Popper's detailed complaint against Freud's own evidential disposal of *prima facie wish-contravening* dreams, such as anxiety dreams and nightmares. After all, on Popper's conception, theories are so-called third-world objects that have a logical life of their own, as it were, apart from the attitudinal scientific integrity of their propounders.*

As we know, Freud relied on the distinction between the manifest and latent content of a dream to claim that a latent wish engenders even those dreams whose manifest content is anything but wish fulfilling. And Popper's critical comment on this distinction is the following:

> Freud repeatedly re-affirms his programme of revealing the latent content of every anxiety dream as a wish-fulfillment.... *Yet Freud never carries out his programme; and in the end he gives it up altogether*—without, however, explicitly saying so.[47]
>
> The reason why Freud does not carry out his original programme of showing (by way of detailed analyses such as he is wont to give) that all anxiety dreams are wish-fulfillments is, clearly, that in the end he no longer believed in it.... I should be the last to criticize such a change of mind. But the change is not a conscious correction, or the admission of a mistake.[48]

---

* *Editor's Note*: Popper's three worlds scheme is composed of the following: (1) the world of physical objects and events, including biological, that is, organic entities; (2) the "mental or psychological world, the world of our feelings of pain and of pleasure, of our thoughts, of our decisions, of our perceptions and our observations"; and (3) "the world of the products of the human mind, such as languages; tales and stories and religious myths; scientific conjectures or theories, and mathematical constructions; songs and symphonies; paintings and sculptures. But also aeroplanes and airports and other feats of engineering." He developed this idea in detail in his *Tanner Lecture on Human Values*, titled "Three Worlds," delivered at the University of Michigan, April 7, 1978. On the topic see also J. C. Eccles, "The World of Objective Knowledge," in: P. A. Schilpp, *The Philosophy of Karl Popper*, LaSalle, IL: Open Court, 1974, pp. 349–370; and the summarized reply of Popper, ibid., pp. 1048–1080.

As we learn from Popper's Postscript,[49] the psychoanalytic theory of dreams in particular was historically a major inspiration of his falsifiability criterion of demarcation. Moreover, though Popper hails Freud's interpretation of dreams as "a great achievement," he nonetheless considers it to be his own center-piece for leveling the following reproach: *"But it is a fundamental mistake to believe that, because it is constantly being 'verified,' it must be a science, based on experience."*[50]

Yet, as we know from the old inductivist tradition of Francis Bacon and John Stuart Mill, Popper's admonition here just carries coals to Newcastle by echoing one of the salutary injunctions familiar from precisely that legacy of elimina-tive inductivism. Freud's wish-fulfillment theory of dream production is clearly a causal hypothesis. And I have argued on earlier occasions[51] that the demands made by that inductivist patrimony for validating such a strong hypothesis are clearly not met by the so-called verifications adduced by Freudians: As I recalled already from my *Foundations* (1984), *it is precisely Freud's theory that furnishes poignant evidence that Popper has caricatured the inductivist tradition by his the-sis of easy inductive confirmability of nearly every theory.* Hence, inductivistically untutored claims of confirmations made by psychoanalytic zealots cannot serve Popper's advocacy of his falsifiability criterion of demarcation as a replacement for the received Bacon-Mill inductivist criterion, which was en vogue when he came upon the philosophic scene.

Thus, malgré lui, the upshot of Popper's account of the psychoanalytic theory of dreams seems to be as follows: Since this part of Freud's corpus is known to be false, it is falsifiable. Hence, I submit, precisely because—qua "third-world" object—the dream theory itself is falsifiable—it is *scientific* by Popper's criterion of demarcation. This conclusion is, of course, compatible with Popper's distinct claim that Freud was aware of the falsity of the dream theory but was unwilling to admit it explicitly.

In the face of his proclamation of unfalsifiability, Popper has a highly incongru-ous attitude toward psychoanalytic explanations. If, as he says, psychoanalysis is indeed unfalsifiable, then it cannot make any risky empirical predictions, because such predictions run the risk of turning out to be false. Hence, if Popper is to be believed, psychoanalysis cannot be corroborated observationally by success-ful risky predictions, as demanded by his standards of corroboration. Yet, in his view, explanations based on a completely uncorroboratable theory are presumably *pseudo*-explanations. Indeed, if psychoanalysis does yield the surfeit of explana-tions deplored by him, then its explanations cannot be genuine, precisely because they explain too much, as it were.

Unabashed by such apparent corollaries of his account, Popper not only enlists psychoanalytic theory to provide several explanations but also even explicitly declares his belief in its essential substantive correctness. Let me document these very odd twists:

(1) In his *The Open Society and Its Enemies* (1950), Popper invokes Freud's unconscious defense mechanism of "reaction–formation" to "explain," as he claims, how Heraclitus embraced two *prima facie* contradictory ideas.[52] But if such Freudian hypotheses as the existence of conflict, actuated by unconscious fear and resistance, are untestable, as Popper thinks, how can they have enough *empirical import* to explain Heraclitus's utterances by Popper's standards of bona fide scientific explanations?

(2) In his *Conjectures and Refutations* (1962), Popper wishes to explain the adoption by some people of a "dogmatic attitude," as distinct from a "critical attitude."[53] In the quest for such an explanation, he registers "a [purported] point of agreement with psychoanalysis" when suggesting the following causal hypothesis: "Most neuroses may be due to a partially arrested development of the critical attitude ... [due] to resistance to demands for the modification and adjustment of certain ... responses. This instance in its turn may perhaps be explained, in some cases, as due to an injury or shock, resulting in fear." Yet, according to psychoanalytic theory, dogmatic rigidity may be seen as a manifestation of neurotic conflict, but hardly as a contributory *cause* of the neurosis itself. To this extent, Popper's purported "agreement with psychoanalysis" here seems spurious.

(3) Also in his *Conjectures and Refutations*,[54] Popper tries to make diagnostic and explanatory use of Freud's theory of the oedipal conflict, which postulates ambivalent feelings toward the father that include hostile, aggressive, and even murderous impulses. Thus, Popper declares: "One need not believe in the 'scientific' character of psycho-analysis (which, I think, is in a metaphysical phase) in order to diagnose the anti-metaphysical fervour of positivism as a form of Father-killing." But, on his account of psychoanalysis, how can the putative desire for father killing yield anything but an ad hoc *pseudo*-explanation of the positivist antipathy to metaphysics? What competing explanations, if any, did Popper canvass? He makes no mention of any.

(4) Let us be mindful of Popper's complaints against the spurious verifications of Freud's dream theory. Then one is simply dumbfounded to learn that he has no hesitation to declare his belief in its essential correctness. As he puts it amazingly:

> In spite of severe shortcomings ... it [Freud's dream theory] contains, beyond any reasonable doubt, a great discovery. I at least feel convinced that there is a world of the unconscious, and that Freud's analyses of dreams given in his book are fundamentally correct, though no doubt incomplete.[55]

But if the dream theory—though purportedly falsified by anxiety dreams—is unfalsifiable, then there can be no corroborating evidence for it by Popper's standards. Therefore, we are left to wonder how he can know "beyond any reasonable doubt" *without corroborating evidence*, that Freud's dream theory "contains ... a great discovery." And if the distinction between the manifest and latent content of a dream is discredited by anxiety dreams, and Freud's wish-fulfillment hypothesis is false, just what *is* the purported great discovery? Indeed, since Popper believes that (the bulk of) Freud's dream analyses are essentially correct, if incomplete, how can the pivotal distinction between the manifest and latent dream content fail to be sound?

▪ 3. THE DEGENERATION OF POPPER'S CRITERION OF DEMARCATION

In the mentioned Postscript, Popper's avowed purpose of treating the dream theory is to show the following: "that the problem of demarcation is not merely one of classifying theories into scientific and nonscientific ones, but that its solution is urgently needed for a critical appraisal of scientific theories, or allegedly scientific theories."[56] Yet ten pages later—at the end of the section on Freud's theory—Popper tells us amazingly enough that "from the beginning" his "problem of demarcation ... certainly was not a problem of classifying or distinguishing some subject matters called 'science' and 'metaphysics.'"[57] But this concluding disclaimer *flatly contradicts* the first conjunct of his initially stated motive for dealing with the psychoanalytic dream theory. By the same token, this concluding disclaimer here is patently incompatible with two of his earlier landmark reports, as we shall now see.

Thus, in his 1953 Personal Report on the development of his entire philosophy of science, which is reprinted in his *Conjectures and Refutations*, he informs us that, beginning in 1919, when he was seventeen years of age, his problem was "to distinguish between science and pseudo-science ... or ... 'metaphysics.'"[58] And, in his Reply to Critics in the Schilpp volume, he explains that his criterion of demarcation

> is more than sharp enough to make a distinction between many physical theories on the one hand, and metaphysical theories, such as psychoanalysis, or Marxism (in its present form), on the other. This is, of course, one of my main theses; and nobody who has not understood it can be said to have understood my theory.[59]

Evidently, in 1983, he insouciantly repudiates just this major, central tenet of his whole philosophy without ado. By the same token, he now denies his repeated other historical accounts of the aim that inspired his criterion of demarcation, starting in 1919. And, as we shall now see, this 1983 retraction occurs on the heels of a qualified, hedged disavowal that plainly serves as an immunizing stratagem for his prior vindication of his notion of demarcation by reference to psychoanalysis and Marxism. Thus, that stratagem in the Postscript runs as follows:

> It hardly matters whether or not I am right concerning the irrefutability of any of these three theories [i.e., psychoanalysis, Adlerian psychology, and Marxism]: here they serve merely as examples, as illustrations. For my purpose is to show that my "problem of

demarcation" was from the beginning the practical problem of assessing theories, and of judging their claims. It [the problem of demarcation] certainly was not a problem of classifying or distinguishing some subject matters called "science" and "metaphysics." It was, rather, an urgent practical problem: under what conditions is a *critical appeal to experience* possible—one that could bear some fruit?[60]

But if—as Popper tells us here—the stated three theories are not essential to the vindication of his falsifiability criterion of demarcation and are mere illustrations, then I must ask: What *other* theories for which scientificity has been wrongly claimed can be adduced to furnish such a vindication vis-à-vis the much older criterion of evidential support, which he wants to replace as unduly permissive? And as for stating conditions under which a "critical appeal to experience" is possible, Francis Bacon had demanded—three centuries before Popper—that the validation of a given theory *T* requires data that are *contrary* to *T's rivals* or competitors while being positive instances of *T*.

Several conclusions seem to me to emerge from my scrutiny of Popper's portrayal of psychoanalytic theory vis-à-vis his falsifiability criterion of demarcation:

(1) Popper's indictment of the Freudian corpus—or of its unspecified "basic" portion—as inherently unfalsifiable has fundamentally misdiagnosed its failure as a scientific theory. More often than not, the intellectual defects of psychoanalysis are too subtle to be detected by his criterion of demarcation. For example, there is no systematic published critique by him of Freud's method of free association, qua purported method of causal validation. Yet just that method is *the* method of clinical investigation in psychoanalysis. Alas, Popper's myth of nonfalsifiability has entrenched itself in current philosophic folklore. Examples of its inveterate enunciation abound. Thus, in a very recent issue of the *New Republic,* we are told that "Freudian explanations" are "beyond the reach of empirical falsification."[61]

(2) As a perspective on the nature of Freud's theory, Popper's portrayal is not viable. Yet perhaps it has had some value sociologically by putting psychoanalysts on notice to become more accountable scientifically. But it is unclear whether it actually had that effect on their attitudes.

(3) Popper's notion of demarcation in his *Realism and the Aim of Science* has become a kind of degenerative philosophical research program in Imre Lakatos's sense of that term.

This said, let me gladly make an acknowledgment: My own scrutiny of psychoanalysis was prompted by my initial doubt as to the soundness of Popper's portrayal of it.

■ 4. OBITER DICTA ON POPPER'S CRITERION OF DEMARCATION

1. As John Watkins has correctly noted, the actual falsifiability of psychoanalysis does not entail that Popper's "demarcation criterion was in trouble [merely] because

it actually included something [psychoanalytic theory] which Popper himself had mistakenly excluded."[62] Nor have I ever suggested such a moral.[63] Instead, I have now furnished further evidence for the falsifiability of psychoanalysis in order to undermine the following pillar of Popper's theory of science: the falsifiability criterion of demarcation is superior to the received inductivist standard, because the former yields a negative verdict on the scientific status of Freud's theory whereas the latter permissively accords such status to it. As I have argued, it is just the other way around!

But as for the merits of Popper's criterion itself, it is instructive to appraise his attempt to defend it[64] against Grover Maxwell's objection that it is *too restrictive*.[65] Maxwell had adduced mixed, quantified statements, for example, "All men are mortal," as hypotheses that belong to empirical science but that cannot be deductively falsified by basic statements. And in my contribution to the Lakatos *Festschrift*,[66] I had countenanced Maxwell's objection. Yet Bartley complained: "Adolf Grünbaum ignores Popper's reply [to Maxwell] and repeats this worn old criticism."[67]

Hence, let me quote Popper's reply and explain why I consider it unavailing. He writes:

> Whenever a pure existential statement [or mixed quantified statement] by being empirically "confirmed" appears to belong to empirical science, it will in fact do so *not on its own account*, but *by virtue of being a consequence of a corroborated falsifiable theory*.

As in Maxwell's example:

> When existential statements are verified, this is done by means of *stronger falsifiable* (though perhaps still existential) statements. "Jones dies," for example, might be verified by "Jones dies before his 150th year."

Thus:

> The totality of evidence "confirming" "All men are mortal" in fact confirms [in the first instance] a much stronger, and *falsifiable*, statement; for example, "All men die before their 150th year."[68]

In short, "All men are mortal"—though not deductively falsifiable—is authenticated as scientific by following deductively from the corroborated falsifiable statement, "All men die before their 150th year," which is logically stronger. Popper is satisfied that this amplified criterion "completely clears up" the scientific status of Maxwell's counterexamples. I claim, however, that it hardly does so and that, contrary to Bartley, "All men are mortal" is anything but a hoary canard.

Popper had argued that (relative degree of) corroboration is indeed preserved under logical entailment.[69] And, as he would have it, the unfalsifiable statement "All men are mortal" derives its scientific status from the (degree of) *corroboration* that it inherits deductively from a stronger hypothesis that is both falsifiable and corroborated. By contrast, there is obviously no deductive inheritance of falsifiability from the stronger hypothesis.

Two serious difficulties, however, vitiate Popper's reply to Maxwell:

(1) The invocation of corroboration to confer scientific status derivatively on an unfalsifiable statement constitutes a *major switch*, rather than an extension, of Popper's demarcation enterprise: He is abandoning his exclusive reliance on potential contrary evidence in favor of the inductivist's dependence on actual supporting evidence. And this abandonment is not lessened by the fact that the corroboration to which Popper appeals pertains, in the first instance, to a *falsifiable*, logically stronger hypothesis.

(2) Popper tells us that the relative degree of corroboration "says nothing whatever about future performance ... *of a theory*."[70] But, as I have shown elsewhere,[71] this anti-inductivist denial is logically incompatible with two other features that he attributes to corroboration: (a) degree of corroboration is preserved under logical entailment; and (b) "the degree of corroboration is a means of stating preference with respect to truth."[72] Indeed, given (b), (a) contravenes Popper's denial of the corroborability of future-tensed statements because at least some corroborated theories have *future-tensed entailments* (e.g., theories whose laws are universally quantified with respect to time).

Evidently, by invoking the deductive inheritance of corroboration along with its other avowed properties, Popper's attempt to accommodate the scientific status of "All men are mortal" runs afoul of his central anti-inductivist plank: degree of corroboration "*says nothing whatever about future performance ... of a theory*." Therefore, Popper can neutralize Grover Maxwell's counterexamples to the falsifiability criterion only on pain of fatal damage to his contra-inductivist construal of corroboration or to its vital feature (b). And I conclude, pace Popper and Bartley, that Grover Maxwell's counterexamples stand.

2. Popper offers two illustrations[73] of his contention that "scientific theories originate from myths and that a myth may contain important anticipations of scientific theories." The myths, though originally unfalsifiable, can become testable. One of his purported examples is:

> Parmenides' myth [i.e., *proto*-scientific world picture] of the unchanging block universe in which nothing ever happens and which, if we add another dimension, becomes Einstein's block universe (in which, too, nothing ever happens, since everything is, four-dimensionally speaking, determined and laid down from the beginning).

But here, Popper commits an error to which I called attention twenty years ago: the conflation of an absence of objective coming-*into*-being (becoming), on one hand, with timelessness or lack of change, on the other.[74] Though Einstein's space–time is devoid of becoming, it hardly features a timeless or unchanging universe. And it is therefore misleading to characterize it as a "block universe." By contrast, Parmenides' universe *is* timeless and—contrary to Popper—does not admit a consistent addition of a *time* dimension. To say, for example, that the career of a radioactive atom is not characterizable objectively by the attribution of the coming *into* being of events on its world line is *not* at all to say that this atom does not *change its state* when it emits an α-particle and thereby loses mass. By the same token, to deny that the events of an Einsteinian space–time themselves change ontologically by becoming present or past is hardly tantamount to declaring that change in such a space–time is illusory in Parmenides' sense. Popper seems to overlook that whereas Parmenides did deny the objective existence of change and hence of becoming, Einstein denied that there is becoming while asserting that there is indeed change. Precisely by being the *time* dimension in more than name, Einstein's fourth dimension cannot be "added" to Parmenides' unchanging world, as Popper would have it, without generating a contradiction.

I conclude that the exclusion of objective becoming from Einstein's space–time does not warrant Popper's belief that this universe originated conceptually in an interesting sense from Parmenides'. Thus, the theories of Parmenides and Einstein apparently do not illustrate the transformation of untestable myths into testable scientific theories, as envisioned by Popper.

■ NOTES

1. K. R. Popper, "Replies to My Critics," in: P. A. Schilpp (ed.), *The Philosophy of Karl Popper*, 1974, Book II, 1004.

2. Popper, *Conjectures and Refutations*, 1962, pp. 33–38, 255–258; "Replies to My Critics," pp. 984–985. See also Grünbaum, *The Foundations of Psychoanalysis: A Philosophical Critique*, 1984, pp. 103–107.

3. Popper, *Conjectures and Refutations*, p. 256.

4. Grünbaum, *Foundations of Psychoanalysis*, p. 280 (original italics).

5. Popper, *Realism and the Aim of Science*, 1983, p. 164, n. 1 (italics added).

6. Grünbaum, "Author's Response," *Behavioral and Brain Sciences* 9 (1986), pp. 266–269.

7. Jaspers, *Allgemeine Psychopathologie*, 1974; Habermas, *Knowledge and Human Interests*, translated by J. J. Shapiro, 1971, ch. 11; Ricoeur, *Freud and Philosophy*, 1970; *Hermeneutics and the Human Sciences*, translated by J. B. Thompson, 1981.

8. Grünbaum, "Why Thematic Kinships Between Events Do *Not* Attest Their Causal Linkage," in: J. R. Brown and J. Mittelstrass (eds.), *An Intimate Relation: Studies in the History and Philosophy of Science*, 1989, pp. 477–493.

9. Cioffi, "Psychoanalysis, Pseudo-Science and Testability," in: G. Currie and A. Musgrave (eds.), *Popper and the Human Sciences*, 1985, pp. 13–44; Cioffi, "Exegetical Myth-Making in Grünbaum's Indictment of Popper and Exoneration of Freud," in: P. Clark and C. Wright (eds.), *Mind, Psychoanalysis and Science*, 1988, pp. 61–87.

10. Grünbaum, "Author's Response," pp. 271–273.

11. Grünbaum, "The Role of the Case Study Method in the Foundations of Psychoanalysis," in: L. Nagl and H. Vetter (eds.), *Die Philosophen und Freud*, 1988; reprinted in *Canadian Journal of Philosophy* 18 (1988), 623–658.

12. Popper, "Autobiography of Karl Popper," in: P. A. Schilpp (ed.), *The Philosophy of Karl Popper*, 1974, Book I, p. 144 (italics in original).

13. Freud, *Standard Edition of the Complete Psychological Works of Sigmund Freud*, cited as S.E. by volume number and year of the original publication of the referenced text. Here: S.E. 20 (1926), pp. 108–109.

14. Popper, "Replies to My Critics," p. 985.

15. Ibid.

16. Popper, *Conjectures and Refutations*, p. 35.

17. Freud, S.E. 14 (1915), pp. 263–272; S.E. 18 (1922), pp. 225–230; earlier Freud gave a *more tentative* statement in S.E. 12 (1911), pp. 59–65.

18. Freud, S.E. 7 (1905), pp. 127, 157–159.

19. Freud, S.E. 14 (1915), pp. 263–272.

20. Ibid., p. 265.

21. Ibid.

22. Ibid.

23. Ibid., p. 268.

24. Ibid., p. 271.

25. S. Fisher and R. P. Greenberg, *The Scientific Credibility of Freud's Theory and Therapy*, 1977, p. 259.

26. H. S. Zamansky, "An Investigation of the Psychoanalytic Theory of Paranoid Delusions," *Journal of Personality* 26 (1958), pp. 410–425.

27. Grünbaum, 'Authors Response' ibid., pp. 269–270.

28. Freud, S.E. 14 (1915), pp. 265–266.

29. Glymour, "Freud, Kepler and the Clinical Evidence," in: R. Wollheim (ed.), *Freud*, 1974, pp. 285–304.

30. Grünbaum, "The Role of the Case Study Method in the Foundations of Psychoanalysis," section 2.B.

31. Popper, *Realism and the Aim of Science*, p. 169.

32. Freud, S.E. 12 (1911), pp. 65–68.

33. Popper, *Realism and the Aim of Science*, p. 169.

34. Popper, "Replies to My Critics," p. 1004.

35. Ibid.

36. Ibid., p. 1005.

37. Popper, *Realism and the Aim of Science*, p. 173.

38. Popper, "Replies to My Critics," p. 1004.

39. Freud, S.E. 14 (1914), p. 16.

40. Popper, *Realism and the Aim of Science*, ibid., p. 163.

41. Popper, "Replies to My Critics," p. 1004.

42. Freud, S.E. 14 (1915), p. 265; S.E. 18 (1922), p. 228.

43. Carnap, "Testability and Meaning," *Philosophy of Science* 3 (1936), pp. 419–471; 4 (1937), pp. 2–40.

44. Salmon, "Psychoanalytic Theory and Evidence," in: S. Hook (ed.), *Psychoanalysis, Scientific Method and Philosophy*, 1959, pp. 252–267.

45. Grünbaum, *Foundations of Psychoanalysis*, pp. 121–123.

46. Popper, *Realism and the Aim of Science*, p. 173 (italics in original).

47. Ibid., p. 165 (italics in original).

48. Ibid., p. 167.

49. Ibid., pp. 164, 172.

50. Ibid., p. 172 (italics in original).

51. Grünbaum, "Is Freudian Psychoanalytic Theory Pseudo-Scientific by Karl Popper's Criterion of Demarcation?," *American Philosophical Quarterly* 16 (1979), pp. 131–141; Grünbaum, *Foundations of Psychoanalysis*, p. 280.

52. Popper, *The Open Society and Its Enemies*, 1950, pp. 16–17.

53. Popper, *Conjectures and Refutations*, p. 49.

54. Ibid., p. 275, n. 52.
55. Popper, *Realism and the Aim of Science*, ibid., p. 164.
56. Ibid., pp. 163–164.
57. Ibid., p. 174.
58. Popper, *Conjectures and Refutations*, p. 33 (italics in original).
59. Popper, "Replies to My Critics," p. 984.
60. Popper, *Realism and the Aim of Science*, p. 174.
61. Robert Wright, "Why Men Are Still Beasts," *New Republic* 3834 (July 11, 1988), p. 30.
62. Watkins, "Corroboration and the Problem of Content-Comparison," in: G. Radnitzky and G. Andersson (eds.), *Progress and Rationality in Science*, 1978, p. 351.
63. See Grünbaum, *Foundations of Psychoanalysis*, p. 105.
64. Popper, "Replies to My Critics," p. 1038.
65. Maxwell, "Corroboration Without Demarcation," in: P. A. Schilpp (ed.), *The Philosophy of Karl Popper*, 1974, p. 294.
66. Grünbaum, "Is Falsifiability the Touchstone of Scientific Rationality? Karl Popper Versus Inductivism," in R. S. Cohen, P. K. Feyerabend, and M. W. Wartofsky (eds.), *Essays in Memory of Imre Lakatos*, 1976, p. 227. This essay is included as chapter 1 in the present volume.
67. Bartley, "The Philosophy of Karl Popper," *Philosophia* (Israel) 11 (1982), p. 221, n. 125.
68. Popper, "Replies to My Critics," p. 1038.
69. Popper, *Objective Knowledge*, 1973, pp. 19–20.
70. Ibid., p. 18.
71. Grünbaum, "Is Falsifiability the Touchstone of Scientific Rationality?" pp. 244–247.
72. Popper, *Objective Knowledge*, pp. 19–20.
73. Popper, *Conjectures and Refutations*, p. 38.
74. Grünbaum, *Modern Science and Zeno's Paradoxes*, 1968, pp. 22–23.

3

The Falsifiability of Theories: Total or Partial? A Contemporary Evaluation of the Duhem-Quine Thesis

It has been maintained that there is an important *asymmetry* between the *verification* and the *refutation* of a theory in empirical science. Refutation has been said to be conclusive or decisive while verification was claimed to be irremediably inconclusive in the following sense: If a theory T_1 entails observational consequences O, then the *truth* of T_1 does *not*, of course, follow *deductively* from the truth of the conjunction

$$(T_1 \to O) \cdot O.$$

On the other hand, the *falsity* of T_1 is indeed *deductively inferable* by *modus tollens* from the truth of the conjunction

$$(T_1 \to O) \cdot \sim O.$$

Thus, F. S. C. Northrop writes, "We find ourselves, therefore, in this somewhat shocking situation: the method which natural science uses to check the postulationally prescribed theories ... is absolutely trustworthy when the proposed theory is not confirmed and logically inconclusive when the theory is experimentally confirmed."[1]

Under the influence of the physicist, philosopher of science, and historian of science Pierre Duhem,[2] this thesis of asymmetry of conclusiveness between verification and refutation has been strongly denied as follows: If "T_1" denotes the kind of individual or *isolated* hypothesis H whose verification or refutation is at issue in the conduct of particular scientific experiments, then Northrop's formal schema is a misleading oversimplification. Upon taking cognizance of the fact that the observational consequences O are deduced *not* from H alone but rather from the conjunction of H and the relevant body of *auxiliary* assumptions A, the refutability of H is seen to be no more conclusive than its verifiability. For now it appears that Northrop's formal schema must be replaced by the following:

(1) $[(H \cdot A) \to O] \cdot O$ (verification)

and

(2) $[(H \cdot A) \to O] \cdot \sim O$ (refutation)

The recognition of the presence of the auxiliary assumptions A in both the verification and refutation of H now makes apparent that the *refutation of H itself* by *adverse* empirical evidence ~*O* can be no more decisive than its *verification* (confirmation) by *favorable* evidence *O*. What can be inferred deductively from the refutational premise (2) is *not* the falsity of *H* itself but only the much weaker conclusion that *H* and *A* cannot both be true. It is immaterial here that the *falsity* of the *conjunction* of *H* and *A* can be inferred *deductively* from the refutational premise (2) while the truth of that conjunction can be inferred only *inductively* from the verificational premise (1). For this does *not* detract from the fact that there is parity between the refutation of *H* *itself* and the verification of *H* itself in the following sense: (2) does *not* entail (deductively) the falsity of *H* itself, just as (1) does not entail the truth of *H* by itself. In short, isolated component hypotheses of far-flung theoretical systems are not separately refutable but only contextually disconfirmable. And Northrop's schema is an adequate representation of the actual logical situation only if "T_1" in his schema refers to the entire theoretical *system* of premises which enters into the deduction of *O* rather than to such mere *components H* as are at issue in specific scientific inquiries.

Under the influence of Duhem's emphasis on the confrontation of an entire theoretical system by the tribunal of experience, writers such as W. v. O. Quine have gone further to make what I take to be the following claim: No matter what the specific content *O'* of the *prima facie* adverse empirical evidence ~ *O*, we can always justifiably affirm the truth of *H* as part of the theoretical *explanans* of *O'* by doing two things: (1) blame the falsity of *O* on the falsity of *A* rather than on the falsity of *H;* and (2) so modify *A* that the conjunction of *H* and the *revised* version *A'* of *A* does entail (explain) the actual findings *O'*. Thus, in his *Two Dogmas of Empiricism,* Quine writes: "Any statement can be held true come what may, if we make drastic enough adjustments elsewhere in the system."[3] And one of Quine's arguments in that provocative essay against the tenability of the analytic–synthetic distinction is that a supposedly synthetic statement, no less than a supposedly analytic one, can be claimed to be true "come what may" on Duhemian grounds.

The aim of my present paper is to establish two main conclusions:

1. Quine's formulation of Duhem's thesis—hereafter called the "*D*-thesis"—is true *only* in various *trivial* senses of what Quine calls "drastic enough adjustments elsewhere in the system." And no one would wish to contest any of these thoroughly uninteresting versions of the *D*-thesis.

2. In its *non*trivial, exciting form, the *D*-thesis is untenable in the following fundamental respects:

   2.1 *Logically*, it is a non sequitur. For *independently* of the particular empirical context to which the hypothesis *H* pertains, there is no logical guarantee at all of the existence of the *required kind* of revised set *A'* of auxiliary assumptions such that

   $$(H \cdot A') \rightarrow O'$$

for any one component hypothesis H and any O'. Instead of being guaranteed logically, the existence of the required set A' needs *separate* and *concrete* demonstration for each particular context. In the absence of the latter kind of *empirical* support for Quine's unrestricted Duhemian claim, that claim is an unempirical dogma or article of faith which the pragmatist Quine is no more entitled to espouse than an empiricist would be.

2.2. The D-thesis is not only a non sequitur but also is actually *false*, as shown by an important counterexample, namely, the *separate* falsifiability of a particular component hypothesis H.

To forestall misunderstanding, let it be noted that my rejection of the very strong assertion made by Quine's D-thesis is *not* at all intended as a repudiation of the following far weaker contention, which I believe to be eminently sound: the logic of every disconfirmation, no less than of every confirmation of an isolated scientific hypothesis H is such as to *involve at some stage or other* an entire network of interwoven hypotheses in which H is ingredient rather than in every stage merely the separate hypothesis H. Furthermore, it is to be understood that the issue before us is the *logical* one whether *in principle* every component H is unrestrictedly preservable by a suitable A', not the *psychological* one whether scientists possess sufficient ingenuity at every turn to propound the required set A', *if it exists*. Of course, *if* there are cases in which the requisite A' simply *does not even exist logically*, then surely no amount of ingenuity on the part of scientists will enable them to ferret out the nonexistent required A' in such cases.

■ 1. THE TRIVIAL VALIDITY OF THE D-THESIS

It can be made evident at once that unless Quine restricts in very specific ways what he understands by "drastic enough adjustments elsewhere in the (theoretical) system" the D-thesis is a thoroughly unenlightening truism. For if someone were to put forward the false empirical hypothesis H, "Ordinary buttermilk is highly toxic to humans," this hypothesis could be saved from refutation in the face of the observed wholesomeness of ordinary buttermilk by making the following "drastic enough" adjustment in our system: changing the rules of English usage so that the intension of the term "ordinary buttermilk" is that of the term "arsenic" in its customary usage. Hence, a *necessary* condition for the nontriviality of Duhem's thesis is that *the theoretical language be semantically stable* in the relevant respects.

Furthermore, it is clear that if one *were* to countenance that O' itself qualifies as A', Duhem's affirmation of the existence of an A' such that

$$(H \cdot A') \to O'$$

would hold trivially, and H would not even be needed to deduce O'. Moreover, the D-thesis can hold trivially even in cases in which H is required in addition to A' to deduce the *explanandum* O': an A' of the trivial form

$$\sim H \vee O'$$

requires H for the deduction of O', but no one will find it enlightening to be told that the D-thesis can thus be sustained.

I am unable to give a formal *and* completely general *sufficient* condition for the *non*triviality of A'. And, so far as I know, neither the originator nor any of the advocates of the D-thesis have even shown any awareness of the need to circumscribe the class of *non*trivial revised auxiliary hypotheses A' so as to render the D-thesis interesting. I shall therefore assume that the proponents of the D-thesis intend it to stand or fall on the kind of A' which we would all recognize as *non*trivial in *any given case*, a kind of A' which I shall symbolize by "A'_{nt}." And I shall endeavor to show that such a *non*trivial form of the D-thesis is indeed untenable after first commenting on the attempt to sustain the D-thesis by resorting to the use of a *nonstandard logic.*

The species of drastic adjustment consisting in recourse to a *non*standard logic is specifically mentioned by Quine. Citing a hypothesis such as, "There are brick houses on Elm Street," he claims that even a statement so "germane to sense experience … can be held true in the face of recalcitrant experience by pleading hallucination or by amending certain statements of the kind called logical laws."[4] I disregard for now the argument from hallucination. In the absence of *specifics* as to the ways in which alterations of logical laws will enable Quine to hold in the face of *recalcitrant* experience that a statement H like, "There are brick houses on Elm Street," is *true*, I must conclude the following: The invocation of nonstandard logics either makes the D-thesis *trivially* true or turns it into an interesting claim which is an unfounded dogma. For suppose that the nonstandard logic used is a three-valued one. Then even if it were otherwise feasible to assert within the framework of such a logic that the particular statement H is "true," the term "true" would no longer have the meaning associated with the two-valued framework of logic within which the D-thesis was enunciated to begin with. It is not to be overlooked that a form of the D-thesis which allows itself to be sustained by alterations in the meaning of "true" is no less trivial *in the context of the expectations raised by the D-thesis* than one which rests its case on calling arsenic "buttermilk." And this triviality obtains *in this context*, notwithstanding the fact that the two-valued and three-valued usages of the word "true" share what Hilary Putnam has usefully termed a common "core meaning."[5] For suppose we had two particular substances I_1 and I_2 which are isomeric. That is to say, these substances are composed of the same elements in the same proportions and with the same molecular weight but the arrangement of the atoms within the molecule is different. Suppose further that I_1 is not at all toxic while I_2 is highly toxic, as in the case of two isomers of trinitrobenzene.[6] Then if we were to call I_1 "duquine" and asserted that "duquine is highly toxic," this statement H could also be trivially saved from refutation in the face of the evidence of the wholesomeness of I_1 by the following device: only *partially* changing the meaning of "duquine" so that its intension is the second, highly toxic isomer I_2, thereby leaving the chemical "core meaning" of "duquine" intact. To avoid misunderstanding of my charge of triviality, let me point out precisely what I regard as trivial here. The preservation of H from refutation in the face of the evidence by a *partial* change in the meaning of "duquine" is trivial in the sense of being only a *trivial* fulfillment of *the expectations raised by the D-thesis*. But, in my view, the possibility *as such* of preserving H by *this particular kind of change in meaning* is not at all trivial. For this possibility as such reflects a fact about the world: the existence of isomeric substances of radically different degrees of toxicity (allergenicity)!

Even if one ignores the change in the meaning of "true" inherent in the resort to a three-valued logic, there is no reason to think that the D-thesis can be successfully upheld in such an altered logical framework: The arguments which I shall present against the *non*trivial form of the D-thesis within the framework of the standard logic apply just as much, so far as I can see, in the three-valued and other nonstandard logics of which I am aware. And if the reply is that there are *other* nonstandard logics which are both viable for the purposes of science and in which my impending polemic against the nontrivial form of the D-thesis does *not* apply, then I retort: As it stands, Quine's assertion of the feasibility of a change in the laws of logic which would thus sustain the D-thesis is an unempirical dogma or at best a promissory note. And until the requisite collateral is supplied, it is not incumbent upon anyone to accept that promissory note.

■ 2. THE UNTENABILITY OF THE NONTRIVIAL D-THESIS

2.1 The Nontrivial D-Thesis Is a Non Sequitur

The nontrivial D-thesis is that for every component hypothesis H of any domain of empirical knowledge and for any observational findings O',

$$(\exists A'_{nt})\left[\left(H \cdot A'_{nt}\right) \to O'\right].$$

But this claim does *not* follow from the fact that the falsity of H is *not* deductively inferable from premise (2), that is, from

$$[(H \cdot A) \to O] \cdot \sim O.$$

For the latter premise utilizes *not* the full empirical information given by O' but only the part of that information which tells us that O' is logically incompatible with O. Hence, the *failure* of $\sim O$ to permit the deduction of $\sim H$ does *not* justify the assertion of the D-thesis that there always *exists* a nontrivial A' such that the conjunction of *H* and that *A'* entails *O'*. In other words, the fact that the falsity of *H* is *not* deducible (by *modus tollens*) from premise (2) is quite insufficient to show that *H* can be preserved nontrivially as part of an *explanans* of *any* potential empirical findings *O'*. I conclude, therefore, from the analysis given so far that in its *non*trivial form Quine's D-thesis is *gratuitous* and that the existence of the required nontrivial *A'* would require *separate* demonstration for each particular case.

2.2 Physical Geometry as a Counterexample to the Nontrivial D-Thesis

Einstein has articulated Duhem's claim by reference to the special case of testing a hypothesis of physical geometry. In opposition to the Carnap-Reichenbach conception, Einstein maintains[7] that no hypothesis of physical geometry by itself is falsifiable even though all of the terms in the vocabulary of the geometrical theory, including the term "congruent" for line segments and angles, have been given a specific physical interpretation. And the substance of his argument is briefly the following: In order to follow the practice of ordinary physics and use rigid solid rods as the physical standard of congruence in the determination of the geometry, it is essential to make computational allowances for the thermal, elastic, electromagnetic, and other deformations exhibited by solid rods. The introduction of these corrections is an essential part of the logic of testing a physical geometry. For the presence of inhomogeneous thermal and other such influences issues in a dependence of the coincidence behavior of transported solid rods on the latter's *chemical composition*. Now, Einstein argues that the geometry itself can never be accessible to experimental falsification *in isolation from* those other laws of physics which enter into the calculation of the corrections compensating for the distortions of the rod. And from this he then concludes that you can always preserve any geometry you like by suitable adjustments in the associated correctional physical laws. Specifically, he states his case in the form of a dialogue in which he attributes his own Duhemian view to Henri Poincaré and offers that view in opposition to Hans Reichenbach's conception. But I submit that Poincaré's text will *not* bear Einstein's interpretation. For in speaking of the variations which solids exhibit under distorting influences, Poincaré says "we neglect these variations in laying the foundations of geometry, because, besides their being very slight, they are irregular and consequently seem to us accidental."[8] I am therefore taking the liberty of replacing the name "Poincaré" in Einstein's dialogue by the term "Duhem and Einstein." *With this modification*, the dialogue reads as follows:

> Duhem and Einstein: The empirically given bodies are not rigid, and consequently cannot be used for the embodiment of geometric intervals. Therefore, the theorems of geometry are not verifiable.
>
> Reichenbach: I admit that there are no bodies which can be *immediately* adduced for the "real definition" (i.e., physical definition) of the interval. Nevertheless, this real

definition can be achieved by taking the thermal volume dependence, elasticity, electro- and magneto-striction, etc., into consideration. That this is really and without contradiction possible, classical physics has surely demonstrated.

Duhem and Einstein: In gaining the real definition improved by yourself you have made use of physical laws, the formulation of which presupposes (in this case) Euclidean geometry. The verification, of which you have spoken, refers, therefore, not merely to geometry but to the entire system of physical laws which constitute its foundation. An examination of geometry by itself is consequently not thinkable. Why should it consequently not be entirely up to me to choose geometry according to my own convenience (i.e., Euclidean) and to fit the remaining (in the usual sense "physical") laws to this choice in such manner that there can arise no contradiction of the whole with experience?[9]

By speaking here of the "real definition" (i.e., the coordinative definition) of "congruent intervals" by the corrected transported rod, Einstein is ignoring that the actual and potential physical meaning of congruence in physics *cannot* be given exhaustively by any *one* physical criterion or test condition. But here we can safely ignore this open cluster character of the concept of congruence. For our concern as well as Einstein's is merely to single out *one* particular congruence class from among an infinitude of such alternative classes. And as long as our specification of that one chosen class is *unambiguous*, it is wholly immaterial that there are also *other* physical criteria (or test conditions) by which it could be specified.

Einstein is making two major points here: (1) In obtaining a physical geometry by giving a physical interpretation of the postulates of a formal geometric axiom system, the specification of the physical meaning of such theoretical terms as "congruent," "length," or "distance" is *not* at all simply a matter of giving an operational definition in the strict sense. Instead, what has been variously called a "rule of correspondence" (Margenau and Carnap), a "coordinative definition" (Reichenbach), an "epistemic correlation" (Northrop), or a "dictionary" (N. R. Campbell) is provided here *through the mediation of hypotheses and laws* which are collateral to the geometric theory whose physical meaning is being specified. Einstein's point that the physical meaning of congruence is given by the transported rod *as corrected theoretically* for idiosyncratic distortions is an illuminating one and has an abundance of analogues throughout physical theory, thus showing, incidentally, that strictly operational definitions are a rather simplified and limiting species of rules of correspondence. In particular, we see that the physical interpretation of the term "length," which is often adduced as the prototype of all "operational" definitions in Bridgman's sense, is *not* given operationally in any *distinctive* sense of that ritually invoked term. (2) Einstein's second claim, which is the cardinal one for our purposes, is that the role of collateral theory in the physical definition of congruence is such as to issue in the following *circularity*, from which there is no escape, he maintains, short of acknowledging the existence of an a priori element *in the sense of the Duhemian ambiguity*: the rigid body is not even defined without first *decreeing* the validity of Euclidean geometry (or of some other particular geometry). For *before* the *corrected* rod can be used to make an empirical determination of the *de facto* geometry, the required corrections must be computed via laws, such as those of elasticity, which involve Euclideanly calculated areas and volumes.[10] But clearly the warrant for thus introducing Euclidean geometry *at this stage* cannot be empirical.

In the same vein, Hermann Weyl endorses Duhem's position as follows: "Geometry, mechanics, and physics form an inseparable theoretical whole ... Philosophers have put forward the thesis that the validity or nonvalidity of Euclidean geometry cannot be proved by empirical observations. It must in fact be granted that in all such observations essentially physical assumptions, such as the statement that the path of a ray of light is a straight line and other similar statements, play a prominent part. This merely bears out ... that it is only the whole composed of geometry and physics that may be tested empirically."[11]

I now wish to set forth my doubts regarding the soundness of Einstein's geometrical form of the *D*-thesis.[12] And I shall do so in two parts, the first of which deals with the special case in which effectively no deforming influences are present in a certain region whose geometry is to be ascertained.

If we are confronted with the problem of the falsifiability of the geometry ascribed to a region which is effectively free from deforming influences, then the *correctional* physical laws play no role as auxiliary assumptions, and the latter reduce to the claim that the region in question is, in fact, effectively *free* from deforming influences. And *if* such freedom can be affirmed *without* presupposing collateral theory, then the geometry alone rather than only a wider theory in which it is ingredient will be falsifiable. On the other hand, if collateral theory *were* presupposed here, then Duhem and Einstein might be able to adduce its modifiability to support their claim that the geometry *itself* is *not* separately falsifiable.

Specifically, they might argue then that the collateral theory could be modified such that the region then turns out *not* to be free from deforming influences with resulting inconclusive falsifiability of the geometry. The question is therefore whether freedom from deforming influences can be asserted and ascertained independently of (sophisticated) collateral theory. My answer to this question is yes. For quite independently of the conceptual elaboration of such physical magnitudes as temperature, whose constancy would characterize a region free from deforming influences, the absence of perturbations is certifiable for the region as follows: Two solid rods of very *different* chemical constitution which coincide at one place in the region will also coincide everywhere else in it (independently of their paths of transport). It would *not* do for the Duhemian to object here that the certification of two solids as quite *different chemically* is theory laden to an extent permitting him to uphold his thesis of the inconclusive falsifiability of the geometry. For suppose that observations were so ambiguous as to permit us to assume that two solids which appear strongly to be chemically *different* are, in fact, chemically identical in all relevant respects. If so rudimentary an observation were thus ambiguous, then no observation could ever possess the required univocity to be incompatible with an observational consequence of a *total theoretical* system. And if that were the case, Duhem could hardly avoid the following conclusion: "Observational findings are always so unrestrictedly ambiguous as not to permit even the refutation of any given *total theoretical* system." But such a result would be tantamount to the absurdity that *any* total theoretical system can be espoused as true a priori. By the same token, incidentally, I cannot see what methodological safeguards would prevent Quine from having to countenance such an outcome within the framework of his *D*-thesis. In view of his avowed willingness to "plead hallucination" to deal with observations *not* conforming to the hypothesis that "there are brick houses on Elm Street," one wonders whether he would be prepared to say that *all* human observers who make disconfirming observations on Elm Street are hallucinating. And, if so, why not discount all observations incompatible with an *arbitrary total theoretical system* as hallucinatory? Thus, it would seem that if Duhem is to maintain, as he does, that a *total theoretical system is* refutable by confrontation with observational results, then he must allow that the coincidence of diverse kinds of rods at different places in the region (independently of their paths of transport) is certifiable observationally. Accordingly, the absence of deforming influences is ascertainable *independently* of any assumptions as to the geometry and of other (sophisticated) collateral theory.

Let us now employ our earlier notation and denote the geometry by "*H*" and the assertion concerning the freedom from perturbations by "*A*." Then, once we have laid down the congruence definition and the remaining semantical rules, the physical geometry *H* becomes separately falsifiable as an explanans of the posited empirical findings *O*'. It is true, of course, that *A* is only more or less highly confirmed by the ubiquitous coincidence of chemically different kinds of solid rods. But the inductive risk thus inherent in affirming *A* does not arise from the alleged inseparability of *H* and *A*, and that risk can be made exceedingly small without any involvement of *H*. Accordingly, the actual logical situation is characterized not by the Duhemian schema but instead by the schema

$$[\{(H \cdot A) \to O\} \cdot \sim O \cdot A] \to \sim H.$$

We now turn to the critique of Einstein's Duhemian argument as applied to the empirical determination of the geometry of a region which is subject to deforming influences.

There can be no question that, when deforming influences *are* present, the laws used to make the corrections for deformations involve areas and volumes in a fundamental way (e.g., in the definitions of the elastic stresses and strains) and that this involvement presupposes a geometry, as is evident from the area and volume formulae of differential geometry, which contains the square root of the determinant of the components g_{ik} of the metric tensor.[13] Thus, the empirical determination of the geometry involves the joint assumption of a geometry and of certain collateral hypotheses. But we see already that this assumption *cannot* be adequately represented by the conjunction $H{\cdot}A$ of the Duhemian schema, where H represents the geometry.

Now suppose that we begin with a set of Euclideanly formulated physical laws P_0 in correcting for the distortions induced by perturbations and then use the thus Euclideanly corrected congruence standard for *empirically* exploring the geometry of space by determining the metric tensor. *The initial stipulational affirmation of the Euclidean geometry G_0 in the physical laws P_0 used to compute the corrections in no way assures that the geometry obtained by the corrected rods will be Euclidean!* If it is *non*-Euclidean, then the question is: What will be required by Einstein's fitting of the physical laws to preserve Euclideanism and avoid a contradiction of the theoretical system with experience? Will the adjustments in P_0 necessitated by the retention of Euclidean geometry entail merely a change in the dependence of the length assigned to the transported rod on such *nonpositional* parameters as temperature, pressure, and magnetic field? Or could the putative empirical findings compel that the length of the transported rod be likewise made a nonconstant function of its *position* and *orientation* as *independent* variables in order to square the coincidence findings with the requirement of Euclideanism? The possibility of obtaining *non*-Euclidean results by measurements carried out in a spatial region uniformly characterized by standard conditions of temperature, pressure, electric and magnetic field strength, and so forth shows it to be *extremely doubtful*, as we shall now show, that the preservation of Euclideanism could *always* be accomplished short of introducing *the dependence of the rod's length on the independent variables of position or orientation.*

But the introduction of the latter dependence is none other than so radical a change in the meaning of the word "congruent" that this term now denotes a class of intervals *different* from the original congruence class denoted by it. And such tampering with the semantical anchorage of the word "congruent" violates the requirement of semantical stability, which is a necessary condition for the *non*-triviality of the D-thesis, as we saw already.

Suppose that, relatively to the customary congruence standard, the geometry prevailing in a given region when *free* from perturbational influences is that of a strongly *non*-Euclidean space of spatially and temporally constant curvature. Then what would be the character of the alterations in the *customary* correctional laws which Einstein's thesis would require to assure the *Euclideanism* of that region relatively to the customary congruence standard under *perturbational* conditions? The required alterations would be *independently falsifiable*, as will now be demonstrated, because they would involve affirming that such coefficients as those of linear thermal expansion *depend on the independent variables of spatial position*. That such a space dependence of the correctional coefficients might well be necessitated by the exigencies of Einstein's Duhemian thesis can be seen as follows by reference to the law of linear thermal expansion. In the usual version of physical theory, the first approximation of that law[14] is given by

$$L = L_0 \, (1 + \alpha{\cdot}\Delta T).$$

If Einstein is to guarantee the Euclideanism of the region under discussion by means of logical devices that are consonant with his thesis, and if our region is subject only to *thermal* perturbations for some time, then we are confronted with the following situation: Unlike the customary law of linear thermal expansion, the revised form of that law needed by Einstein will have to bear the *twin* burden of effecting *both* of the following two kinds of superposed corrections: (1) the *changes* in the lengths ascribed to the transported rod in different positions or orientations which would be required even if our region *were* everywhere at the standard temperature, merely for the sake of rendering Euclidean its otherwise non-Euclidean geometry; and (2) corrections compensating for the effects of the de facto deviations from the standard temperature, these corrections being the sole onus of the usual version of the law of linear thermal expansion. What will be the consequences of requiring the revised version of the law of thermal elongation to implement the first of these two kinds of corrections in a context in which the deviation ΔT from the standard temperature is the same at different points of the region, that temperature deviation having been measured in the manner chosen by the Duhemian? Specifically, what will be the character of the coefficients α of the revised law of thermal elongation under the posited circumstances, if Einstein's thesis is to be implemented by effecting the first set of corrections? Since the new version of the law of thermal expansion will then have to guarantee that the lengths L assigned to the rod at the various points of equal temperature T differ appropriately, it would seem clear that logically possible empirical findings could compel Einstein to make the coefficients α of solids depend on the space coordinates.

But such a spatial dependence is independently falsifiable: Comparison of the thermal elongations of an aluminum rod, for example, with an invar rod of essentially zero α by, say, the Fizeau method might well show that the α of the aluminum rod is a characteristic of aluminum which is not dependent on the space coordinates. And even if it were the case that the α's are found to be space dependent, how could Duhem and Einstein assure that this space dependence would have the particular functional form required for the success of their thesis?

We see that the required resort to the introduction of a spatial dependence of the thermal coefficients might well *not* be open to Einstein. Hence, in order to retain Euclideanism, it would then be necessary to *remetrize* the space in the sense of abandoning the customary definition of congruence, entirely apart from any consideration of idiosyncratic distortions and even after correcting for these in some way or other. But this kind of remetrization, though entirely admissible in *other* contexts, does *not* provide the requisite support for Einstein's Duhemian thesis! For Einstein offered it as a criticism of Reichenbach's conception. And hence it is the *avowed onus* of that thesis to show that the geometry by *itself* cannot be held to be empirical, that is, separately falsifiable, even when, with Reichenbach, we have sought to assure its empirical character by choosing and then adhering to the usual (standard) definition of spatial congruence, which *excludes* resorting to such remetrization.

Thus, there may well obtain observational findings O', expressed in terms of a particular definition of congruence (e.g., the *customary* one), which are such that there does *not* exist any nontrivial set A' of auxiliary assumptions capable of preserving the Euclidean H in the face of O'. And this result alone suffices to invalidate the Einsteinian version of Duhem's thesis to the effect that any geometry, such as Euclid's, can be preserved in the face of any experimental findings which are expressed in terms of the customary definition of congruence.

It might appear that my geometric counterexample to the Duhemian thesis of unavoidably contextual falsifiability of an explanans is vulnerable to the following criticism: To be sure, Einstein's geometric articulation of that thesis does not leave room for saving it by resorting to a remetrization in the sense of making the length of the rod *vary* with position or orientation even *after* it has been corrected for idiosyncratic distortions. But why saddle the Duhemian thesis as such with a restriction peculiar to Einstein's particular version of it? And thus why not allow Duhem to save his thesis by countenancing those *alterations in the congruence definition* which are *remetrizations*?

My reply is that to deny the Duhemian the invocation of such an alteration of the congruence definition *in this context* is *not* a matter of gratuitously requiring him to justify his thesis within the confines of Einstein's particular version of that thesis; instead, the imposition of this restriction is entirely legitimate here, and the Duhemian could hardly wish to reject it as unwarranted. For it is of the essence of Duhem's contention that H (in this case, Euclidean geometry) can always be preserved *not* by tampering with the principal semantical rules (interpretive sentences) linking H to the observational base (e.g., specifying a particular congruence class of intervals) but rather by availing oneself of the alleged *inductive latitude* afforded by the ambiguity of the experimental evidence to do the following: (a) leave the factual commitments of H *essentially unaltered* by retaining both the statement of H and the *principal* semantical rules linking its terms to the observational base; and (b) replace the set A by A' such that A and A' are logically incompatible under the hypothesis H. The qualifying words "principal" and "essential" are needed here in order to obviate the possible objection that it may not be logically possible to supplant the auxiliary assumptions A by A' *without also* changing the factual content of H in *some* respect. Suppose, for example, that one were to abandon the optical hypothesis A that light will require equal times to traverse *congruent* closed paths in an inertial system in favor of some rival hypothesis. Then the semantical linkage of the term "congruent space intervals" to the observational base is changed to the extent that this term no longer denotes intervals traversed by light in equal round-trip times. But such a change in the semantics of the word "congruent" is innocuous in this context, since it leaves wholly intact the membership of the class of spatial intervals that is referred to as a "congruence class." In this sense, then, the modification of the optical hypothesis leaves intact both the "*principal*" semantical

rules governing the term "congruent" *and* the "*essential*" factual content of the geometric hypothesis *H*, which is predicated on a particular congruence class of intervals. That "essential" factual content is the following: Relatively to the congruence
specified by unperturbed transported rods—*among other things*—the geometry is
Euclidean.

Now, the essential factual content of a geometrical hypothesis can be changed
either by preserving the original statement of the hypothesis while changing
one or more of the principal semantical rules or by keeping all of the semantical
rules intact and suitably changing the statement of the hypothesis. We can see,
therefore, that the retention of a Euclidean *H* by the device of changing through
remetrization the semantical rule governing the meaning of "congruent" (for line
segments) effects a retention *not* of the essential *factual commitments* of the original Euclidean *H* but only of its *linguistic trappings*. That the thus "preserved"
Euclidean *H* actually repudiates the essential factual commitments of the *original*
one is clear from the following: The *original* Euclidean *H* had asserted that the
coincidence behavior common to all kinds of solid rods is Euclidean *if* such transported rods are taken as the physical realization of congruent intervals, but the
Euclidean *H* which survived the confrontation with the posited empirical findings only by dint of a *remetrization* is predicated on a denial of the very assertion
that was made by the original Euclidean *H*, which it was to "preserve." It is as if
a physician were to endeavor to "preserve" an a priori diagnosis that a patient
has acute appendicitis in the face of a negative finding (yielded by an explor-
atory operation) as follows: He would redefine "acute appendicitis" to denote the
healthy state of the appendix!

Hence, the confines within which the Duhemian must make good his claim
of the preservability of a Euclidean *H* do *not* admit of the kind of change in the
congruence definition which alone would render his claim tenable under the
assumed empirical conditions. Accordingly, the geometrical critique of Duhem's
thesis given in this paper does *not* depend for its validity on restrictions peculiar
to Einstein's version of it.

Even apart from the fact that Duhem's thesis precludes resorting to an alternative metrization to save it from refutation in our geometrical context, the very
feasibility of alternative metrizations is vouchsafed *not* by any general Duhemian
considerations pertaining to the logic of falsifiability but by a property peculiar to
the subject matter of geometry (and chronometry): the latitude for *convention* in
the ascription of the spatial (or temporal) *equality* relation to intervals in the con-
tinuous manifolds of physical space (or time).

It would seem that the least we can conclude from the analysis of Einstein's geometrical *D*-thesis given in this paper is the following: Since empirical findings can
greatly narrow down the range of uncertainty as to the prevailing geometry, there
is no assurance of the *latitude* for the choice of a geometry which Einstein takes for
granted in the manner of the *D*-thesis.

■ NOTES

1. Notes. F. S. C. Northrop, *The Logic of the Sciences and the Humanities*, 1947, p. 146.

2. Cf. Pierre Duhem, *The Aim and Structure of Physical Theory*, 1954, esp. pp. 183–190.

3. W. v. O. Quine, "Two Dogmas of Empiricism," in his: *From a Logical Point of View*, 2d ed., 1961, p. 43. Cf. also p. 41, n. 17.

4. Quine, "Two Dogmas of Empiricism," p. 43.

5. H. Putnam, "Three-Valued Logic," *Philosophical Studies* 8 (1957), 74.

6. Cf. H. L. Alexander, *Reactions With Drug Therapy*, 1955, p. 14. Alexander writes: "It is true that drugs with closely related chemical structures do not always behave clinically in a similar manner, for antigenicity of simple chemical compounds may be changed by minor alterations of molecular structures.... 1, 2, 4-Trinitrobenzene ... is a highly antigenic compound.... 1, 3, 5-Trinitrobenzene is allergenically inert." (I am indebted to Dr. A. I. Braude for this reference).

7. Cf. A. Einstein, "Reply to Criticisms," in: P. A. Schilpp (ed.), *Albert Einstein: Philosopher-Scientist*, 1949, pp. 676–678.

8. H. Poincaré, *The Foundations of Science*, 1946, p. 76.

9. Cf. Einstein, "Reply to Criticisms," 676–678.

10. Cf. I. S. Sokolnikoff, *Mathematical Theory of Elasticity*, 1946; S. Timoshenko and J. N. Goodier, *Theory of Elasticity*, 1951.

11. H. Weyl, *Space-Time-Matter*, 1950, pp. 67, 93.

12. I draw here on my more detailed treatment of this and related issues in Grünbaum, "Geometry, Chronometry and Empiricism," *Minnesota Studies in the Philosophy of Science* 3, 1962, pp. 510–521.

13. L. P. Eisenhart, *Riemannian Geometry*, 1949, p. 177.

14. This law is only the first approximation, because the rate of thermal expansion varies with the temperature. The general equation giving the magnitude mt (length or volume) at a temperature t, where m_0 is the magnitude at 0° C, is $m_t = m_0(1+\alpha t+\beta t^2+\gamma t^3+\ldots)$, where α, β, γ etc., are empirically determined coefficients (cf. *Handbook of Chemistry and Physics*, 1941, p. 2194). The argument which is about to be given by reference to the approximate form of the law can be readily generalized to forms of the law involving more than one coefficient of expansion.

■ APPENDIX: A LETTER BY WILLARD VAN ORMAN QUINE

Editorial Note:

In a letter to Adolf Grünbaum dated June 1, 1962, W. v. O. Quine responded to the preceding paper. We include this letter here on special request by Professor Grünbaum. It was previously published in: Sandra G. Harding (ed.), *Can Theories Be Refuted? Essays on the Duhem-Quine Thesis*, Dordrecht, Holland: D. Reidel Publishing 1976, p.132. We thank W. v. O. Quine for permission to reprint the letter here. The letter reads:

Dear Professor Grünbaum:

I have read your paper on the falsifiability of theories with interest. Your claim that the Duhem-Quine thesis, as you call it, is untenable if taken nontrivially, strikes me as persuasive. Certainly it is carefully argued.

For my own part I would say that the thesis as I have used it *is* probably trivial. I haven't advanced it as an interesting thesis as such. I bring it in only in the course of arguing against such notions as that the empirical content of sentences can in general be sorted out distributively, sentence by sentence, or that the understanding of a term can be segregated from collateral information regarding the object. For such purposes I am not concerned even to avoid the trivial extreme of sustaining a law by changing a meaning; for the cleavage between meaning and fact is part of what, in such contexts, I am questioning.

Actually my holism is not as extreme as those brief vague paragraphs at the end of "Two Dogmas of Empiricism" are bound to sound. See sections 1–3 and 7–10 of *Word and Object*.

Sincerely yours,

W. V. Quine

PART II

Determinism and the Human Condition

4   Free Will and Laws of Human Behavior

■ 1. INTRODUCTION

Is man's possession of free will compatible with causal and/or statistical laws of human behavior? This problem has been one of the perennial issues of modern philosophy.[1] Its ramifications include the applicability of the scientific method to the study of our voluntary behavior, the logical consistency of *advocating* specific social changes while also predicting their occurrence, and the justice of making moral evaluations of human agents as a basis for meting out punishment or rewards to them. Thus, in his recent *Action and Purpose*, Richard Taylor has tried to show, by some new arguments, that if volitions are the causes of our actions, then we are not free but are compelled to act as we do. He therefore claims that our free actions must be exempt from the causal sphere.[2] And in a paper entitled "Some Limitations of Science," Thomas Murray of the United States Atomic Energy Commission wrote as follows:

> However useful science is to investigate the privacy of tiny chambers called atoms, it is all but useless to investigate the inner and higher life of man. You can't examine free-will in a test tube. Yet, much of what man does for weal or woe springs from this inner life of free choice.[3]

We are told that while causal and statistical laws characterize the physical world and thus make possible predictions and retrodictions of the careers of physical processes, the consciousness characteristic of man in some sense intrinsically defies any such characterization by laws. And since scientific mastery of a domain requires successful predictions, retrodictions, and/or explanations which only the existence of laws makes possible, it is claimed that consciously directed human behavior is beyond the scope of scientific comprehensibility.

It is ironic that this claim often finds adherents among executives like Murray, whose every step in the management of people is based on the unwitting assumption of *causal* connections between *influences* on men and their *responses*. But it is precisely this causal or *deterministic* conception of man which is repudiated by Murray. And in view of that conception's pivotal role in our inquiry, we shall need to make a careful assessment of its credentials. The deterministic conception of the *inorganic* sector of nature found its modern prototype in classical Newtonian mechanics, and in classical physics generally. These classical theories feature exceptionless functional dependencies relating the states of physical systems as follows: Given the state of a physical system at one or more times, its state at other times is uniquely determined. But, of course, these particular theories are not asserted by determinism as such and are only exemplifications of a deterministic kind of theory. Moreover, there is a notorious vagueness in determinism as a general thesis about the world and even as a regulative principle of scientific research. Yet the problems which we are about to discuss have significance, I believe, because they would confront any reasonably satisfactory explication of the notion of determinism which has evolved in modern science.

The deterministic conception of *human behavior* is inspired by the view that man is an integral part and product of nature and that his behavior can reasonably be held to exhibit scientifically ascertainable regularities just as any other *macroscopic* sector of nature. Determinism must be distinguished from predictability, since there are at least two kinds of situations in which there may be no predictability for special *epistemic* reasons even though determinism is true: (i) Though determinism may hold in virtue of the existence of one-to-one functional dependencies between specifiable attributes of events, some such attributes *may* be "emergent" in the sense of being unpredictable relatively to any and all laws that could possibly have been discovered by us humans in advance of the first occurrence of the attribute(s) in question; and (ii) there can be persons who choose among alternative courses of action *not* in the light of the benefits that may accrue from the action but with a view to assuring that someone else's prediction of their choice behavior turns out to be false. And there may be conditions under which such persons are bound to succeed in behaving *contra*-predictively.[4] For there may be conditions under which these persons can make it *physically impossible* for the predictor to gain access to some of the information essential to the success of the prediction. Thus, there may be cases in which determinism is true even though prediction is not epistemically feasible even in principle. Indeed, it *may* be that for *external* predictors, either humans or computers, it is not physically possible to predict certain deterministic processes in the interior of the Schwarzschild radius of a Schwarzschild gravitational field of general relativity theory, and *perhaps* certain *deterministic* processes inside the so-called black holes of astrophysics are thus unpredictable for *outside* predictors.

Furthermore, as Popper has argued, no deterministically operating computing machine can calculate *its own future* in full detail, and hence the career of a deterministic system will *not* be completely predictable by a human or a computer if they themselves are constituents of that system.[5] Mindful of this result concerning self-referential or auto-predictability, Wilfrid Sellars has written as follows:

> Conceptual difficulties do arise about universal predictability if we fail to distinguish between what I shall call *epistemic* predictability and *logical* predictability. By epistemic predictability, I mean predictability by a predictor *in the system*. The concept of universal epistemic predictability does seem to be bound up with difficulties of the type explored by Gödel. By logical predictability, on the other hand, is meant that property of the process laws governing a physical system which involves the derivability of a description of the state of the system at a later time from a description of its state at an

earlier time, without stipulating that the latter description be obtained by operations within the system. It can be argued, I believe, with considerable force, that the latter is a misuse of the term "predictability," but it does seem to me that this is what philosophers concerned with the free will and determinism issue have had in mind, and it simply muddies up the waters to harass these philosophers with Gödel problems about epistemic predictability.[6]

Thus, there are several weighty reasons for not identifying or equating determinism with universal predictability, that is, with the physical possibility of universal prediction by predictors both inside and outside the system to which the prediction pertains.

The moral opponent of determinism, or "indeterminist libertarian," maintains that determinism is *logically incompatible* with the known fact that people respond meaningfully to moral imperatives. Specifically, this indeterminist says: If each one of us makes decisions which are determined by the sum total of all the relevant influences upon us (e.g., heredity, environmental background, the stimuli affecting us at the moment), then no man *can help* doing what he does. And then the consequences are allegedly as follows: (a) It is impossible to allow for our feeling that we are able to act freely except by dismissing it as devoid of any factual foundation. (b) It is useless to try to choose between good and bad courses of action. (c) It is ill conceived to hold people responsible for their acts. (d) It is unjust to punish people for wrongdoing or reward and praise them for good deeds. (e) It is mere self-delusion to feel remorse or guilt for past misdeeds. (Hereafter the exponent of these views will be called "indeterminist" for brevity.)

Furthermore, the indeterminist sometimes makes the ominous declaration that, if determinism became known to the masses of people and was accepted by them, moral chaos would result, because—so he claims—everyone would forthwith drop his inhibitions. The excuse would be that he cannot help acting uninhibitedly, and people would fatalistically sink into a state of futility, laziness, and indifference. Moreover, we are told that, if determinism were believed, the great fighters against injustice in human history would give up raising their voices in protest, since the truth of determinism would allegedly make such efforts useless.

Thus, the indeterminist goes on to contend that there is a basic *inconsistency* in *any* deterministic *and* activistic sociopolitical theory. The alleged inconsistency is the following: to *advocate* a social activism with the aim of thereby bringing about a future state whose eventuation the given theory regards as assured by historical causation. This argument is applied to any kind of deterministic theory independently of whether the explanatory variables of the historical process are held by that theory to be economic, climatic, sexual, demographic, geopolitical, or the inscrutable will of God. Accordingly, the indeterminist objects to such diverse doctrines as (a) Justice Oliver Wendell Holmes's dictum that "the mode in which the inevitable comes to pass is through effort"[7]; and (b) St. Augustine's (and Calvin's) belief in divine foreordination, when coupled with the advocacy of

Christian virtue. And correlatively, the indeterminist claims that if determinism is true, it is futile for men to discuss how to optimize the achievement of their ends by a change in personal or group behavior.

I do not regard the persistence of the polemics associated with these contentions as betokening the futility of the controversy. Instead, I believe that a good deal has already been clarified by the literature on it and that further progress is possible. Indeed, the present essay is an attempt to improve substantially upon my earlier defense of the thesis that man's "inner life of free choice" in principle no more eludes scientific intelligibility than does "the privacy of tiny chambers called atoms," to use Murray's parlance.[8]

■ 2. THE ARGUMENT FROM MORALITY

To introduce the objections to the argument from morality given by *some* of the indeterminists, let us suppose, for argument's sake, that the truth of determinism would actually render moral imperatives ill conceived by entailing that moral appraisals and exhortations rest on an illusion. This would indeed be tragic. But it could hardly be claimed that determinism is false on the mere grounds that its alleged consequences would be terrible. We would show concern for the sanity of anyone who would say that his house could not have burned down because this fact would make him unhappy. And it is a stubborn fact that no amount of Norman Vincent Peale's "positive thinking"* can assure that the sun will warm the earth forever so that the human race will avoid the calamity of ultimate extinction.

But is it actually the case that there are data from the field of human responses to moral rules or from the phenomena of reflective action which refute the deterministic hypothesis? I shall argue that the answer is decidedly negative. For I shall maintain that in important respects the data are *not* what they are alleged to be. And insofar as they are, I shall argue that they are not evidence against determinism. Nay, I shall claim that in part, these data are first rendered intelligible by determinism. Furthermore, I shall show what precise meaning must be given to certain moral concepts like responsibility, remorse, and punishment within the context of a deterministic theory. Of course, determinism does exclude, as we shall see, *some* of the moral conceptions entertained by philosophical indeterminists like Charles Arthur Campbell. But I shall maintain that this involves no actual loss for ethics. And we have already seen that even if it did, this would not constitute evidence against the causal character of human behavior, which is asserted by determinism. I wish to emphasize, however, that the truth of that deterministic assertion can be established inductively *not* by logical analysis alone but requires the working psychologist's empirical discovery of specific causal laws.

* *Editor's Note*: Norman Vincent Peale was an influential Methodist minister and author in the United States in the 1960 and 1970s. His most prominent book was *The Power of Positive Thinking*, Random House, 1955.

To establish the invalidity of the moral argument of the indeterminist, I shall now try to show that there is no incompatibility between the existence of either causal or statistical laws of voluntary behavior, on one hand, and the feelings of freedom which we actually do have, the meaningful assignment of responsibility, the rational infliction of punishment, and the existence of feelings of remorse or guilt, on the other.

2.1 The Fallacious Identification of Determinism With Fatalism

In some cases, the charge that determinism rules out meaningful ethical injunctions springs from the indeterminist's confusion of determinism with one or both of two versions of fatalism. Let us now distinguish determinism from each of these versions in turn.

The first of these two forms of fatalism is the appallingly primitive prescientific doctrine that in every situation, regardless of what we do, the outcome will be unaffected by our efforts. We can grant at once that if we were to learn that the sun will become a supernova during the latter part of this century or that the earth will soon collide with a giant astronomical body, we would be helpless to avert total catastrophe. But if a diabetic is in glucose shock on a certain day, it is plainly empirically false to say that it is immaterial whether he is then administered insulin or sugar. The fatalist tells us that if the diabetic's time is up on that day, he will die then in either case but that if he is destined to live beyond that day then he will survive it in either case. That a person dies at the time of his death is an utter triviality. And it is banal that for each of us there is a time at which we die. But these truisms do *not* lend any credence to the fatalist, who tells us that a man dies when his time is up in the sense that human effort to postpone death is *always* futile.

The latter false thesis of fatalism does not follow at all from determinism. The determinist believes that specifiable causes determine our actions and that these, in turn, determine the effects that will ensue from them. But this doctrine allows that human effort be efficacious in *some* contexts while being futile in others. Thus, determinism allows the existence of situations which are correctly characterized by Justice Oliver Wendell Holmes's cited epigram that the inevitable comes to pass through effort. The mere fact that both fatalism and determinism affirm the fixity or determinedness of future outcomes has led some indeterminists to infer fallaciously that determinism is committed to the futility of *all* human effort. The determinist maintains that existing causes determine or fix whether certain efforts will in fact be made at certain times while allowing that future outcomes are indeed dependent on our efforts in particular contexts. By contrast, the fatalist holds falsely that such outcomes are always independent of all human efforts. But the determinist's claim of the fixity of the outcome does not entail that the outcome is independent of our efforts. Hence, determinism does not allow the deduction that human intervention or exertion is futile in every case. One might as well deduce the following absurdity: Determinism guarantees that explosions are always independent of the presence of detonating substances because determinism asserts that, in specified contexts, the effects of the presence of explosives are determined!

The misidentification of determinism with fatalism can now be seen to underlie two of the contentions by indeterminists, which were stated in the Introduction.

The predictions that might be made by contemporary historical determinists concerning the social organization of industrial society, for example, pertain to a society of which these forecasters are themselves members. Hence, such predictions are self-referential. But these predictions are made by social prophets who, qua deterministic forecasters, consider their own society externally, from *without* rather than as active contributors to its destiny. And the predictions made from that theoretically external perspective are *predicated* on the prior fulfillment of certain initial conditions. These conditions include the presence in that society of people—among whom they themselves may happen to be included—who are dissatisfied with the existing state of affairs and are therefore actively seeking the future realization of the externally predicted social state. To ignore that the determinist rests his social prediction in part on the existence of the latter initial conditions, just as much as a physicist makes a prediction of a thermal expansion conditional upon the presence of heat, is to commit the fallacy of equating determinism with fatalism. Thus, a person's role as *predictor* of social change from an "external" perspective, as it were, is quite compatible logically with his belief in the necessity of his being an *advocate* of social change internal to society. We see that the indeterminist has no valid grounds for the following objection of his: It is logically inconsistent for an historical determinist, qua participating citizen, to advocate that action be taken by his fellow citizens to create the social system whose advent he is predicting on the basis of his theory. For it is now plain that the indeterminist's charge derives its semblance of plausibility from his confusion of determinism with fatalism in the context of self-referential predictions. This confusion is present, for example, in Arthur Koestler's claim, in his *Darkness at Noon*, that the espousal of historical determinism by Marxists is *logically inconsistent* with their reproaching the labor movement in capitalist countries for insufficient effort on behalf of socialism.

Equally fallacious is the indeterminist's claim that it is practically *futile* for a determinist to weigh alternative modes of social organization with a view to optimizing the organization of his own society. For the determinist does *not* maintain, in fatalist fashion, that the future state of society is independent of the decisions which men make in response to (a) facts (both physical and social), (b) their own *interpretation* of these facts (which, of course, is often false), and (c) their value objectives. It is precisely because, on the deterministic theory, human decisions *are* causally dependent upon these factors that deliberation concerning optimal courses of action and social arrangements can be reasonably expected to issue in successful action rather than lose its significance by adventitiousness. In short, the causal determinedness of the outcome of a process of human deliberation does not at all render futile those deliberations which issue in true beliefs about the efficacy of specified actions.

There is another version of fatalism which Gilbert Ryle has articulated[9] and rightly criticized as a non sequitur.[10] If Ryle went to bed at a particular time on a certain Sunday, then it was true a thousand years before then that he would go to bed a millennium later as stated. Indeed in the posited case, it was true forever beforehand that he would be going to bed on Sunday, January 25, 1953. But given this fact, it was impossible for Ryle not to have gone to bed at the specified time on that date. For it would be self-contradictory to assert that although it was true beforehand that Ryle would do something at a certain time t, he did not do it. Whatever it was to be. Thence the fatalist believes he can conclude the following: *Irrespective of the antecedent circumstances*, "nothing that does occur could have been helped and nothing that has not actually been done could possibly have been done."[11]

My concern here is *not* with the fallaciousness of this argument, which Ryle has no difficulty demonstrating. Instead, I wish to point out that determinism as such is irrelevant to its truistic premises and even denies its fatalist conclusion. Note that these premises are compatible with indeterminism no less than with causal or merely statistical determinism.[12] And we already saw that far from asserting the independence of events from any and all circumstances prevailing at other times, the determinist explicitly affirms that kind of dependence.

The illicitness of saddling determinism with either version of fatalism is now apparent.

2.2 The Confusion of Causal Determination With Compulsion

Some of those indeterminists who do not equate determinism with fatalism fail to see, however, that psychological laws do *not* force us to do or desire anything against our will. These laws merely state what, as a matter of fact, we do or desire under certain conditions. Thus, if there were a psychological law enabling us to predict that under certain conditions a man will desire to commit a certain act, this law would not be making him act in a manner *contrary* to his own desires, for the desire would be his. It follows that neither the causes of our desires nor psychological laws, which state under what conditions our desires arise and issue in specified kinds of behavior, compel us in any way to act in a manner contrary to our own will. There is in the indeterminist's thinking a confusion of physical and psychological law, on one hand, with statutory law, on the other. As Moritz Schlick emphasized decades ago, psychological laws do not coerce us against our will and do not *as such* make for the frustration or contravention of our desires. By contrast, statutory laws do frustrate the desires of some and are passed only because of the need to do so. Such laws are violated when they contravene powerful desires. But natural laws (as distinct from erroneous guesses as to what they are) cannot be broken. Anyone who steps off the top of the Empire State Building shouting defiance and insubordination to the law of gravitation will not break that law but rather will give a pathetic illustration of its applicability.

We act under *compulsion*, in the literal sense relevant here, *when we are literally being physically restrained from without in implementing the desires which we have upon reacting to the total stimulus situation in our environment and are physically made to perform a different act instead*. In that case, our desires are essentially causally irrelevant to the act which we are being compelled to perform. For example, if I am locked up and therefore cannot make an appointment, then I would be compelled to miss my appointment. Or, if a stronger man literally forces my hand to press a button which I do not wish to press, then I would be compelled to blow up a bridge. The meaning of "compulsion" intended here should not, of course, be identified with the meaning of that term familiar to students of neuroses. In the case of neurotic compulsion, the compulsive person does *unreflectively* what he wishes, although his behavior is inspired by unwarranted anxiety and hence is insensitive to normally deterring factors, as in the case of an obsession with germs issuing in hand washing every time a doorknob is touched.

In using the locution "act under compulsion," I employ the term "act" in a somewhat Pickwickian sense. For I agree with Wilfrid Sellars' remark that "if an action is not voluntary ... then it is not really an action, but rather behavior of a sort which *would* be an action if it *were* voluntary. An involuntary wink is not a wink at all, but rather a blink.... To go out the window propelled by a team of professional wrestlers is not to *do* but to *suffer*, to be a patient rather than an agent."[13]

To emphasize the meaning of "compulsion" relevant to the issue before us, I wish to point out that when a bank teller hands over cash during a robbery upon feeling the revolver pressing against his ribs, he is *not* acting under compulsion in my literal sense, any more than you and I act under compulsion when deciding not to go out to play tennis during a heavy rain. When handing over the money in preference to being shot, the bank teller is doing what he genuinely wants to do *under the given conditions*. Of course, in the absence of the revolver, the teller would not have desired to surrender the cash in response to a mere request. By the same token, it was the heavy rain that induced the hopeful tennis player to wish to stay indoors. But both the bank teller and the frustrated tennis player are doing what they wish to do in the face of the existing conditions.

The relevant similarity of the bank teller case to a case of genuine compulsion in my literal sense lies only in the fact that our legal system does *not* decree punishment either in a case of genuine compulsion, like having one's hand literally forced to blow up a bridge by pressing a button, or in a cases like that of the bank teller. For although the bank teller is actually physically free to hold on to the money and sound the alarm, he is not punished for surrendering the money because the alternative to such surrender would be to sound the alarm at the cost of his own life. The armed bank robber can therefore be said to have "compelled" the teller to surrender the money *not* in our literal sense but only in the sense that the robber's threat was the decisive determinant of the *particular kind of action* that was taken by the teller. Similarly, when deference to their duties under the law "compels" a judge to sentence a dear friend to imprisonment or "compels" a policeman to arrest his own kin, the behaviors of the judge and of the policeman are each voluntary. Their behavior is compelled only in the nonliteral sense of springing from motives which had to overcome their affection for the culprits. Thus, what is common to genuine compulsion and deciding to hand over money in preference to dying is that both of them are treated as *excusing conditions*.

But *causal determination of voluntary behavior is not identical with what we have called literal compulsion!* For voluntary behavior does not cease to be voluntary and become "compelled" in our literal sense just because there are causes for that behavior. The indeterminist needs to show that responsibility is rendered meaningless in every case of causation no less than in the case of literal compulsion. Unless he does, he is not entitled to assert that if determinism were true, the assignment of responsibility to people for their acts would be ill conceived, just because such assignment is inappropriate in the case of literal compulsion, as we shall see.

There are borderline cases between acting freely and acting under compulsion in the aforementioned literal sense. An example of such a borderline case is being under posthypnotic suggestion. Although people will not do things under posthypnotic suggestion which run very strongly counter to powerfully ingrained dispositions of theirs, they will experience a strong sense of urgency to do what the hypnotist told them to do, even though they have no conscious awareness of any particular motivation for doing so. The interference with the person's freedom by the posthypnotic suggestion is constituted by the fact that, when he acts under the suggestion, he is not responding to motivational stimulae from outside on the basis of his normally operative dispositional makeup. It is true that he is doing what the hypnotist instructed him to do quasi-voluntarily in the sense that his hand is not being physically forced from outside, as it were. Therefore, the case of posthypnotic suggestion is in some ways a borderline case between acting freely and acting under compulsion in my literal sense, but perhaps more predominantly a case of acting freely.

It should not be thought that the indeterminist is now prepared to surrender, for he has yet to use the most plausible of his arguments. Says he:

> We are all familiar with the fact that when we look back upon past conduct, we frequently feel very strongly that we could have done otherwise. For example, someone might have chosen to come to work today via a route different from the one he actually did use. If the determinist is right in saying that our behavior was unavoidably determined by earlier causes, this retrospective feeling of freedom either should not exist or else it is fraudulent. In either case, the burden of proof rests upon the determinist.

The determinist gladly accepts this challenge, and his reply is as follows: Let us carefully examine the content of the feeling that on a certain occasion we could have acted other than the way we did, in fact, act. What do we find? Does the feeling we have inform us that we could have acted otherwise *under exactly the same kinds of relevant external and internal motivational conditions? No* says the determinist, even if we take this feeling at face value.

What this feeling discloses is that (1) we were able to act in accord with our strongest desire at that time instead of acting under compulsion; and (2) we not only might but also could and indeed would have acted otherwise if a suitably different motive had prevailed at the time. And this state of affairs is entirely in accord with determinism. For the absence of freedom obtains in cases of literal compulsion and does not depend on the strength of mere psychological causation. Thus, the determinist answer is that the content of this "consciousness of freedom" consists in our awareness that we were able to act in response to our strongest motive at the time and that we were not "under compulsion" in that sense. We were able to do what we wanted. Moritz Schlick has explained in his classic treatment of the free will problem in chapter 2 of *Problems of Ethics* (New York, 1939) that it is *not* a covert tautology to assert, as I have done, that we act in response to our strongest (conscious or unconscious) motive. Thus, the determinist reminds us that our feeling of "freedom" does *not* disclose that, given the motives which acted on us at the time and given their relative strength and temporal distribution, we could have acted differently from the way in which we did, in fact, act.

Neither do we feel that we could have responded to the weaker of two contending motives, or acted without a cause or motive or could have chosen the motives which acted upon us. I never wake up totally devoid of any content of consciousness and then ask my *blank* self: With what motives shall I populate my consciousness this morning? Shall I have the aspirations of Al Capone or those of Albert Schweitzer? Nor do I even know what the indeterminist means by the supposition that we could have chosen our own character *from scratch*: Every decision to shape or choose one's character must be one's own, that is, must be made by an already existing personality, constituted by a set of dispositions. The notion of choosing one's character involves an infinite regress, because an initial "I" is presupposed in the making of the choice.[14] Choices which mold the *subsequent* development of one's character are fully compatible with determinism. Since the retrospective feeling of freedom that we have does *not* report any ability of making the kinds of choices envisioned by the indeterminist, its deliverances contain no facts incompatible with the claim of the determinist. The compatibility of scientific determinism with the ability to do otherwise inherent in freedom of action is exhibited illuminatingly in a partially formalized way in Wilfrid Sellars's more detailed analysis of the practical concepts of ability to do and ability to will.[15]

Thus, according to determinism, freedom of action is rooted in the fact that one or more causal laws of voluntary human behavior do allow specifiable alternative actions with respect to suitably different initial motivational conditions. But given a sufficient set of initial motivational conditions, determinism asserts that we cannot act differently from the way in which we actually do act, under these given circumstances. Yet libertarians such as Charles A. Campbell* have contended that a person's will is free only if he could have acted differently under the given circumstances rather than merely under suitably different conditions. To this conception of freedom of action I say: (i) I, for one, simply do not find introspectively that I possess freedom of action in this Campbellian sense; and (ii) if appropriate causal laws of human behavior do hold, then the universe just does not accommodate Campbell's demand for his kind of freedom, since it simply does not exist in that case. Indeed, as I shall now argue, this kind of freedom is unavailable even on the assumption of merely statistical rather than causal laws of voluntary human behavior. My reason for dealing with this point is that many indeterminists have claimed that only deterministic laws of human behavior and not statistical ones are incompatible with freedom and hence with the meaningful assignment of moral responsibility. In this way, they sought to acknowledge the incontestable existence of an impressive measure of lawful regularity in human behavior (witness the insights of Shakespeare!).

Thus, we now inquire: Are the empirical claims made by the philosophical indeterminist as part of his doctrine of moral responsibility compatible with the assumption of the existence of laws of human behavior which are *irreducibly statistical* rather than deterministic? Irreducibly statistical laws are laws whose statisticality does *not* arise from human ignorance of relevant hidden parameters. Statisticality of laws may be due to human ignorance, since there may be cases in which the possession of more information on our part regarding hidden attributes of the entities in question could serve to disclose the existence of a substratum of deterministic laws masked by the incompleteness of our knowledge. Thus, classical physicists believed that the statisticality of the regularities governing the results of tossing unbiased coins is attributable to human ignorance of the underlying initial conditions and deterministic dynamics of the throws: In principle, the result of each throw was presumed to be uniquely determined by the combination of these factors. By contrast, the majority of present-day quantum physicists believe that the statistical character of the linkages between events involving the microcosm does not mask a crypto-determinism but is irreducible in the sense of not being removable by the possession of more complete information pertaining to the micro-entities in question. Since the philosophical indeterminist's doctrine of freedom and moral responsibility has denied the existence of deterministic laws of human behavior, our problem now is whether that doctrine can allow that there are irreducibly statistical laws of human behavior.

My answer is that it cannot. Consider a statistical law (whose statisticality we presume to be irreducible) stating with near certainty that, of all the people belonging to a certain reference class, in the long run 80 percent will commit a certain

---

* *Editor's Note*: Reference here is to Campbell's inaugural address at Glasgow University: *In Defence of Free Will*, 1938 (published in the anthology C. A. Campbell, *In Defence of Free Will*, London: Allen & Unwin, 1967, pp. 35–55).

kind of crime at some time during their adulthood. To be sure, this statistical law would not entitle us to say that any particular individual(s) belonging to the reference class will become criminal; hence, it does not preclude the possibility that some particular person (or persons) who did, in fact, commit the crime, might have been among the 20 percent whose conduct is legal and that, *to this extent*, the person in question be regarded as having acted "freely" in the indeterminist sense. Insofar as responsibility is an individual matter, it might even *seem* that our statistical law would permit the indeterminist to claim that his conditions for being held morally responsible are met by each member of the reference class who does commit the crime in question. For the cardinal one of these conditions is that each person guilty of the crime "could have done otherwise" under the given kinds of conditions.

But *in what sense* does the statistical law under consideration entitle one to say that each criminal *could have done otherwise* when it tells us that, in the long run, one of every five members of the reference class does avoid the crime, that is, that the probability of crime avoidance is 1/5? As Kurt Baier has helpfully put it, the irreducibly statistical law entitles us to say, with regard to any particular individual, A, who commits the crime, that it might well have been A who would not commit the crime. But the law does not entitle us to say that the individual who commits the crime could have done otherwise. The former says merely that on the basis of the law it is not specified *who* would commit the crime. The latter asserts that those who we know committed the crime could have refrained from committing it. The former expresses nonspecification of what a *particular* person *would* do, the latter claims specification of what he *could have done instead* of what he *did* do. Hence, the indeterminist is not even entitled to invoke the statistical law to assert that any one of the criminals could have done otherwise in the merely distributive sense of "anyone." A fortiori, he cannot appeal to the statisticality of the law to make the following claim, as required by his cardinal condition for ascribing moral responsibility: in the *collective* sense of "anyone," each and every one of the culprits in the reference class could have avoided the crime under the given circumstances in both the short and long run. Nay, if the statistical law is to have empirical content by having finitary significance, this contention *contradicts* the statistical law for any typical long finitary run of cases.

We see that the statistical law under consideration, whose statisticality we are presuming to be irreducible, assures sufficient causality to preclude the kind of freedom required by the philosophical indeterminist for the assignment of moral responsibility. If the indeterminist denies the existence of freedom and the justice of punishment on the assumption of deterministic laws of human behavior, then he cannot assent to the punishment of individuals belonging to reference classes concerning which statistical laws assert a behavioral regularity. Accordingly, the indeterminist must have moral objections to deterministic and statistical laws of human behavior alike. And this means that he is logically committed to the futility of the behavioral scientist's quest even for statistical laws.

The analysis we have offered is applicable at once to the case of remorse, regret, or guilt. We sometimes experience remorse over past conduct when we reconsider that conduct in the light of different motives or of a new awareness of its consequences. Once we bring a different awareness of consequences and/or a new set of motives to bear on a given situation, we may feel that a different decision is called for. If our motives or appraisals do not change, we do not regret a past deed no matter how reprehensible it would otherwise appear. Nathan Hale told a British court that he wished he had more than one life to give for the American Revolution.* And many a killer has honestly declared that if he had to choose again, he would do again precisely what he had done. In that case, the relevant motives had not changed. Regret is an expression of our emotion toward the disvalue and injustice which issued from our past conduct, as seen in the light of new motives. The regret we experience can then act as a *deterrent* against the repetition of past behavior which issued in disvalue. If the determinist expresses regret concerning past misconduct, he is applying motives of self-improvement to himself but not indulging in retroactive self-blame. Now answering for one's past misdeeds may redirect later conduct. Thus, when ascribing responsibility for misdeeds the determinist does *not* mean retroactive blameworthiness but rather that being held responsible is tantamount to *liability to reformative or educative punishment*. Punishment can be educative in the sense that, when properly administered, it institutes counter-causes to the repetition of injurious conduct. For the determinist who wishes to spare culprits no less than others *gratuitous* suffering, punishment is never an end in itself. It is *not* intended as a revenge catharsis. The humane determinist rejects as barbarous the primitive vengeful idea of retaliatory, retributive, or vindictive punishment. He condemns hurting a man simply because the man has hurt others, for the same reason that he would condemn stealing from a thief or cheating a swindler. He fails to see how the damage done by the wrongdoer is remedied by the mere infliction of pain or sorrow on the culprit, *unless* such infliction of pain promises to act as a causal deterrent against the repetition of evil conduct. For the humane determinist, the decision whether pain is to be inflicted on the culprit and, if so, to what extent is governed solely by the conduciveness of such punishment to the reform and re-education of the culprit and to repairing his damage, where possible, or to the deterrence of other potential criminals. The requirement that punishment can be expected to be reformatory does not itself specify what choice is to be made among two equally effective punishments of differing severity. But the use of the moral requirement that *gratuitous* suffering be avoided as a principle of justice does indeed make this choice unique. The mathematics professor at the University of Pennsylvania whose three-year-old daughter was murdered by

---

* *Editor's Note*: Nathan Hale (1755–1776) was a member of the Continental Army during the American Revolution and was captured by the British. In the United States he is perhaps best known and praised for his purported last words before being hanged: "I only regret that I have but one life to give for my country."

an adolescent in Philadelphia nobly disavowed all caveman revenge. But he did ask for greater preventive efforts in diagnosing potentially homicidal but seemingly exemplary adolescents.

The implementation of this conception requires psychological and sociological research into *causal* connections and the institution of a *rational prison system*. If kindness rather than punishment were to deter the recidivist criminal, then it is clearly rational to be kind. The design and organization of the penal system must be set up accordingly, and the social cost would probably be less. Revenge seekers do not care whether prison hardens criminals further or whether the social cost of protecting society increases further. Their motto is that of the English schoolmaster: Be pure in heart boys, or I'll flog you until you are.

On the determinist view of punishment as educative, "punishment" of the criminally insane automatically takes the form of treatment since their insanity makes for their lack of a unified point, as it were, for applying a counter-motive by the usual kinds of punishment. Indeed, modern brain research indicates that felons committing wanton crimes of violence from irresistible impulses they themselves cannot understand often suffer from temporal lobe brain disease and other lesions of the brain. In a number of cases, this damage can be traced to head injuries, accident during birth, and poor nutrition in the early years of life. Instead of counseling the infliction of futile vindictive reprisal on such culprits from anger and fear, the humane determinist enjoins the courts to screen all such offenders so that they may receive appropriate neurological therapy, if at all possible.

Insanity and punishment should *not* be judged by the McNaughton criterion of English law: Did the wrongdoer "really know" the difference between right and wrong at the time of the act? For this criterion does *not* suffice to determine whether a man is sane in the sense of being capable of being deterred from repeating his crime by the same punishment that would deter the rest of us. Often a man who is judged sane by the *traditional* criterion does *not* respond to the standard punishment and may even be attracted by the prospect of such punishment.

Our legal and penal system is to a certain extent an inconsistent and unresolved compromise between the revenge philosophy of medieval spiritism which prescribed tortures for the insane and a grudging recognition of some of the findings of modern science concerning the conditions breeding criminal behavior. For the humane determinist, there is no "tempering of justice with mercy": The punishment is never made more severe than is believed necessary to reform the criminal or prevent him from continuing his destructive behavior. "Tempering justice with mercy" is the philosophy of either a revenge seeker who has qualms or a man who is torn between the revenge conception and the reformatory one.

The *New York Sun* editorially supported revenge in the Leopold parole case and said that reform is not enough, although it was not against throwing in a touch of mercy for good measure. On the other hand, the great attorney Clarence Darrow, who was as warmhearted as he was mistaken, used a revenge conception of punishment in conjunction with a belief in determinism to plead with his juries that criminals should be excused for their misdeeds since they did not "choose their own character." But he and most of his juries failed to realize that the notion of choosing one's own character is self-contradictory, and hence this notion can hardly be used to characterize the conditions under which revengeful or vindictive punishment would be just.

Contrary to Darrow, we see that a humane determinism does *not* entail the doctrine that "tout comprendre, c'est tout pardonner," that is, does *not* claim that *to understand all is to forgive all*: Punishment that prevents or deters human beings from committing acts issuing in much greater pain than is inflicted by the punishment is the lesser evil. But in the case of injurious conduct which qualifies as compelled in the aforementioned literal sense, the determinist does not administer punishment. For the dispositions of a person who acted injuriously under literal compulsion were *irrelevant* causally to his so acting. And since the volitional dispositions of such a person require no reforming on this score, punishment would be completely misplaced, gratuitous, *and* unjust in such a case. Thus, if an act which is socially injurious is committed under posthypnotic suggestion, it would be misplaced to punish the individual in the sense of attempting to use means of punishment usually appropriate to the deterrence of unhypnotized individuals who commit the same injurious act. For the individual who underwent posthypnotic suggestion would presumably not be disposed to repeat the act when not hypnotized even though the other relevant circumstances are the same. As we noted earlier, the interference with the person's freedom by the posthypnotic suggestion is constituted by the fact that, when he acts under the suggestion, he is not responding to motivational stimulae from outside on the basis of his normally operative dispositional makeup, although it is true that he is doing what the hypnotist instructed him to do quasi-voluntarily in the sense that his hand is not being physically forced from outside, as it were. Hence, the only punishment appropriate in the case of such a posthypnotically induced misdeed would be deterrence of the individual against resubmitting to that kind of risky hypnosis.

For the determinist, therefore, the theoretical edifice of responsibility for misdeeds and liability to punishment rests importantly on his distinction between voluntary and literally compelled behavior. And he emphasizes that the pertinent crucial difference between these two kinds of behavior is not gainsaid in the least by his contention that there are causes for both kinds of behavior. Our focus has been on responsibility for misdeeds. But our account of the humane determinist's position is readily extended to the case of ascribing responsibility for *desirable* deeds. In the latter case, to assign responsibility is to give credit with a view to reinforcing the desirable behavior patterns.

It is apparent that the entire problem of responsibility can be solved within the domain of deterministic assumptions. Thus, the issue is not *whether* conduct is determined but rather *by what factors* it is determined, when responsibility is to be assigned: specified kinds of insanity, compulsion in the literal sense, ordinary volition, and so forth. And it is clear now why the determinist does not see any merit in the claim that culprits should be exempt from punishment merely because they could not have acted differently under the given conditions. Far from facing insuperable difficulties with the problem of responsibility, the determinist and the scientific psychologist now challenge the indeterminist to provide a logical foundation for a penal system. We recall that the indeterminist accused the determinist of cruelly punishing people who, if determinism is true, cannot help acting as they do. The determinist now turns the tables on his antagonist and accuses him of being gratuitously vengeful, on the grounds that the indeterminist is committed by his own theory to a retaliatory theory of punishment. The indeterminist cannot consistently expect to achieve anything better than retaliation by inflicting punishment, for were he to admit that punishment will causally influence all or some of the criminals, then he would be abandoning the basis for his entire argument against the determinist. As Leszek Kolakowski has rightly remarked: "It is an obvious truth ... that if one believes punishment ... can be effective, then one posits by that very fact some kind of determinism of human behavior."[16]

2.3 Other Arguments of the Indeterminist

It is sometimes said that, when applied to man, the deterministic doctrine becomes untenable by virtue of becoming self-contradictory. This contention is often stated as follows: The determinist, by his own doctrine, must admit that his very acceptance of determinism was causally conditioned or determined. Since he could not help accepting it, he cannot argue that he has chosen a true doctrine. To justify this claim, it is first pointed out rightly that determinism implies a causal determination of its own acceptance by its defenders. Then it is further maintained, however, that, since the determinist could not, by his own theory, help accepting determinism under the given conditions, he can have no confidence in its truth. Thus, it is asserted that the determinist's acceptance of his own doctrine was forced upon him. I submit that this inference involves a radical fallacy. The proponent of this argument is gratuitously invoking the view that if our beliefs have causes, these causes *force* the beliefs in question upon us, against our better judgment, as it were. Nothing could be further from the truth; this argument is another case of confusing *causation* with *compulsion*. Its proponent fails to allow that the decisive cause of the acceptance of determinism by one of its adherents may have been his belief that the available evidence supports this doctrine. And evidence is a uniquely distinguished touchstone for gaining true knowledge of the world. For what makes a factual claim true is that it asserts what is the case. And the very existence of the state of affairs truly affirmed by the claim will then manifest itself to us under appropriate conditions in the form of evidence for it.

The causal generation of a belief does not, of itself, detract in the least from its truth. My belief that I address a class at certain times derives from the fact that the presence of students in their seats is causally inducing certain images on the retinas of my eyes at those times and that these images, in turn, then cause me to infer that corresponding people are actually present before me. The reason that I do not suppose that I am witnessing a performance of *Aïda* at those times is that the images which Aïda, Radames, and Amneris would produce are not then in my visual field. The causal generation of a belief in no way detracts from its veridicality. In fact, if a given belief were not produced in us by definite causes, we should have no reason to accept that belief as a correct description of the world rather than some other belief arbitrarily selected. Far from making knowledge either adventitious or impossible, the deterministic theory about the origin of our beliefs alone provides the basis for thinking that our judgments of the world are or may be true. Knowing and judging are indeed causal processes in which the facts we judge are determining elements along with the cerebral mechanism employed in their interpretation. It follows that although the determinist's assent to his own doctrine is caused or determined, the truth of determinism is not jeopardized by this fact; if anything, it is made credible.

More generally, both true beliefs and false beliefs have psychological causes. The difference between a true or warranted belief and a false or unwarranted one must therefore be sought *not* in *whether* the belief in question is caused. Instead, the difference must be sought in the particular *character* of the psychological causal factors which issued in the entertaining of the belief; *a warrantedly held belief, which has the presumption of being true, is one to which a person gave assent in response to awareness of supporting evidence.* Assent in the face of awareness of a *lack* of supporting evidence is irrational, although there are indeed psychological causes in such cases for giving assent. Thus, one person may be prompted to give assent to a certain belief solely because this belief is wish fulfilling for him, while another may accept the same conclusion in response to his recognition of the existence of strong supporting evidence.[17]

I hope that these considerations have shown, therefore, that it is entirely possible to give a *causal* account of both rational and irrational beliefs and behavior. And since a causal account is based on principles and regularities which are based on evidence, it follows that we can indeed give a rational explanation of why it is that people do behave irrationally under certain conditions, no less than we can provide a causal account of their rational behavior.

Last, a remark on the belief that determinism would lead to moral cynicism: Why should my belief that my motives for wishing to help someone are caused lessen my readiness to implement my desire to help that person? Lincoln's view that his own beliefs (ethical and other) were causally determined did not weaken in the least his desire to abolish slavery, as demanded by his ethical theory, and similarly for Augustine, Calvin, Spinoza, and hosts of lesser men. Furthermore, it is inconsistent for an indeterminist to invoke a *causal connection* between the espousal of determinism and moral cynicism.

■ 3. IS LOVE MEASURABLE?

The phenomena involving human affection have sometimes been adduced as concrete evidence against the existence of laws of human behavior. Human love, we are told, is in principle "intangible" or not amenable to measurement. Hence, it is claimed that love does not lend itself to description by laws and thereby eludes scientific understanding. The proponents of this indeterminist view do not, of course, mean mere intangibility in the purely etymological sense that would accord to the sense of touch a preferred status as an avenue of scientific knowledge. But awareness of our reasons for *not* regarding that sense as having unique reliability will throw light on the issue of the alleged intangibility or nonmeasurability of love. These reasons emerge from the following twofold considerations:

(a) As far as vision alone is concerned, we can interpret the visually bent appearance of a stick partially immersed in water as follows: The stick bends when partially immersed and then unbends when taken out. By means of pure vision, we are quite ignorant of any optical theories of refraction. But the sense of touch is invoked by some to declare that sight deceived us as to the condition of the stick: Touch tells us that the partially immersed stick is straight. Hence, here touch presumably corrects sight.

(b) But the inverse order of trustworthiness prevails among the senses of sight and touch in regard to ascertaining the dimensions of a distressing cavity in one's tooth: tactile exploration with the tongue suggests a huge crater, but the dentist's mirror is held to tell the true story visually.

How can we oscillate like that in accepting the verdicts of a given sense organ in one case and not in the other? The answer was given by Immanuel Kant when he said that percepts without concepts are blind. *Theories* are used to assess the *significance* of observational findings. And the same holds true for *measures*, whether in physics or psychology. For some measures are totally devoid of theoretical significance, while other measures do indeed possess such significance. A given kind of measure has theoretical significance if it is explanatorily and/or predictively fruitful by virtue of being lawfully linked to other quantities or attributes, as we shall now see.

Suppose that I were to define the "Mathew measure" of a person at a given time as follows: blood count, multiplied by the number of hairs, divided by the square root of the height, the measure being plus or minus depending on whether the person is Rh+ or −. I venture to say that although this is a perfectly well-defined measure, no physiologist would be interested in it. Again, suppose that I were to define the "George index of a piece of metal" as follows: 3/2 π, times its mass, times the square root of its electrical conductivity, divided by 45 times its volume. The George index plays no role in metallurgy. Neither is the Mathew measure of interest to the physiologist. And why not? Because these respective measures or indices are not lawfully (predictively) related to *other* properties whose occurrence is of concern to the physiologist or metallurgist. Similarly, there are a host of perfectly trivial measures of love that one could introduce and which are unavailing, *not* because they fail to generate numbers but because they are explanatorily sterile. Thus, consider the following measures of love: Disregarding ethical or legal complications, how much electric shock or how much imprisonment up to ten years would a man be willing to stand to marry the woman he loves? Apart from its impracticality, the sterility of this measure lies in the fact that it is not lawfully related to whether a man will exhibit *other* types of behavior which his wife may expect from him by virtue of his love. Thus, there is no "intangibility" of love in the sense of lack of measures per se, that is, of procedures which generate numbers nonarbitrarily. On the other hand, if the "intangibility" of love is held to lie in the purely inferential, indirect knowledge we have of *other* people's feelings of affection toward us, then indeed neutrinos, nuclear forces, and the interiors of stars would also have to be deemed "intangible" and scientifically intractable, which they are surely not.

But, you might ask, do we not have reason to think that in the human domain there simply are no theoretically significant fruitful measures *at all* and, correlatively, no regularities, whereas in physical science there are some celebratedly successful measures and laws? This seems to me a misleading oversimplification. First of all, there are a host of reliable indices of the presence or absence of love which we generally regard as obvious because of their familiarity. It is banal to say that the willful and knowing throwing of sulfuric acid by a man into the face of another for pay is a reliable indication of deficient affection and even callous indifference to human suffering. And it is equally trite to predict that under the usual assumed initial conditions, the vast majority of undergraduates will stay away from their college campuses during vacations. Similarly, people are unimpressed when a businessman makes dependable forecasts of the economic fortunes of neighborhood grocery stores in small communities. But they ask economists to predict the state of the U.S. national economy for the next ten years. Why? Because the exigencies of life confer great urgency on the immediate solution of highly complex problems by the social scientist at times when he is no more ready to cope with *these* particular problems than Galileo was to discover general relativity or Pascal was to design the rocketry of planetary probes. To be sure, there are social demands made on the physical scientist also. But almost any discovery by him issuing in a change in the externals of life will be hailed as an indication of the power of the scientific method to deal with inanimate nature and will be contrasted with the failure of the psychiatrist to find the cause and cure of schizophrenia and of the political scientist to devise an immediately workable disarmament agreement. Psychoanalysis was no sooner born when it was confronted with the demand to administer successful therapy. Yet no one complained in 1905 that the special theory of relativity failed to provide an engineering recipe for a controlled nuclear chain reaction,[18] although its possibility was foreseen. When people say that the physical sciences are more successful than the social sciences, what measure of failure do they use? I believe that the implicit measure is very often the relative social urgency of the problems which these disciplines leave *unsolved*. It is clear that in such comparison the social or behavioral sciences are bound to end up on the bottom. But, this purely *pragmatic* measure has none of the derogatory implications for the scientific tractability of man which indeterminists claim.

■ 4. QUANTUM PHYSICS AND THE FREEDOM OF THE WILL

In the sense of section 2.2 let us hereafter assume the irreducibility of the *statistical* links asserted by quantum physics between the attributes of individual events in physical space and time. And let us examine the import of this assumed kind of quantum mechanical indeterminism for the possibility of the scientific study of man by stating what conclusions, if any, seem to be warranted by the logic of the situation concerning free will.[19]

I have argued that our retrospective feeling of freedom that we could have acted otherwise does *not* tell us that our decisions are uncaused in the sense that they could have been different in the face of the same kinds of relevant circumstances and desires. Furthermore, so far as I can see, our feeling of freedom merely discloses often that we can do what we wish but *not* that we can will what desires we shall have. As we saw, the very concept of "I" or of self already involves a set of dispositions which come into play when this self finds itself with desires that it has not chosen from scratch and deliberates in response to them. In short, I have maintained that the kind of feeling of freedom which the indeterminist takes for granted does not, in fact, exist. But I defer until section 5 a discussion of the interesting *new gloss* which the brain physicist Donald MacKay has put upon the feeling of freedom.

I therefore regard as wholly ill conceived the quest for indeterministic neurological correlates of the nonexistent kind of feeling of freedom postulated by the indeterminist free willist. For the nonexistence of the latter kind of feeling seems to me to make it idle to try to find quantum processes in the nervous system which are its supposed physical correlates. Moreover, in the present state of neurophysiology, it would seem to be quite unclear what is the physico-neurological correlate of the feeling of freedom which I claim we do experience, namely, the feeling or informational awareness state that we can often do what we wish under given circumstances and that under *other* circumstances we might well both have different desires and act differently. By the same token, even if the retrospective feeling of freedom *did* disclose anything incompatible with the causal generation of our decisions—which I claim it does not—it is altogether unclear in the present state of neurophysiology how the discoveries of quantum physics could be adduced to show *on the level of the human organism* that this feeling *must* have a foundation in fact and should be taken at face value. Nay, I now want to argue that such an inference must be a non sequitur for the following reason: an assumedly irreducible quantum mechanical indeterminism is itself fully compatible with a kind of determinism on the level of human behavior that denies epiphenomenalism and is coupled with the denial of the existence of any soul substance or disembodied consciousness.

Let us denote the predicate variables used in the physicalistic vocabulary of neurophysiology and biochemistry by letters P_1, P_2, P_3.... And let α and t be variables ranging over real numbers so that, in general, any one of the variables P_n can assume one of the values α at some one time t. The neurophysiological state S_p of an organism at some time t_0 would therefore be given by an n-tuplet of specifications $P_{n\alpha t_0}$ in which only the value of t_0 would have to be the same. It may well be that at a given time t_0 one or more of the attributes P_n cannot be specified by a single real number but requires a real-valued *function*. Thus, if one of the P_n were the electric field intensity E, for example, E might vary with spatial position in the body. Hence, it might be that at a given time t_0 the specification of S_p would require an n-tuplet of functions rather than of mere real numbers. But this complication is not problematic for our purposes.

Analogously, let the predicate variables of a *mentalistic* psychological vocabulary be denoted by $\psi_1, \psi_2,...\psi_m$, where it is left open whether the individual attributes over which these variables can range are specified by means of real numbers or integers or in some other way. Thus, one of the ψ_m might range over the awareness states covering the spectrum between manic elation and deep depression, while another of the ψ_m might range over the various visually discriminable color contents of awareness. In this noncommittal sense, let particular values of K represent the individual attributes in the range of any one of the predicates ψ_m, while t is the time variable as before. Then the *mental* state S_ψ of a human organism at any given time t_0, as distinct from its *physical* (neurophysiological) state S_p, is given by an m-tuplet of specifications Ψ_{mKt_0}. And the total P-cum-ψ state of the organism at the given time t_0 would then be given by an $(n + m)$-tuplet of specifications $(P_{n\alpha t_0}, \Psi_{mKt_0})$.

Then, to say that the organism behaves deterministically with respect to this set of attributes of state is to say the following: There is one or more laws such that the organism's total P-cum-ψ state at any one time t is uniquely specified by its corresponding total P-cum-ψ state at one *or more* other times.

Now assume, merely for the sake of argument, that the S_p description itself can be furnished by means of quantum mechanics, supplemented by some explicit definitions of neurophysiological and/or biochemical terms. And let us be mindful of our initial assumption of the irreducibility of the statistical character of quantum mechanics in the sense of section 2.2. It is thus being assumed now that, in general, the lawful links between the various S_p states obtaining at different times are irreducibly statistical with respect to the attributes of quantum mechanical entities. One of the logical consequences of this assumption is the following: There does not exist any *physical* mode of preparing systems so as to yield in *every* case (i.e., with certainty) a preassigned pair of exact measured values (x, p) for position *and* momentum or for other conjugate attributes. It is crucial for our particular purposes here to note that this negative existential statement of quantum theory is predicated on the *exclusion* of at least a large proper subset of the mentalistic predicates ψ_m from the descriptions of the physical modes of preparing systems on which measurements are then made.

Given these considerations, we see that there is logical compatibility between (1) an irreducibly statistical temporal evolution of the human organism's S_p states, taken by themselves, and (2) a deterministic temporal evolution of the total P-cum-ψ states whose specification draws on the *full* range of the ψ_m. And this logical compatibility exhibits the non sequitur in the inference that a presumed indeterminist feeling of freedom *must* have a foundation in quantum mechanical fact. Needless to say, nothing in this demonstration of a non sequitur depends on whether there are, in fact, good empirical grounds for a P-cum-ψ determinism. Hence, this charge of non sequitur is not erected on begging the question. Indeed, it does not gainsay my charge of non sequitur that I allow, of course, for the future discovery of empirical evidence that would lead psychophysics to postulate the irreducible statisticality of the P-cum-ψ evolution.

The compatibility of S_p statistico-indeterminism with P-cum-ψ determinism which I have claimed presupposes as a necessary condition that any given S_p substate of the organism at a time t_0 can coexist at t_0 with any one of *many different mental states S_ψ* rather than uniquely giving rise to only one particular S_ψ. Otherwise, the assumed P-cum-ψ determinism would make for a deterministic linkage among the S_p states themselves, in contravention of quantum mechanics. Indeed, the range of different S_ψ, any one of which can coexist with any given S_p at any one time, must be sufficiently great to make for the quantum mechanical probabilistic spread among the subsequent S_p in the context of the P-cum-ψ determinism.

For precisely this reason, our putative P-cum-ψ determinism constitutes a *denial* of *epiphenomenalism*: The mental states S_ψ have causal efficacy precisely because their copresence with the physical states S_p transforms the merely probabilistic linkages among the latter alone into deterministic ones! To take an absurdly oversimplified example, let the role of the human soma be played by an electron, and let its S_p state at t_0 be an energy eigenstate, which is followed by a position measurement at a later time t_1. A repeated repreparation of the same energy state S_p will then issue in a dispersion of subsequently measured position values. But if this free electron could play the role of the soma in the human organism and the aforementioned energy S_p were to coexist with some one of the many mental S_ψ assumedly compatible with it, then the position *and* S_ψ at any subsequent time would be uniquely determined on the strength of the assumed P-cum-ψ determinism.

In characterizing the postulated P-cum-ψ determinism, I stated that it can also be coupled with the naturalistic denial of the existence of any soul substance or disembodied consciousness. In our context, this denial takes the form of asserting that the occurrence of the ψ-attributes requires a neurophysiological (biochemical) material base or perhaps a cybernetic-hardware base. And this existential dependence *may* take the form that the coexistence relation between the S_ψ and S_p is many–one rather than many–many. But see the Postscript at the end of this paper for an important quantum mechanical *caveat*.

Perhaps there are cases in which the human eye responds to as little energy as a single photon or in which the triggering of neurons is a statistical affair in the sense of being a physical process sufficiently microscopic to be subject to quantum indeterminacies. If these processes of vision or neural excitation then issue in particular responses and actions or inactions on the part of humans in whose bodies they transpire, then one can say that quantum indeterminacies enter into human macroconduct and one can speak of a corresponding reduction in predictability in *principle*. Hence, the most drastic kind of revision that quantum physics itself could possibly compel in the deterministic conception of human behavior is that at least some of the laws governing human behavior are irreducibly statistical. But we saw in section 2.2 that such statisticality can lend no support at all to the particular empirical claims of freedom made by the philosophical indeterminist as part of his moral doctrine. And thus, irreducibly statistical laws of human behavior do *not* satisfy the cardinal condition laid down by the philosophical indeterminist for the assignment of moral responsibility. Moreover, unless it is shown that a significant number of human responses are indeed subject to quantum indeterminacies, it would seem that the vast bulk if not all human responses and acts involve physical agencies of such magnitude that quantum indeterminacies become irrelevant to them and classical deterministic characterization holds to all intents and purposes for these physical agencies.

It is hoped that this discussion of the alleged intangibility of man's inner life and of the bearing of quantum indeterminacy on human freedom has served to support the view that there are very important respects in which science *can and ought* to deal cognitively with man—as with the rest of nature.

▪ 5. AUTO-PREDICTABILITY AND FREEDOM OF ACTION IN A DETERMINISTIC UNIVERSE

In section 1, we noted Popper's result concerning the limitations governing *auto*-predictability of a *deterministically* evolving system. In his 1967 Eddington Memorial Lecture "Freedom of Action in a Mechanistic Universe" in Cambridge (hereafter cited as FAMU), D. M. MacKay offers an account of the import of this *limited* auto-predictability for freedom of action in a deterministic world. MacKay likewise appeals to Niels Bohr's 1948 thesis that the subjective and causal explanations of voluntary acts are *complementary* in Bohr's technical sense of that term (FAMU, pp. 25, 39).

I shall summarize MacKay's interesting account in order to make some comments on it and relate it to the preceding analysis.

MacKay invites us to suppose that the human "brain were as mechanical as clockwork, and as accessible to deterministic analysis" (FAMU, p. 8). Then he calls attention to "the change that must take place in the physical brain (according to mechanistic brain theory itself) when knowledge is acquired by a human agent" (p. 9). And he argues that "even then—even with this wildly idealized pre-Heisenberg conception of a physically determinate brain—the denial of human freedom in general would be not only unfounded but demonstrably mistaken" (p. 8). Let us see more specifically how MacKay construes the application of Popper's limited auto-predictability to the cognitive brain mechanism and how he reasons that consequently there is a brain-theoretic confirmation of our subjective feeling of freedom to choose.

MacKay writes: "Now we have granted for the sake of argument the assumption that what a man believes, sees, feels, or thinks is rigorously represented by the state of his brain, so that someone who 'knew the code' could 'read off' what goes on in his mind from the state of his brain. (In practice, of course, this is ludicrously beyond the bounds of possibility, but that is beside the present point)" (FAMU, pp. 27–28). Thus, he contends that "if we suppose that what a person believes is rigorously represented by the state of what we may call his cognitive mechanism, then discrete, nonnegligible changes in what he believes must entail discrete nonnegligible changes in that state" (p. 11). In terms of the notation of section 4, the mechanistic brain theory whose import MacKay proposes to develop presumably asserts the following: There is a single-valued function having the set of brain state types S_p as domain and the set of mental state types S_ψ as range, so that *distinct* belief-state types S_ψ *cannot* be associated with one and the same brain state type S_p, although it *might be* (p. 25, footnote) that different kinds of S_p are associated with *one* kind of S_ψ.

On the basis of this version of mechanistic brain theory, MacKay's dialectic concerning limited auto-predictability becomes specific:

> It follows that even if the brain were as mechanical as clockwork, no completely detailed present or future[20] description of a man's brain can be equally accurate whether the man believes it or not. (*a*) It may be accurate *before* he believes it, and then it would automatically be rendered out of date by the brain-changes produced by his believing it; or (*b*) it might be possible to arrange that the brain-changes produced by his believing it would bring his brain into the state it describes, in which case it must be inaccurate *unless* he believes it, so he would not be in error to *disbelieve* it. (Pp. 11–12)

Accordingly, MacKay immediately draws the following fundamental distinction between auto-predictability (i.e., W. Sellars' universal "epistemic" predictability of our sec. 1), on one hand, and hetero-predictability, on the other:

> In either case, then, the brain-description lacks the "take it or leave it" character of scientific descriptions of the rest of the physical world, since its validity depends precisely upon whether the subject takes it or leaves it! True, any number of *detached* observers

could predict whether the subject will "take it" or "leave it"; but this prediction in turn, though valid for the observers in detachment, would lack any "take it or leave it" validity for the subject. It would still be true that for such brain-states, and future events causally dependent upon them, no *universally valid (pre)determination exists*: no complete and certain prediction waits undiscovered upon which the subject and his observers would be correct to agree.

Notice that we are not saying only that the subject cannot *make* or *discover* a prediction of his future brain-states, but that there *exists* no definitive prediction that could claim his assent. There is thus no question here of "ignorance of the truth" on the subject's part, nor are we concerned with what he subjectively *feels* about the future event. What we are saying is that as a matter of objective fact, whatever he feels about it, the event in question is one about which for him *no definitive information yet exists*.

It would be nonsensical, and not just ineffectual, for him to "wish he knew" what form the event would take; for on that matter there is nothing final for him to know. Even though the whole physical process concerned were as mechanical as clockwork, his special relation to it makes an observer's view for him not just unknowable but logically invalid and inconclusive. For him, strange as it may seem, the only logically admissible view is that this particular future event is as yet *indeterminate*. (FAMU, pp. 12–13)

MacKay is clearly introducing a special technical use of the terms "determinate" (or "determined") and "indeterminate" (or "undetermined"). For he assumes rather than denies that future brain events are *causally* determined physically. But he is concerned to distinguish this established sense of the term "determined" from the special, different sense of having a determinate specification with a demonstrable claim to unconditional assent (FAMU, pp. 12, 17, 20, 21, 22, 24). For in view of the limitations governing auto-predictability in the face of presumedly universal hetero-predictability, MacKay deems it essential to render the fact that a prediction's possession of an unconditional claim to assent is *relative to the person* in a very special way: In the case of at least some future states of the brains of human agents, there is no unique, fully detailed predictive specification which can demonstrably command the unconditional assent of everyone, since the subjective assent of the agent would, in general, materially affect the accuracy of the specification. MacKay understands a predictive specification's claim to assent from a person P to be "unconditional" iff its accuracy does not depend on its being believed by P. Hence, a specification whose accuracy depends on the agent's *not* assenting to it is ipso facto deprived of any unconditional claim to his assent. Thus, MacKay considers it important to point out that brain states, and actions flowing from them which are causally determined, may not be determined for the agent in MacKay's special technical sense.

Employing his technical sense of "determined," MacKay now proceeds to offer his account of freedom of action at the psychological as well as at the physical (brain-theoretic) level. His aim is to show that a cognitive agent A's subjective feeling of freedom of choice rests on the fact, confirmable by outside observers, that at least before *A* makes up his mind no complete prediction of his *brain state* exists with an unconditional, take-it-or-leave-it logical claim to *A's own* assent. At the same time, MacKay wants it to be understood that if *A* is not at all made privy in advance to a hetero-prediction of the outcome of his choice, then *A* has no reason to deny that there exists a prediction, both as to the outcome of his choice and as what he will believe at the time, which the *outside* observers are correct to believe. MacKay writes:

> Let us now see how what we call a free choice appears in the light of our discussion. Suppose you have to decide at 7 p.m. whether to leave for the 7.15 p.m. train, or to wait for a later one. This means that just before 7 p.m. there are two possible descriptions of the future event at 7 p.m: either (*G*) you decide to go, or (*W*) you decide to wait. Until 7 p.m. you contemplate *G* and *W* as *undetermined* alternatives. At 7 p.m. you decide to go, thus making description *G* true and *W* false; and you act freely as far as you can tell. But next day a Determinist calls with a ciné film of data taken from your brain before 7 p.m., from which he can prove that the outcome was calculable in advance to be *G*. Does this evidence retrospectively refute your belief that your action was free?
>
> If "free" were defined to mean "unpredictable by anyone," then of course in that sense your belief would be falsified. I suggest, however, that this definition, though true perhaps of the "freedom" of *caprice*, begs too many questions to be adopted uncritically for the "freedom" of responsible action. (FAMU, p. 16)

Of freedom in the sense relevant to responsible action, MacKay says:

> To call an action "free" in this sense is therefore to deny the existence of any determinate specification that is binding on (valid and definitive for) everyone, *including the agent*, before he makes up his mind. It is this kind of freedom that I suggest underlies human responsibility. (FAMU, p. 17)

And furthermore, "some of your most responsible actions may be highly predictable by people who know you well, without any examination of your brain" (p. 17).

MacKay's characterization of action that is *not* free has very strong affinities with my category of action which is "compelled" in the special sense of section 2.2. For he explains:

> If, however, before you made up your mind, there existed a determinate and unique specification of the outcome which could command the assent of everyone, including yourself, whether you knew it or not or liked it or not, then the case would be different. Here we would have to say that the outcome was not only predictable by others, but also *inevitable by you*, and you could not be held to have acted "freely." (FAMU, p. 17)

Let us now distinguish the account of freedom of action given by MacKay at the purely psychological level from the one he gives at the brain theoretic level with a view to evaluating critically how he relates the two to one another. The former holds "regardless of any brain theory" (FAMU, p. 18). To state it more fully, he refers to the above case of the agent who has to decide at 7 p.m. whether to leave for the 7:15 p.m. train or wait for a later one. And MacKay says:

> Consider the two situations (i) before 7 p.m. and (ii) at 7 p.m., and the possibilities (a) and (b) discussed earlier. First, regardless of any brain theory, if a prediction specifies that you will take the decision at (*and not before*) 7 p.m. to go for your train, then clearly you must be wrong to believe it as *certain and inevitable* beforehand, since believing it as certain and inevitable would amount to deciding the outcome *before* 7 p.m.—which would refute the prediction!
>
> ... At 7 p.m. on the other hand, you are ready to decide. It is no longer self-contradictory for you to accept either G or W; but you feel (indeed you might claim to know) that both are *open* to you. The crucial question is whether our mechanistic brain theory implies that you are mistaken. As we might by now expect, it does not. (ibid.)

We see that, on the psychological level, *before* 7 p.m. the indeterminateness-for-the-agent of the outcome of a decision among specified alternatives actually first made by him *at* 7 p.m. is guaranteed by the following *logical* fact: If the agent affirms that he will not decide until 7 p.m., then it would be self-contradictory for the agent to assent to the proposition that he has already made the pertinent decision *before* 7 p.m. Thus, on the purely psychological level, the advance indeterminateness for the agent does *not* depend on the putative *empirical* law codified by the aforementioned functional claim that $S_\psi = f(S_p)$ and, a fortiori, does not depend on the *specific assertive content* of that law.

By contrast, let us see how MacKay argues his view of the bearing of mechanistic brain theory on freedom of action:

> Before 7 p.m., since we are assuming that what you believe is rigorously represented by the state of your brain, your brain is in a state of type (a), and its future state at 7 p.m. cannot be fully defined in any way that could be believed by you without self-falsification.
>
> ... Your situation at 7 p.m. fits precisely the case we labeled (b), where there exists a definitive description that can be believed by you without self-falsification—but it is one that could also be rejected by you without any inconsistency, because it will be correct *only* if you accept it. It has no "take it or leave it" inevitability for you, even on mechanistic assumptions. All that our mechanistic theory can do is to confirm your subjective feeling that your choice of G or W is genuinely undetermined: an event which depends on your cognitive attitude, as distinct from the myriad other bodily happenings (for example in your digestive system) which are unaffected by your attitude to them and in that sense outwith [*sic*! without] your control. (FAMU, pp. 18–19)

To my mind, the key question raised by this statement is the following: Does the mechanistic hypothesis $S_\psi = f(S_p)$ of the assumed brain theory have enough specificity *concerning the causal coupling between the brain mechanism and the choosing process* to sustain MacKay's contention here that the brain theory *confirms* the agent's subjective feeling of freedom of action? MacKay points out convincingly enough that the mechanistic brain theory as sketched does *not* imply even retrospectively the *mistakenness* of the agent's belief that beforehand his choice between G and W is genuinely undetermined for him on the psychological level. But the failure to disconfirm a belief is *not* tantamount to *confirming* or *demonstrating* the correctness of the belief. That MacKay's use of the term "confirm" is not an isolated slip of the pen is apparent from his repeated use in this connection of the verb "demonstrate" (FAMU, p. 20) and of its adjectival and adverbial cognates. Thus, when he speaks of the mechanistic evidence furnished by the "imaginary Determinist and his ciné film" (p. 19), he concludes the following, among other things:

> Moreover, (4) when you came to make up your mind at 7 p.m., the different possibilities you contemplated were still demonstrably open to you on his mechanistic assumptions; so that (5) even if you were disposed to deny that you were free to choose between them, his mechanistic evidence could in principle demonstrate the reality and extent of your freedom. (P. 20)

Granted that the mechanistic theory does not imply the mistakenness of the feeling of freedom, let me state my reasons for thinking that the brain theory likewise does *not* underwrite or justify it. We are assuming with MacKay that brain theoretic warranted auto-predictability of one's brain states is limited at least to the following extent: Such auto-prediction is *not* accurate to within differences induced by one's own belief states about present or future physical states of one's own brain. Does this entail that there is no auto-predictability of brain states to within the degree of detail causally relevant to what the agent will *choose* or *decide* to do at 7 p.m. in regard to something like taking a train? In view of the absence of *specifics* as to the particulars of the causal coupling $S_\psi = f(S_p)$ between the brain mechanism and *the choosing process*, it would appear to be a non sequitur to infer the *specific* latter kind of limited auto-predictability from the former. Note that what is at issue is the auto-predictability of those features of the brain state which are causally relevant to choice behavior. We know that on the *purely psychological* level choices are causally influenced by *beliefs* pertaining to the consequences of our actions and to some other related matters. But neither this fact nor MacKay's *general* assumption $S_\psi = f(S_p)$ rules out the following possibility: Two kinds of brain state which differ only with respect to the modifying effects corresponding to the agent's believing or not believing a statement as to the physical state of his brain at 7 p.m. may well *each* be causally coupled to the decision outcome G, instead of being respectively associated with the different outcomes G and W among which the choice was made! For the specific content of the law $S_\psi = f(S_p)$ may be such as to permit or even entail this state of affairs. And indeed, as we remarked earlier, MacKay explicitly countenances (FAMU, p. 25, footnote) the possibility that different kinds of S_p are associated with *one* kind of S_ψ.

Hence, unless I misunderstood MacKay, it would appear that he erred in maintaining that the mechanistic brain theory demonstrates or underwrites the correctness of the agent's subjective feeling of freedom.

Finally, a comment on the bearing of MacKay's valuable and stimulating analysis on the soundness of my criticism of Campbell's libertarianism in section 2.2. There I was concerned to argue that there is no incompatibility between the content of our feeling of freedom of action, even if taken at face value, and the thesis that the outcome of our deliberation or decision is causally determined. MacKay shares my view that our feeling of freedom does not gainsay the following claim of causal determinism: *Given* the relevant internal and external motivational conditions, it could *not* have turned out otherwise. But MacKay emphatically distinguishes the latter aloof or detached kind of impossibility from the state of affairs which he would describe by the sentence: "Under the given conditions, *the agent* could not have *done* otherwise." For MacKay is concerned to reserve the affirmative counterpart of the latter denial, viz. the sentence: "Under the given conditions, *the agent* could have *done* otherwise," to render what he perhaps rightly deems to be at least an important facet of the subjective feeling of freedom: The compound fact that (i) by MacKay's unconditional assent criterion, the outcome of the agent's decision was indeterminate *for him* beforehand; and (ii) in retrospect, nothing contradicts this kind of advance indeterminateness.

I take it that Campbell's libertarianism was intended as a denial of causal determinism on the basis of the feeling of freedom rather than merely as an implicit endorsement of the kind of innovative articulation of the feeling of freedom which MacKay has given us. Hence it seems to me that there is only a terminological and substantively spurious inconsistency between the following two assertions: (i) MacKay's avowal that if the agent acted freely, then it must be the case that "*he* could have *done* otherwise *under the given conditions*"; and (ii) my denial of this statement in section 2.2.[21]

Postscript for Section 4

Arthur Fine and Michael Gardner have remarked that if the S_p states are quantum mechanical *observables*, then the speculative P-cum-ψ determinism of Section 4 may run afoul of a 1967 finding by Kochen and Specker (K&S)* to the following effect: There is no "hidden variable" state or property of a quantum mechanical system S at a time t which determines what a measurement of any observable of S would yield at t. But, as Gardner then pointed out, I can contest the free willist's invocation of quantum indeterminacy without relying on any assumptions disallowed by the stated result of K&S. For, as will be recalled, in opposition to epiphenomenalism and mind–body identity theories, I specifically countenanced

---

* *Editor's Note*: Reference is to Simon B. Kochen & Ernst Specker, "The Problem of Hidden Variables in Quantum Mechanics," *Journal of Mathematics and Mechanics* 17 (1967), 59–87.

the possibility that *one* kind of S_p may be associated with *many* kinds of S_ψ. And if there is no single-valued function from the S_p to the S_ψ, then the indeterministic evolution of the S_p does *not* imply any such indeterminism of the S_ψ. Indeed, there is then no implication of indeterminism for *any* attributes with which the S_p *fail* to be uniquely correlated.

■ NOTES

1. For a historical and critical treatment of some of the earlier literature on this problem, see Ernst Cassirer, *Determinism and Indeterminism in Modern Physics*, 1956. A very useful recent anthology, B. Berofsky (ed.), *Free Will and Determinism*, 1966, presents four fundamentally different approaches to the issue. Other valuable collections are Sidney Hook (ed.), *Determinism and Freedom in the Age of Modern Science*, 1961; and K. Lehrer (ed.), *Freedom and Determinism*, 1966. See also H. Margenau, *Scientific Indeterminism & Human Freedom*, 1968; Corliss Lamont, *Freedom of Choice Affirmed*, 1967. For a contemporary account from the standpoint of dialectical materialism, see Tamas Földesi, *The Problem of Free Will*, 1966.

2. Richard Taylor, *Action and Purpose*, 1966.

3. Thomas E. Murray, *Chemical Engineering Progress*, vol. 48 (1952), p. 22.

4. Cf. M. Scriven, "An Essential Unpredictability in Human Behavior," in: *Scientific Psychology*, ed. B. B. Wolman and E. Nagel (1965), pp. 411–425. For a criticism of Scriven's interpretation of the significance of contra-predictive behavior, see P. Suppes, "On an Example of Unpredictability in Human Behavior," *Philosophy of Science* 31 (1964), pp. 143–148.

5. K. R. Popper, "Indeterminism in Quantum Physics and in Classical Physics," *British Journal for the Philosophy of Science* 1 (1950), pt. I, pp. 117–133; pt. II, pp. 173–195. In section 5, I shall comment on the account of freedom of action which was proposed on the basis of Popper's result by D. M. MacKay in his 1967 Eddington Memorial Lecture "Freedom of Action in a Mechanistic Universe."

6. Wilfrid Sellars, "Fatalism and Determinism," in: Lehrer, op. cit., *Freedom and Determinism*, pp. 143–144.

7. O. W. Holmes, "Ideas and Doubts," *Illinois Law Review* 10 (1915), p. 2.

8. A. Grünbaum, "Science and Man," *Perspectives in Biology and Medicine* 5 (1962), pp. 483–502; reprinted in Louis Z. Hammer (ed.), *Value and Man*, 1966, pp. 55–66. There was also a shorter version of the present essay in *L'Age de la Science* 2 (1965), pp. 105–127.

9. G. Ryle, *Dilemmas*, 1954, p. 28.

10. Ibid., pp. 15–35 (chapter 2: "It Was To Be").

11. Ibid., p. 15.

12. For details, see A Grünbaum, *Modern Science and Zeno's Paradoxes*, 1967, ch. 1, §5.

13. Cf. Sellars, "Fatalism and Determinism," 159–160.

14. For an account of the role of the "I" in volition and choice behavior, see Sellars, "Metaphysics and the Concept of a Person," in: *The Logical Way of Doing Things*, ed. J. F. Lambert, 1969, 219–252. In that publication, Sellars also gives a valuable critique of Richard Taylor's argument in *Action and Purpose* that a person's free actions must be exempt from the causal sphere.

15. See Sellars, "Fatalism and Determinism."

16. L. Kolakowski, "Determinism and Responsibility," in: *Toward a Marxist Humanism*, 1968, p. 188.

17. Cf. Erich Fromm's useful comments on Sigmund Freud's views on this issue in *Psychoanalysis and Religion*, 1950, p. 12, footnote.

18. Cf. Philipp Frank, *Einstein: His Life and Times*, 1947, pp. 66–67, 173–174.

19. This supplanting of classical determinism by the mere statistico-determinism of quantum physics need *not* be predicated on a version of the latter which precludes the *simultaneous* measurability of exact values of the conjugate variables of Heisenberg's indeterminacy relations: see James L. Park and Henry Margenau, "Simultaneous Measurability in Quantum Theory," *International Journal of Theoretical Physics* 1 (1968), pp. 211–283. Cf. also A. Grünbaum, "Determinism in the Light of Recent Physics," *Journal of Philosophy* 54 (1957), sect. 2, pp. 713–715.

20. "Some future states may of course be only trivially dependent on the present state. In general the effect referred to will increase as the future state approaches."

21. The exposition in the present essay has had the benefit of helpful comments from Kurt Baier, Richard Gale, Allen Janis, Donald MacKay, Nicholas Rescher, and Gordon Welty for which I wish to thank them.

5 Historical Determinism, Social Activism, and Predictions in the Social Sciences

In a recent issue of the *British Journal for the Philosophy of Science*, Mordecai Roshwald maintained that it was *self-contradictory* for Marx to combine a belief in the need for exhorting men to establish socialism with the thesis of the historical inevitability of its establishment.[1] Presumably, Roshwald's criticism of this feature of Marx's theory does not depend on the merits of Marx's particular prognosis of the career of industrialism but derives from Roshwald's view that it is inconsistent for *any* deterministic sociopolitical theory to *advocate* a social activism with the aim of thereby bringing about a future state whose eventuation the theory in question regards as assured by historical causation. Since Roshwald's thesis does not depend on whether the explanatory variables of the historical process are held to be economic, climatic, demographic, geopolitical, or the inscrutable will of God, it applies not only to the Marxian view but also, mutatis mutandis, to each of the following: To Augustine's (and Calvin's) belief in divine fore-ordination, when coupled with the advocacy of Christian virtue; to the advocacy on the part of a British economist, who pays taxes in England, of the passage of a certain tax law by Parliament for the purpose of assuring a specified income distribution, when combined with a deterministic *prediction* by him of the passage of that tax law and of the ensuing desired income distribution; and to Justice Holmes's dictum that the inevitable comes to pass through effort.*

In the present note, I wish to challenge (1) Roshwald's contention that historical determinism is logically incompatible with the advocacy of social activism. In addition, I shall deal critically with the following claims: (2) Roshwald's assertions that "an ethics ... combined with a strictly deterministic philosophy would have no practical significance as a motivating force" and that determinism "implies the practical futility of discussion about the best way to be chosen by men in forming their social relations;"[2] and (3) Robert Merton's contention that the "self-stultifying" and "self-fulfilling" predictions encountered in the social sciences are "peculiar to human affairs" and are "not found among predictions about the world of nature."[3] The inclusion of a rebuttal of Merton's statement by means of counterexamples is prompted by the fact that his statement might otherwise be adduced in support of Roshwald's cardinal tenet that "social science ... seems as radically different from natural science as man is from inanimate objects."[4]

---

* *Editor's Note*: The reference is to Oliver Wendell Holmes, "Ideas and Doubts," *Illinois Law Review* 10 (1915), p. 2. See also Essay 4, n. 7, and accompanying text.

(1) Although the predictions made by a contemporary (Marxian or anti-Marxian) historical determinist concerning the social organization of industrial society and those made by our British economist pertain to a society of which these forecasters are members and are thus *self-referential*, they are made by social prophets who, qua deterministic forecasters, consider their own society ab extra rather than as active contributors to its destiny. But the predictions made from that theoretically external perspective are predicated on the prior fulfilment of certain initial conditions which include the presence in that society of men who are dissatisfied with the existing state of affairs and are therefore actively seeking the future realization of the predicted social state. To ignore that the determinist rests his social prediction in part on the existence of the latter initial condition, just as much as a physicist makes a prediction of a thermal expansion conditional upon the presence of heat, is to commit the fallacy of equating determinism with fatalism.[5] Now, in actual fact, not only are the social forecasters whom we are considering spectatorial theoretical analysts of the society whose future they are predicting but, qua members of that society, they also participate in the fashioning of its destiny. On what grounds then does Roshwald feel entitled to maintain that it is logically inconsistent for them, *qua* participating citizens, to advocate that action be taken by their fellow-citizens to create the social system whose advent they are predicting on the basis of their theory? Is it not plain now that Roshwald's charge derives its semblance of plausibility from his confusion of determinism with fatalism in the context of self-referential predictions?

(2) Roshwald's denial of the practical significance of an ethics which is coupled with determinism is tantamount to asserting that a belief in determinism is *causally* incompatible with the determinist's possession of the psychological incentives necessary for taking the action required to implement the moral directives of his ethics. This contention is both empirically false and inconsistent with Roshwald's own premises. For Abraham Lincoln's view that his own beliefs (ethical and other) were causally determined did not weaken in the least his desire to abolish slavery, as demanded by his ethical theory, and similarly for Augustine, Calvin, Spinoza, and a host of lesser men. Moreover, Roshwald's mention of "practical significance as a motivating force" shows that he is actually affirming a *causal* connection between two kinds of psychic states: a belief in determinism and indolent futilitarianism. But this thesis not only contradicts his correct observation that "*psychologically* the necessity of the 'final victory' moved the adherents of the [Marxist] creed to participate in its realisation and served [as] a powerful source of revolutionary activity"[6] but also constitutes a covert and unwitting invocation of psychological determinism to which Roshwald himself is hardly entitled.

Equally untenable is his claim that it is practically futile for determinists to weigh alternative modes of social organization with a view to optimizing their own social arrangements. For the determinist does *not* maintain, in fatalist fashion, that the future state of society is independent of the decisions which men make in response to (i) facts (both physical and social), (ii) their own *interpretation* of these facts (which, of course, is often false), and (iii) their value objectives. It is precisely because, on the deterministic theory, human decisions *are* causally dependent upon these factors that deliberation concerning optimal courses of action and social arrangements can be reasonably expected to issue in successful action rather than lose its significance by adventitiousness. Roshwald's objection here springs from the false supposition that if our beliefs and decisions have causes, these causes force the beliefs in question upon us, against our better judgment, as it were, thus rendering the attempt to exercise that judgment futile.

(3) It remains to consider Robert Merton's interpretation of what John Venn has called "suicidal prophecies" in the social sciences. As an example of such a prophecy, Merton cites a government economist's distant forecast of an oversupply of wheat which, upon becoming publicly known, induces individual wheat growers so to curtail their initially planned production as to invalidate the economist's forecast.[7] Since failure to take cognizance of the possible "perturbational" influence of the dissemination of a social forecast may issue in that prediction's *spurious disconfirmation*, Merton rightly points out that the possibility that a social prophecy stultify itself by becoming known creates a problem for the reliable testing of social predictions. It would, of course, be an error to infer that the phenomenon of suicidal prophecies encountered in the social studies constitutes evidence against determinism, since the dissemination of these prophecies alters the initial conditions on which the forecasts were predicated and would not prevent another forecaster, at least in principle, from correctly predicting the outcome on the basis of the actual, modified initial conditions.[8] Merton does not commit *this* error. But he does make the equally vulnerable claim that self-stultifying and self-fulfilling prophecies are endemic to the domain of human affairs and are "not found among predictions about the world of nature." As evidence, he cites the fact that a "meteorologist's prediction of continued rainfall has until now not perversely led to the occurrence of a drought" and that "predictions of the return of Halley's comet do not influence its orbit."[9] To be sure, these particular predictions of purely physical phenomena are not self-stultifying any more than those social predictions whose success is essentially independent of whether they are made public. But instead of confining ourselves to commonplace meteorological and astronomical phenomena, consider the goal-directed behavior of a servo-mechanism like a homing device which employs feedback and is subject to automatic fire control. Clearly, every phase of the operation of such a device constitutes an exemplification of one or more purely *physical* principles. Yet the following situation is *allowed* by these very principles: A computer predicts that, in its present course, the missile will miss its target, and the communication of this information to the missile in the form of a new set of instructions induces it to alter its course and thereby to reach its target, contrary to the computer's original prediction. How does this differ, in principle, from the case in which the government economist's forecast of an oversupply of wheat has the effect of instructing the wheat growers to alter their original planting intentions?

Corresponding remarks can be made concerning self-fulfilling prophecies, which Merton likewise believes to be confined to the realm of human affairs. Such prophecies are characteriszd by the fact that an *initially false* rumour is believed and thereby creates the conditions of its own fulfilment and spurious confirmation: When the depositors of a bank whose assets are initially entirely adequate give credence to a false rumor of insolvency, this belief can inspire a run on the bank with resulting actual insolvency. Merton himself shows illuminatingly how the pattern characteristic of self-fulfilling predictions serves to buttress prejudices against minority groups by the mechanism of spurious confirmation.[10]

Is there any analog to such predictions in the domain of physical phenomena? It is at least physically possible, though not very likely, that the reception of a false message (e.g., a false warning of an impending collision) will actuate a servo-mechanism so as to realize the very conditions which the originally false message predicted.

It would be unavailing to object that the counterexamples which I have adduced against Merton's contention involve physical artifacts which depend for their existence upon having been constructed by human beings. For the phenomena in question fall entirely within the purview of physical laws and thus invalidate the widely held belief, espoused by Merton, that, in principle, self-stultifying and self-fulfilling prophecies are "not found among predictions about the world of nature."

▨ NOTES

1. Roshwald, "Value-Judgments in the Social Sciences," *British Journal for the Philosophy of Science* 6 (1955), pp. 186–208; the reference is to 191, n. 3. A similar contention is found in George L. Kline, "A Philosophical Critique of Soviet Marxism," *Review of Metaphysics* 9 (1955), 100.

2. Roshwald, "Value-Judgments," p. 192.

3. Robert K. Merton, *Social Theory and Social Structure*, 1949, p. 122.

4. Roshwald, "Value-Judgments," pp. 202–203.

5. For a discussion of this fallacy and of related issues, see Grünbaum, "Causality and the Science of Human Behavior," *American Scientist*, 40 (1952), p. 671, reprinted in Feigl and Brodbeck (eds.), *Readings in the Philosophy of Science*, 1953, pp. 766–778; Grünbaum, "Time and Entropy," *American Scientist* 43 (1955), pp. 568–570.

6. Roshwald, "Value-Judgment," pp. 191–192, n. 3.

7. Merton, *Social Theory*, p. 122.

8. The dissemination of the *adjusted* prediction may then, in turn, require that the *latter* forecast be revised to allow for the effects of *its* publication, and so on ad infinitum. Fortunately, successful prediction *is* possible nonetheless in such cases under very general conditions, as has recently been shown by E. Grunberg and F. Modigliani, "The Predictability of Social Events," *Journal of Political Economy* (U.S.A.) 62 (1954), p. 465; to which Professor H. Feigl has kindly called my attention. These authors elaborate their thesis by reference to the public prediction at time t of the price of some commodity that will prevail at time $t + 1$

and then generalize this analysis of the one-variable case to cover the correct public prediction of the values of n variables.

9. Merton, *Social Theory*, p. 181.

10. Ibid., pp. 179–182.

6 In Defense of Secular Humanism

During a period of considerable strife and moral turmoil in society, there is a perennial tendency in some quarters to offer ethical nostrums. Often we are told that the theistic creeds permit the resolution of our moral perplexities, whereas secular humanism only exacerbates them, leaving moral decay and the decline of our civilization in its wake. These claims have also been turned into a political gospel in the United States. Gravely, William A. Rusher, the former editor of the *National Review*, has blamed secular humanism for producing an amoral sort of human being in our inner cities:

> What is happening to us, and what can be done? Simply put, the secular humanists have been gnawing away at the foundations of Western civilization (God, morality, the family) for two centuries, and have finally succeeded in producing, especially in our inner cities, an almost totally amoral kind of human being—a sort of human pit bull. Our country will recover, if at all, only by discovering and recommitting itself to the great salvific truths on which our civilization was founded.[1]

Indeed, as we shall see, our culture is rife with smug and politically coercive proclamations of the *moral superiority* of theism over secular humanism as follows:

(1) Theism is *normatively* indispensable for the acceptability of moral imperatives.

(2) Religious belief in theism is *motivationally* necessary, as a matter of psychological fact, to assure such adherence to moral standards as there is in society at large.

(3) "Secular humanism is brain dead" (Irving Kristol).

(4) "The taking away of God dissolves all. Every text becomes pretext, every interpretation misinterpretation, and every oath a deceit" (Richard John Neuhaus). In the same vein, Dostoyevsky had told us earlier that "if God does not exist, all things are permissible." Just such theses are also espoused in the recent Jewish journal *Ultimate Issues*.[2]

More recently, at the Republican National Convention in Houston, Texas, in August 1992, Pat Buchanan and Pat Robertson declared a religious war on secularism in our society. And George H. W. Bush, standing before a sign reading "GOD," tried demagogically to secure electoral advantage by complaining that the word "God" was absent from the election platform of the Democratic Party!

Even the philosophically trained William Bennett, one-time secretary of education and antidrug czar, pugnaciously intoned the purported religious foundations of democracy.

Alarmed by these untutored, if not malicious, attacks on secular humanism, I shall examine the conceptual relations between the theological and moral components of the relevant religious creeds and enlist my conclusions in the defense of secular humanism.

In a free society, the purveyors of religious nostrums have, of course, every right to preach to their own faithful and, indeed, to make all others aware of their moral injunctions. Thus, the Pope is entitled to condemn the use of so-called artificial birth control, as distinct from the rhythm method. Yet secular humanists claim entitlement to consider that prohibition barbaric, not only sexually but also demographically, if only because it contributes to the population explosion and concomitant ecological ravages, especially in the third-world countries of Latin America and Africa. Alas, in the current Pope's new encyclical *Veritatis Splendor*, John Paul II reaffirms opposition to artificial birth control (and to divorce). But he turns a deaf ear to the plight of the Catholic families for whom the observance of so-called infertile times *fails for biological reasons.*[3]

As has been documented by the Nobelist Max F. Perutz from Pope Pius's 1930 *Casti Connubii* and from Pope Paul VI's 1965 *Humanae Vitae*, "successive popes have ordained that married couples sharing a bed must practice strict chastity unless they desire a child, with the reluctantly conceded exception of the woman's short infertile period before and after menstruation."[4] And Perutz concludes: "Such inhuman demands could only have been conceived in the minds of celibate old men who mistook their own envy of happily married couples for the voice of God." Refreshingly, the Archbishop of Canterbury, George Carey, went to see Pope John Paul II before the Rio de Janeiro Earth Summit in 1992 to urge that the Catholic ban on birth control is bad for the planet and must be abandoned. Carey also blamed "the dominant dogma of the Catholic Church" for excluding population control from the 160-nation summit's agenda.[5] Thus, significantly, even within orthodox Christendom God hardly speaks with a single voice on the morality of artificial birth control.

Yet undaunted, nowadays theistic moral advocacy is again readily turned into political intimidation, designed to browbeat into conformity or silence those who share Sidney Hook's perception: "Whatever is wrong with Western culture, there are no religious remedies for it, for they have all been tried."[6] Such coercive attempts are being made in our society by both Christians and Jews.

The centerpiece of the religious creeds that are purported to be essential to both private morality and good citizenship is theism: The belief in the existence of an omnibenevolent, omnipotent, and omniscient God to whose will the universe owes its existence at all times and who is distinct as well as independent from his own creation. We learn that this theistic doctrine is normatively indispensable as the source of meaningful ethical prescriptions, although the combined attributes of omnipotence and omnibenevolence are *impugned* with the abundant existence of moral evil in the world, which includes evil that is *not* man-made. Thus, in the eighteenth century, Immanuel Kant argued that the realizability of morality, *as construed by him*, requires the God of theism and indeed with immortality of the soul as its underwriter. To boot, often we are also told, without the slightest attempt to supply supporting evidence, that at least for the vast majority of people, such religious belief is actually *motivationally* necessary, in point of empirical fact, to assure such adherence to moral standards as is found in society. In short, the theistic nostrum is that its species of religious belief is normatively, and typically also motivationally, indispensable to moral conduct and good citizenship in our society. My stated concerns here do not, of course, include dealing with the tenets of a completely atheistic yet avowedly religious humanism, as exemplified by classical Buddhism and certain versions, perhaps, of some other Far Eastern religions. Suffice it to say that these tenets are cognate to secular humanism and therefore pose no issues here.

I should call attention to various modifications or purported reconstructions of the classical theism already outlined. Thus, on one reading of the Book of Genesis, it contains no attributions of omnipotence and omnibenevolence to God. And explicit denials of these attributions have been issued by modern religious thinkers such as Hermann Cohen (of the Marburg School), who was the dominant influence in German Jewish philosophy after the turn of the century, and by the American Protestant theologians Edgar Brightman and Charles Hartshorne, for example. On some of these construals, God is powerful but not the "Almighty," and good but not morally perfect. In this way, God's responsibility for the world is considerably curtailed.

Yet if so, then these theists do not give us an inventory of what God can or cannot do or of what virtues he possesses or lacks. For instance, can God cure otherwise fatally ill people, whose loved ones address petitionary prayers to Him for their recovery? If not, are such prayers not a snare and a delusion? And why have the "anti-omni"-theists not issued a sobering caveat to the faithful who say petitionary prayers? It would seem that their modification of classical theism effects an escapist immunizing maneuver. It serves as an *asylum ignorantiae* in the face of the challenge to a theodicy to reconcile the existence of moral evil with the joint divine attributes of omnipotence and omnibenevolence.

Worse still, some proposed reconstructions of theism turn its doctrines into babble. Thus, what is one to make of Paul Tillich's view that the assertion of the existence of God is *meaningless*, rather than false, and of Martin Buber's incoherent claim that God does not exist per se but only in the I–Thou context of human beings? Buber seems to make God a mere figment of the human imagination à la Feuerbach. Indeed, at the hands of Karl Barth's "wholly other" God and of Moses Maimonides's denial that *any* humanly conceivable properties at all can be predicated of God (the *via negativa*), all the inveterate contorted God talk becomes at best a vast circumlocutory sham, if not just gibberish.

What, for example, has thus become of God the creator of the universe in the opening sentence of Genesis? And why should we not regard such a purported reconstruction of the Old and New Testaments as a case of linguistically misleading social engineering or regimentation of the "masses" of the faithful, if not as bordering on thought pathology? Those beset by doubts about the biblical God who turn to Maimonides's *Guide for the Perplexed* for reassurance find their expectations harshly dashed by false advertising. As Freud wrote aptly in another context in *The Future of an Illusion* (1927):[7]

> Philosophers ... give the name of "God" to some vague abstraction which they have created for themselves; having done so they can pose before all the world as deists, as believers in God, and they can even boast that they have recognized a higher, purer concept of God, notwithstanding that their God is now nothing more than an insubstantial shadow and no longer the mighty personality of religious doctrines.

For example, Paul Tillich is seen as a Lutheran, even though for him "God" is just shorthand for a set of human "ultimate" concerns.

Why then not drop all the biblical discourse about a single or trinitarian personal God "above naming," who is the creator of the universe and of man, cares for his creation, and intervenes in history? And why not just preserve a code of social justice as in the prophetic Judaism of the admirable Isaiah? Such "coming clean" would, of course, amount to embracing secular humanism. Just that challenge prompts some theists in each of the mainline denominations to distance themselves explicitly even from "religious humanism." Thus, in an advertisement "Why Are Catholics Afraid To Be Catholics?"[8] the lay Catholic editors of the *New Oxford Review* wrote:

> The Vatican thunders against abortion, tyrants, illicit sex, consumerism, dissenting theologians, disobedient priests and nuns, and more. But walk into your average parish. Where's the beef? We get crumbs—and platitudes. We don't hear much, if anything, about the Church's teachings on abortion, euthanasia, homosexuality, pre-marital sex, pornography, the indissolubility of marriage—"too controversial." Birth control and Hell are taboo subjects. Pop psychology and feel-good theology are "in." Sin is "out," prompting one to wonder why Christ bothered to get crucified.
>
> We at the *New Oxford Review*, a monthly magazine edited by lay Catholics, say: Enough!
>
> We refuse to turn the wine of Catholicism into the water of religious humanism.

Alas, humanism has again become a major target, if not the object of outright slander, by self-declared classical theists. I shall therefore hereafter ignore the merely *nominal* theists who have no quarrel with philosophical naturalism and atheism.

In the latter vein, Henry Grunwald, former editor-in-chief of *Time* and one-time U.S. ambassador to Austria, opined: "Secular humanism (a respectable term even though it became a right-wing swearword) stubbornly insisted that morality need not be based on the supernatural. But it gradually became clear that ethics without the sanction of some higher authority simply were not compelling."[9] And to emphasize the alleged moral anarchy ensuing from secular humanism, Grunwald approvingly quotes G. K. Chesterton's dictum: "When men stop believing in God, they don't believe in nothing; they believe in anything." A like note of moral self-congratulation for theism is struck by Irving Kristol, as we shall see, who opined that "secular rationalism has been unable to produce a compelling, self-justifying moral code,"[10] whereas theism allegedly had done so.

This pejorative attitude toward atheism is even codified in the ethically derogatory *secondary* meaning of the term "atheist" given in the unabridged *Webster's Dictionary*: "a godless person; one who lives immorally as if disbelieving in God."

Furthermore, as reported in an article on "America's Holy War,"[11] it is now being argued that the separation of church and state in the United States has gone too far: "A nation's identity is informed by morality, and morality by faith" (p. 62), "faith" being faith in the God of the mainline theistic religions. This "accommodationist" position is epitomized by Chief Justice Rehnquist of the U.S. Supreme Court, who declared that the wall of separation between church and state is "based on bad history .... It should be frankly and explicitly abandoned" (p. 63, caption).* It is also espoused by Yale law professor Stephen L. Carter, who claimed that this separation was designed "to protect religion from the state, not the state from religion."[12] Relatedly, many devout parents see evil as instantiated *alike* by "sex, drugs or secular humanism" (p. 65).

Indeed, as *Time* tells us further, "such families also believe that faith is central to serious intellectual activity and should not be relegated to Sunday school" (p. 65). One must wonder at once how intellectual titans like Bertrand Russell or Einstein, who rejected theism, ever managed to make their contributions! Fear of the alleged dire consequences of secular humanism may well also animate creationist opposition to the theory of biological evolution, which many creationists see as abetting secular humanism.[13]

For brevity and style, here let the terms "religious" or "religion" refer to the *theistic* species of religion, that is, to theism. This usage is indeed the *primary* one given in Webster's Dictionary. The theistic religions are usually held to comprise Judaism, which is unequivocally monotheistic, trinitarian Christianity, and Islam. Christianity and Islam were successor religions of Judaism.

Yet the term "religion" is employed very ambiguously. For example, John Dewey's notion of "religion" is far wider than the doctrine of *theism*. Sometimes the term is meant to refer to the historical phenomenon of an institutionalized form of social

---

* *Editor's Note*: The original formulation is from Justice Rehnquist's dissenting opinion in *Wallace v. Jaffree*. There it reads: "The 'wall of separation between church and State' is a metaphor based on bad history, a metaphor which has proved useless as a guide to judging. It should be frankly and explicitly abandoned"; 472 U.S. 38 (1985); quotation is from p. 108.

communion involving participation in a set of ritualistic practices, in abstraction from any doctrines that may provide the rationale for them. Yet none other than a Hebrew prophet like Isaiah hailed righteous conduct as far superior to the fulfillment of the traditional rituals and issued a fervent plea for social justice.

The theistic creeds feature claims about the existence of God, his nature, including his causal relations to the world, as well as ethical teachings that are held to codify the divine moral order of the world within the framework of the theological tenets. Yet the appraisal of the complaints made by theists against secular humanism and of the moral worth they avow for theism requires that we distinguish the *theological* from the *moral* components of their creeds in order to clarify the *conceptual relations* between them.

One vital lesson of that analysis will be that, contrary to the widespread claims of moral *asymmetry* between theism and atheism, *neither theism nor atheism as such permits the logical deduction of any judgments of moral value or of any ethical rules of conduct*. Moral codes turn out to be logically extraneous to each of these competing philosophical theories alike. And if such a code is to be integrated with either of them in a wider system, the ethical component must be imported from elsewhere.

In the case of theism, it will emerge that neither the attribution of omnibenevolence to God nor the invocation of divine commandments enables its theology to give a cogent justification for any particular actionable moral code. Theism, no less than atheism, is itself *morally sterile*: Concrete ethical codes are autonomous with respect to either of them.

Just as a system of morals can be *tacked onto* theism, so also atheism may be *embedded* in a secular humanism in which concrete principles of *humane* rights and wrongs are supplied on other grounds. Though atheism itself is devoid of any specific moral precepts, secular humanism evidently need not be. By the same token, a suitably articulated form of secular humanism can *rule out* some modes of conduct while enjoining others, no less than a religious code in which concrete ethical injunctions have been externally adjoined to theism (e.g., "do not covet thy neighbor's wife").

Therefore, it should hardly occasion surprise that theism is not logically *necessary* as one of the premises of a systematic moral code any more than it is *sufficient*. And this *failure* of logical indispensability patently discredits Dostoyevsky's affirmation of it via Smerdyakov's dictum in *The Brothers Karamazov*: "If God doesn't exist, all things are permissible." Indeed, Smerdyakov's epigram boomerangs: Since atheism and theism are alike ethically barren, neither doctrine itself imposes any concrete moral prohibitions on human conduct.

One major conclusion that will emerge from the application of Socrates's insight in the *Euthyphro* is the following: In regard to the theoretical foundation of any and all specific, concrete norms of conduct, all ethical injunctions, whether their *auspices* be theistic or secular, have an *extra*-theological, mundane, and sociocultural inspiration in particular historical contexts. Thus, this moral will be seen to hold, even when the statement of the ethical code and/or its de facto social inculcation invokes the fear or love of God or employs theological language and imagery.

My arguments will also undermine the rather strident attacks leveled against secular humanism in 1991 by Irving Kristol and Richard John Neuhaus as well as those delivered earlier by Alexander Solzhenitsyn.

Some twentieth century theists articulated the notion of divine omnibenevolence with a view to reconciling it with what most civilized people would surely regard as great moral and natural evil. Theological apologetics—or so-called "theodicy"—is designed to vindicate the justice and omnibenevolence of an omnipotent and omniscient God in a world of rampant evil. The pronouncements of some prominent orthodox rabbis will illustrate that the notion of divine omnibenevolence is shockingly *permissive* morally, to the point of sanctioning the justice of the Holocaust. True enough, as we shall see, there are indeed other theists who would reject these fundamentalist biblical theodicies. Yet I shall argue in detail that precisely their divergence will itself be evidence for the moral *hollowness* of theism and for the ubiquitous inter-denominational and intra-sectarian ethical discord among theists!

■ 2. THE PROBLEM OF EVIL AND THE MORAL PERMISSIVENESS OF THEISM

The problem of acknowledged moral evil has perennially bedeviled those who believe in the governance of the world by a just, or even omnibenevolent God. No wonder, therefore, that the influential twentieth-century Jewish theologian Martin Buber saw the Nazi Holocaust as a particularly acute challenge to the doctrine of divine justice. Bewailing the horrors of Auschwitz, Buber acknowledges its moral challenge:

> One asks again and again: how is a Jewish life still possible after Auschwitz? I would like to frame this question more correctly: how is a life with God still possible in a time in which there is an Auschwitz? The estrangement has become too cruel, the hiddenness too deep. [...] Dare we recommend to the survivors of Auschwitz, the Job of the gas chambers: "Give thanks unto the Lord, for He is good; for His mercy endureth forever"?[14]

Paul Edwards explains:[15]

> Phenomena like Auschwitz, according to Buber, do not show that there is no God but rather that there are periods when God is in eclipse. It is not just that modern men, because of their absorption in technology and material progress, have become incapable of hearing God's voice. God himself is silent in our age and this is the real reason why his voice has not been heard.

Actually in an attempt to come to terms with the acute challenge posed by monstrous moral evil to the notion of divine righteousness and omnibenevolence, Buber offers *two distinct* versions of an "eclipse of God" doctrine, one of which is theocentric, while the other is anthropocentrically phenomenological: Citing Isaiah (45:15), Martin Buber tells us that, according to the Hebrew Bible, "the living God is not only a self-revealing but also a self-concealing God."[16] Indeed, he asks rhetorically (1952, p. 66):

> ...whether it may not be literally true that God formerly spoke to us and is now silent, and whether this is not to be understood as the Hebrew Bible understands it, namely, that the living God is not only a self-revealing but also a self-concealing God [reference omitted]. Let us realize what it means to live in the age of such a concealment, such a divine silence....

Buber (p. 105) speaks of this self-concealing God as possessing "unlimited power and knowledge." And he also tells us that the "righteousness" of the "God of Israel" is "the confirmation of what is just and the overcoming of what is unjust" (pp. 103–104). Yet the self-concealment of *such* a God is simply *frivolous*. As Edwards goes on to explain in his 1969 Lindley Lecture, according to Buber's theocentric version of the eclipse-of-God doctrine, "men cannot in our times find God, not or not just because they have become incapable of I-Thou relationships, but rather because God has turned his back on the world. This 'divine silence,' in [Buber's disciple] Fackenheim's words, 'persists no matter how devoutly we listen.'"[17]

Indeed, the theocentric version of the eclipse theory, which focuses on God's *self-concealment* from the world in our age is, "as Buber rightly observes, ... clearly implied in various passages in the Bible." But this doctrine of the Deus absconditus is also espoused by such Christians as Martin Luther.

Yet what of the *merits* of Buber's hypothesis that though God is always very much alive, there are periods when he conceals Himself by withdrawing into silence and inaction? Edwards's devastating reply is right on the mark:[18]

> The obvious retort to it is that God's self-concealment is inconsistent with his perfect goodness or indeed with any kind of goodness on his part. If a child is in terrible trouble and his father knows about it and could come to the child's help but refuses to do so, i.e., begins to "conceal" himself, this would not, surely, be the mark of a perfectly good father. On the contrary, we would regard him as a monster. It is difficult to see what other response could be justified toward a deity behaving in this fashion. If a Jew in Auschwitz desperately needs God's assistance, if God knows about the Jew's need (and he must know it since he is omniscient), if God furthermore is capable of coming to the Jew's assistance (since he is omnipotent he can do this) and if he nevertheless refuses to do so but instead "conceals himself," then this is not simply a deity falling short of complete goodness but a monstrous deity in comparison with whom, as Bertrand Russell once put it, Nero would have to be regarded as a saint.

William Safire is completely unmoved by such considerations in his article "God Bless Us." Thus, Safire opines à propos of Abraham Lincoln's Second Inaugural Address:

> God is not in moral bondage to man. His design is not for us to discern. As the biblical Job learned, God does not have to do justice on earth—nor need He explain the suffering of innocent babes in Somalia, Bosnia or Kurdistan.[19]

Emil Fackenheim gave an elaboration of Buber's view and offered a defense of the theocentric version of Buber's eclipse doctrine.[20] (Fackenheim's paper appeared earlier in *Commentary*, 1964, and in his book *In Quest for Past and Future*, Indiana University Press, 1968). Paul Edwards takes issue with Fackenheim in his afore-cited Lindley Lecture.[21] As Edwards shows there, Fackenheim even elevates the escapist and evasive role of the eclipse doctrine into an *epistemological virtue*: As Fackenheim sees it, whereas the goodness in the world does *verify* the benevolence of God, the evils of the world do not refute it, because the faith of the true believer will not be *psychologically* shaken by the horrors of this world. But, as we know, characteristically the delusions of paranoiacs and of fanatics are likewise not dislodged by adverse evidence! Such is Fackenheim's deplorable slide from epistemological reasoning to the cognitive devices familiar from psychotic behavior: Heads I win, tails you lose.

The upshot of Fackenheim's Buberian stratagem of rendering Judaic theology irrefutable is this: In Fackenheim's view, all that follows from the rampancy of evil is that "God's ways are unintelligible, not that there are no ways of God.... God was even more inscrutable than had hitherto been thought, and His revelations even more ambiguous and intermittent."[22] In short, Fackenheim parries the refuting import of the problem of evil by the twin devices of (i) attributing morally irresponsible absenteeism to God, and (ii) declaring the reason for this irresponsibility to be unfathomable.

The anthropocentric, phenomenological version of Buber's eclipse doctrine pertains to a decline in man's receptivity to the light from God. This formulation makes God's elusiveness into a human artifact. As given on some pages of Buber's *Eclipse of God*,[23] it states, in Edward's (1970, p. 33) words: "Modern man, in Buber's terminology, is so absorbed in I-It dealings that he has lost the capacity for the I-Thou relationship; and this has made it impossible for him to find God." As Edwards notes (p. 34), this phenomenological version is hardly original with Buber. Besides having been held by other theologians, it even resembles "Heidegger's claim that modern man, because of his immersion in beings and his excessive concern with technology, has 'forgotten Being' [whatever *that* is]" (p. 33). In short, the phenomenological version is that "God is not deliberately hiding himself from men—it is they who have become incapable of seeing Him."[24]

For my part, it boggles the mind how Buber's theocentric and biblical doctrine of God as *self-concealing* can be compatible at all with his anthropocentric version, which blames God's elusiveness on us, unless the two versions are restricted as pertaining to different times, or are qualified in some other way. In any case, Buber felt driven to conclude that God temporarily goes into eclipse during such periods as that of the Holocaust. But just why a benevolent God would go into eclipse to accommodate the likes of Adolf Hitler, Buber left glaringly unexplained. After all, as Paul Edwards noted eloquently and cogently, going into such an eclipse would seem to be a case of morally irresponsible absenteeism on God's part. Indeed, if Buber is to be believed, and if one looks at the history of the societies that have embraced theism in one form or another, it is difficult to find any time at all when God was not at least partially in eclipse.

Buber does *not* offer a vindication or theodicy of the Holocaust *as such*. Yet his theocentric eclipse-of-God doctrine is, in effect, a shabby, lame, and evasive gambit, serving to *immunize* the notion of divine benevolence and righteousness against outright refutation by the perennial existence of evil, including not only the Holocaust but also *much natural evil that is not man-made*!

Worse, some recent apologias for the Holocaust from some Jewish religious quarters have been nothing less than *obscene*. In a 1987 article, Lord Immanuel Jakobovitz, the chief Orthodox rabbi of Britain and the Commonwealth, asserted that the Nazi Holocaust was divine punishment for the apostasy of the German Jews who founded assimilationist Reform Judaism. "This idol of individual assimilation," he wrote almost gleefully, "exploded in the very country in which it was invented, to be eventually melted down and incinerated in the crematoria of Auschwitz."[25]

Now, when the SS men who implemented the "final solution" had their reunions, they could say—on the authority of none other than the chief Orthodox rabbi of the United Kingdom—that they were merely the instruments of the God of Moses. Indeed, if Rabbi Jakobovitz is to be believed, the wrath of God is so *indiscriminate* that it prompted the Nazis to incinerate *devoutly orthodox* Jews from all over Central Europe, no less than the supposedly wicked reform Jews of Germany. Moreover, the vindictiveness of this God is such that the punishment for the doctrinal deviance of reform Jews, even *within* a Mosaic theistic framework, had to be nothing short of live incineration rather than some lesser, reversible misfortune. Far from being just, a God who indiscriminately assigns wholesale lethal punishment and allows babies to be killed in front of their mothers by SS guards at extermination camps is a sadistic, satanic monster deserving of cosmic loathing rather than worship and love.

Rabbi Jakobovitz is hardly alone in the view that the Holocaust was divinely sanctioned. As reported by the noted Israeli scholar Amos Funkenstein, the ultra-Orthodox Rabbi Joel Teitelbaum—who lives in Jerusalem but regards the Jewish secular state and government in Israel as sinful—sees the Holocaust as God's punishment for the Zionist founding of a Jewish state *in advance* of the promised arrival of the purported new Messiah. As Avishai Margalit just pointed out:[26]

The ultra-Orthodox did not experience any crisis of faith or of theology when confronted with the absolute evil of the Holocaust. Their ... response to the Holocaust ... was directed, then, not at God for having allowed the Jews to be murdered but at the Zionists.... According to the prominent Orthodox rabbi Moshe Scheinfeld ... the Zionist leaders ... were "the criminals of the Holocaust who contributed their part to the destruction."

Evidently, the ultra-Orthodox (*haredim*) also regard God's justice as morally indiscriminate. After all, many of the European Jews who perished in the crematoria were not even Zionists, let alone participating citizens of the state of Israel, which had not yet been founded at the time of the Holocaust. And it seems to have been lost on all three of the rabbis that the principle of wholesale, *collective* guilt and justice is invoked by Islamic terrorists who attack Israeli citizens no less than others.

Not to be outdone by Rabbis Jakobovitz and Teitelbaum, the ultra-Orthodox Brooklyn rabbi Menachem Mendel Schneerson, who was even hailed as the new Messiah by his disciples, gave his own twist to the vindication of the Holocaust. In his 1980 book *Faith and Science (Emunah v' Madah)*, this revered sage of orthodoxy opined that, in permitting the Holocaust, God cut off the gangrenous arm of the Jewish people. On this basis, this man of God concludes, the Holocaust was a good thing because without it the entire Jewish people would perish in the future. Just why *that* would happen is left unclear.[27] The zealots who proclaim Schneerson to be the new Messiah suggest that the wonders he will enact are imminent. Yet we can be sure that, when these miracles fail to materialize, we will be treated to other, soothing prophecies on the model of the "Barnum statements" found in astrological forecasts or Chinese fortune cookies. Indeed, Schneerson died uneventfully.

Donald J. Dietrich, chairman of the Department of Theology at the Jesuit Boston College, in his 1994 book *God and Humanity at Auschwitz: Jewish–Christian Relations and Sanctioned Murder*, illuminatingly calls attention to those religious factors which created a climate that permitted the Holocaust by being *theologically encultured*.

Sidney Hook explained why he rejects theism, including Judaism, the religion of his ancestors, in favor of atheism. In a response, the orthodox Chicago rabbi Yaakov Homnick (*Free Inquiry*, Fall 1987) indicted Hook's rejection of his heritage as "a far greater tragedy than all of the physically maimed children in the world." Indeed, Rabbi Homnick goes Buber, as well as Rabbis Jakobovitz, Teitelbaum, and Scheinfeld, one better in his discernment of the hand of God, which he deems patent in the Holocaust: "Yes, without a doubt, the guidance of history by G-d is perceptible even to our limited gaze. The sense of justice ... is palpable .... Especially is the Holocaust a proof of G-d's justice, coming as a climax of a century in which the vast majority of Jews, after thousands of years of loyalty in exile, decided to cast off the yoke of the Torah."

The rabbi's deletion of the letter "o" from the spelling of "God" is intended to convey reverence, as if the *word* "God" were God's true, hallowed name. No wonder that, in their prayers, the Orthodox ask: "May His *name* be blessed," in the manner of word magic, although it boggles the mind just what would happen to his "true" name Yahweh (Jehovah) if the blessing were *effective*! No wonder that the kabbalah of Jewish mysticism is replete with abracadabra and numerology.

Rabbi Homnick's veritable paean to divine retribution prompted Sidney Hook[28] to reply: "All apologists, whether Christian or Jewish, for the divine inspiration of the Bible end up justifying … actions that in ordinary moral discourse we should regard as wicked or evil. This would be evidence enough that, in our discussions with them, we are not using terms like *good* and *bad*, *right* and *wrong* in the same sense." After all, Hook points out, these apologists "cannot really share with us a common universe of moral discourse, since they claim that every event inspired or approved by Jehovah [—such as the Holocaust—] is morally good."

In fact, the Bible, though called the "Good Book," features some appalling teachings ranging from genocide in Deuteronomy to slavery and the inferior status of women in the New Testament. Thus, in a barbaric message to male homosexuals in Leviticus, it reads: "If a man also lie with mankind, as he lieth with a woman, both of them have committed an abomination: they shall surely be put to death; their blood shall be upon them" (20:13). Besides: "And if a man take a wife and her mother [sexually], it is wickedness: they shall be burned with fire, both he and they" (20:14). A father who has sex with his daughter-in-law "shall be put to death" (20:12).

James A. Michener cites these passages and adds that Muslim law requires the stoning to death of an adulterous woman, an event he witnessed in Afghanistan in the 1950s in the presence of a cheering crowd.[29] Yet Michener points to the utter unruliness of the ancient Hebrews as justification of the harshness of Leviticus. But even if, as he claims oddly in the title of his Op-Ed piece, "God Is Not a Homophobe," the biblical proscription is still being invoked nowadays in the service of homophobia. Thus, as reported in the *New York Times*:[30] "A Vatican document on homosexuality [dating from 1986/1987] condemned not only the behavior but also the orientation as a 'tendency ordered toward an intrinsic moral evil.'" Besides, as Robin Lane Fox has shown in great detail,[31] the Bible contains massive historical errors and contradictions, which furnish an additional and devastating case against strict biblical fundamentalism.

If theological teachings lend themselves to countenancing the stated enormities, then this unconscionable permissiveness provides strong reason to reject the pertinent creedal systems.

In my 1992 paper "In Defense of Secular Humanism,"[32] I developed some of the above criticisms of the recent rabbinical holocaust theodicies. In an indignant reply, Seymour Cain, a veteran historian of world religions, editor of an anthology on theological responses to the Holocaust, and a Jewish theist, unwittingly supplies further grist to my thesis in the next section concerning the moral sterility and the glaring ethical ambiguity of theism.[33] Cain does acknowledge the genuineness of these rabbinical endorsements of the Holocaust as justifying divine punishment of the Jews for religious backsliding. Yet he goes into high dudgeon, because these apologias are not *statistically representative* of Jewish theological opinion on the Holocaust. As he puts it (p. 56):

> One only has to recall Eliezer Berkovits's *Faith After the Holocaust*, which puts the onus for the Holocaust not on backsliding Jews, but on Western civilization and its religion, Christianity.... I assume that this Orthodox theologian was not mentioned either because Grünbaum was ignorant of his work or because it did not suit the needs of the adversarial argument.
>
> Or why not mention [Rabbi] Richard Rubenstein, who [in his *After Auschwitz*] proclaimed the death of the God who was traditionally believed to be the protector of his chosen people? Rubenstein went acutely to the root of the matter, not merely the general problem of theodicy, but the specific problem of a covenantal God who let his chosen people endure abysmal humiliation, torture, and death—a now absolutely unbelievable God. He even blamed the Chosen People claim for leading ultimately to the Holocaust. Here again Grünbaum makes no mention of an eminent Holocaust theologian who does not blame Jewish backsliding for the cataclysm, again a skewed omission.... We are led to believe that Jakobovits, Schneerson, and Teitelbaum, who interpret the Holocaust as divine punishment for the modern Jews' abandonment of Torah belief and observance, are the representative voices of contemporary Jewish theology. Nothing could be further from the truth. Many Jewish theologians have voiced exactly the same rejection of the idea of the killer-God of Auschwitz in practically the same words as Grünbaum ..., e.g., Eugene Borowitz. Any mention of them would not serve the purpose of Grünbaum's adversarial argument.

But Cain turns a deaf ear to precisely the damaging fact: It is scandalous that Judaism is *sufficiently permissive morally* to enable some world-renowned rabbis to offer a Holocaust theodicy *at all* with theological impunity: It attests to the moral bankruptcy of the notion of a theological foundation of Jewish ethics. Cain (and other apologists for Judaism) *ought to be deeply embarrassed* by this situation instead of offering the witless complaint that the rabbinical Holocaust apologists made "easy targets" for me, like "fish in a barrel." Rabbi Jacobovitz and Rabbi Schneerson, who both vindicated the Holocaust as divine justice, are *world figures* in Orthodox Judaism! Clearly, I submit, precisely the statistics on the depth of the cleavage among the moral verdicts of Jewish theologians on so overarching an occurrence as the Holocaust bespeaks the ethical bankruptcy of their theology. By the same token, Cain's complaint that I made no allowance for that statistical dispersion boomerangs.

William Safire sounds the same note as Cain but in regard to Islam. Thus, recent attacks by Islamic fundamentalists prompted Safire's admonition[34] that Islam is "one of the world's great religions" and that non-Muslims should refrain from "thoughtlessly lumping together the orthodox, the secular and the extremist." And a lead editorial in the *New Republic*[35] went much further, complaining very implausibly that "the mass media, showing its habitual contempt for religion, conflated Islam with the most bizarre of modern cults [in Waco, Texas] and treated the two as almost interchangeable."

No doubt, there are great numbers of Muslims who abhor terrorism and who interpret their religion in a humane way. But Safire's caveat against lumping the orthodox together with the extremist surely makes insufficient allowance for several stubborn facts:

> (1) Shiite clerics have loudly claimed the sanction of Islam for meting out death sentences to apostates for affronts against Allah (God) or against the Prophet. Thus, declaring someone an unbeliever, that is, to engage in *takfir al hakim*, provides religious warrant for killing the infidel.
>
> (2) Notoriously, the Imam Khomeini in Iran issued a *fatwa* (religious ruling), making it the religious duty of any Muslim to assassinate Salman Rushdie for blasphemy. To boot, the Imam's successor, Ayatollah Khamenai ("the leader"), rejected appeals to rescind Rushdie's death sentence and actually doubled the $1 million bounty for carrying it out. Besides, President Rafsanjani of Iran reaffirmed the *fatwa* as irrevocable.
>
> (3) *Fatwas* may also be issued to enjoin a jihad (holy war) or to counteract any perceived threats to Islam.

Fatwas have been and are now being used in some Islamic countries to suppress secularism. And even in Egypt, the Ministry of Culture of the secular government is increasingly yielding to the threats from fundamentalists by permitting them to censor books scheduled for publication by the Ministry. Indeed, one of Egypt's most senior theologians testified in court that secularists are apostates "who should be put to death by the Government" and that "if the Government failed to carry out that 'duty,' individuals were free to do so."[36]

True enough, some Egyptian Sunni clerics have deemed the *fatwa* against Rushdie as less than justified. *But just this elasticity in the conception of theologically sanctioned moral injunctions demonstrates anew the ethical permissiveness that I deplored in Jewish Holocaust theology.* Hence, it is misleading on Safire's part to depict Islam as currently being "under attack from within." And Cain ought to be deeply embarrassed anew by the murderous *fatwas*, precisely because—in Safire's words—"Islam [is] one of the world's great religions."

■ 3. THE MORAL STERILITY OF THEISM

The moral hollowness of the theistic superstructure requires both clarification and argument. Why are theological trappings morally unavailing? It was Socrates who permitted us to realize that if a religious creed is to yield any specific moral prescriptions at all, the ethics must be *extraneously* imported or *tacked on* to theism on *extra*-theological, worldly grounds, being put into the mouth of God by the clergy when asserting His goodness or omnibenevolence. This moral sterility of theism comes into view from the failure of divine omnibenevolence to deal with the challenge posed by a key question from Socrates in Plato's *Euthyphro*: Is the conduct approved by the gods right ("pious") because of *properties of its own* or merely because it pleases the gods to value or command it? In the former case, divine omnibenevolence and revelation are at best ethically superfluous, and in the latter the absolute divine commands fail to provide any *reason* at all for imposing *particular kinds* of conduct.

For if God values and enjoins us to do what is desirable in its own right, then ethical rules do *not* depend for their validity on divine command, and they can then be *independently* adopted. But, on the other hand, if conduct is good merely because God decrees it, then nowadays we also have the morally insoluble problem of deciding, in a multireligious world, *which one* of the conflicting purported divine revelations of ethical commands we are to accept. Indeed, Richard Gale sees the thrust of Plato's *Euthyphro* to be the claim that "ethical propositions are not of the right categorical sort to be made true by anyone's decision [command], even God's."[37]

The plurality of competing revelations is illustrated by those in which Jesus is the Lord and those in which he is not, as in Islam and Judaism. And how are we to resolve *theologically* the basic ethical disagreements existing even *within* the clergy of the same religious denomination, such as the debate on pacifism in times of war or the justice of capital punishment for crime? Just these conflicting moral revelations and intradenominational disagreements spell a cardinal lesson: Even if a person is minded to defer completely to theological authority on moral matters, he or she cannot avoid deciding *which one* of the conflicting religious authorities is to be his or her ethical guide. Thus, try as they may, people cannot abdicate their own responsibility for deciding by what moral norms they are to live. In just this decision-making sense man is inescapably the measure of all things, for better or for worse. And it is quite otiose to speak, as Reinhold Niebuhr obviously did, of "God giving us to see the right."[38]

True enough, assuming divine omnibenevolence, it presumably follows that all divinely ordained conduct is morally right. But that is unavailing, because this much leaves us wholly in the dark as to which moral directives are binding on us or what goals are ethically desirable. How, for example, does divine omnibenevolence tell us whether to share or abhor the Reverend Falwell's and Rabbi Kahane's claim that a nuclear Armageddon is part of God's just and loving plan for us, because only the righteous will be resurrected thereafter? In any case, the existence of states of affairs in the world that theists themselves acknowledge to be morally evil, no less than others do, does indeed impugn the purported omnibenevolence of God. And the existence of evil that is *not* wrought by human volition cannot be explained away by recourse to the so-called free-will defense. That apologia adduces the *value* of human freedom to perpetrate evil deeds no less than to do good ones.

The inability of the theological superstructure to yield a moral code also crops out in Kant's invocation of God (and of immortality of the soul) as underpinnings of his own system of deontological ethics. His argument for such a theological foundation starts out from his moral doctrine that there is a categorical imperative to act only on principles that *everyone* could adopt consistently. But Kant avowedly offered only a formula: Alas, it does not tell us which moral directives to adopt from a set of competing ones. Thus, instead of being a source of concrete ethical injunctions, his formula provides only a necessary condition of their acceptability.

Even at that, Kant's theological underpinning of his ethics loses its force, if only because the required realizability of the highest good is hardly assured. Besides, his case for a divine underwriter founders on its dubious assumption of immortality of the soul. And his argument becomes baseless in the context of such rival conceptions of ethics as are offered by the teleological or self-realization schools. Indeed, even if the philosophical viability of morality were evidence for the existence of God, as claimed by Kant, the ubiquitous reality of evil in the world would be stronger evidence against theism. It would seem that Kant's own special version of a theological foundation for ethics fails, even if one disregards the legitimacy of nondeontological systems of ethics.

Alexander Solzhenitsyn's 1978 Commencement Address at Harvard showed no awareness of the moral sterility of theism:

> There is a disaster which is already very much with us. I am referring to the calamity of an autonomous [despiritualized] and irreligious, humanistic consciousness. It has made man the measure of all things on Earth, imperfect man, who is never free of pride, self-interest, envy, vanity, and dozens of other defects …. Is it true that man is above everything? Is there no Superior Spirit above him?

Prima facie, this declaration may sound ingratiatingly modest. But, as it stands, it is morally hollow and theologically question begging. *Whose revelation*, one must ask, is to supplant man as the measure of all things? That of the Czarist Russian Orthodox Church? Or the edicts of the Ayatollah Khomeini, as enforced by his mullahs? Those of the Dutch Reformed Church in apartheid South Africa? Or the teachings of Pope John Paul II, who—amid starvation in Africa—is getting support from the native episcopate for the prohibition of "artificial" birth control? Or yet those of the Orthodox rabbinate in Israel, which prohibits autopsies, for example? And, if the latter, which of the two doctrinally *competing* chief rabbis is to be believed: the Ashkenazi or the Sephardic one? If the ethical perplexity of modern man is to be resolved by concrete moral injunctions, Solzhenitsyn's jeremiad simply replaces secular man by *selected clergymen*, who become the moral touchstone of everything by claiming revealed truth for their particular ethical directives.

It appears that the moment a theology is to be used to yield ethical prescriptions, these rules of conduct are obtained by deliberations in whose outcome secular aims and thought are every bit as decisive as in the reflections of secular ethicists who deny theism. And the perplexity of moral problems is not lessened by the theological superstructure, which itself leaves us in an ethical quandary.

No wonder that Judeo–Christian theology has been invoked as a sanction for such diverse ethical doctrines as the divine right of kings; the inalienable rights of life, liberty, and the pursuit of happiness; black slavery; "Deutschland über alles"; the social Darwinism of Herbert Spencer; and socialism. Indeed, as Sidney Hook has pointed out in his own critique of Solzhenitsyn: "Neither Christianity nor Judaism, in principle, ever condemned slavery or feudalism. In their modern forms, they have been humanized in consequence of [the challenge from] the rise of secular humanism."[39] As the Roman Catholic judge John T. Noonan Jr. pointed out more specifically most recently, from the time of St. Paul to well beyond the middle of the nineteenth century the Catholic Church taught that slavery was morally acceptable. And it was not until 1890 that Pope Leo XIII finally condemned slavery, but "only after the laws of every civilized land [had] eliminated the practice."[40] At last, Pope John Paul II included slavery among the intrinsic evils in his latest encyclical *Veritatis Splendor*.

Furthermore, Noonan explains, for 1,200 years, "popes, bishops and theologians regularly and unanimously denied the religious liberty of heretics." Indeed, "the duty of a good ruler was to extirpate not only heresy but heretics,"[41] and the Church did all it could to help. Even when the Church came to acquiesce in religious tolerance after compelling orthodoxy by force, its papal advisors continued to uphold the enforcement of orthodoxy by the state as an ideal.[42]

Some religious sects in India would have us abstain from the surgical excision of cancerous growths in man, and Christian Scientists in the West reach somewhat similar conclusions from rather different premises. Roman Catholics, on the other hand, endorse the medical prevention of death but condemn interference with nature in the form of birth control, a position not shared by leading Protestant and Jewish clergymen. Indeed, both Mahatma Gandhi and Hitler saw themselves as serving God. And divine Providence was a frequent feature of Hitler's speeches, illustrating anew that religion can also be the last refuge of the scoundrel. Indeed, one believer's will of God is another's will of Satan, as illustrated by the exchange between Ayatollah Khomeini and President Jimmy Carter, a born-again Christian.

Unfortunately, leading opinion-makers in the United States seem unaware not only of the moral sterility of theism but also of the ethical abominations perpetrated by theocracies, past and present.

Solzhenitsyn's charge of moral inadequacy against an irreligious humanistic consciousness is of-a-piece with the point of his rhetorical questions: "Is it true that man is above everything? Is there no Superior Spirit above him?" Surely the assumption that man may well *not* be above everything hardly requires belief in the existence of God. As we know, NASA has been scanning the skies for signals from extraterrestrial and indeed extra-solar humanoids, whose intelligence may indeed be superhuman.

Nor will it do for clergymen to appeal—as they often do when thus challenged by the stated damaging considerations—to the finitude of our minds or to the inscrutability of God, who is said to transcend human understanding. After all, the clergy is in no better position to transcend that finitude than anyone else! Nor, it must be emphasized, do religious apologists have greater expertise than nonbelievers for discerning the limits of human cognition. Besides, one would expect that the avowed inscrutability of God would induce great modesty in regard to fathoming his purported will and alleged ethical commands.

Those who claim a divine foundation for their *otherwise* favorite moral code, as against its available rivals, compensate for the ethical emptiness of theism by begging the question: They blithely claim *revealed* divine sanction for their own moral code. It was Moses, not God, who issued the Ten Commandments. The famous law code of the Babylonian King Hammurabi was purportedly received by him from the sun god Shamash during prayer, a tale similar to the legend of Moses and the revelation of the Decalogue by Yahweh on Mt. Sinai. Indeed, the theological grounding of ethics is so shaky that the craving for it legitimately calls for *psychological* explanation as part of the psychology of fideist acceptance of theism.[43]

In a recent widely touted plea for the theoretical relevance of religious ethics to U.S. public policy, Yale's law professor Stephen L. Carter inadvertently undermines his basis for just that plea. In his book *The Culture of Disbelief* he writes: "What was wrong with the 1992 Republican convention was not the effort to link the name of God to secular political ends. What was wrong was the choice of secular ends to which the name of God was linked."[44] Anna Quindlen quotes this passage after praising Carter's book as "exceptionally intelligent and provocative."[45]

But clearly, Stephen Carter makes the linkage to God logically irrelevant precisely by assuming that *we must already know, independently* of any purported divine commands, *which* secular political ends are ethically proper and thereby may *properly* be chosen for linkage to the name of God! Otherwise, *any* secular political ends can be given such a linkage with theological impunity, as they have been historically and at the 1992 Republican convention, to Carter's discomfiture.

Thus, George H. W. Bush's avowed belief that Jesus is his Savior understandably did not prevent him from making demagogic use of the "GOD" sign when complaining at the 1992 Houston convention that it was absent from the election platform of the Democratic national convention. Alas, as the *New York Times* reported[46] Bush's Democratic successor, President Bill Clinton, predicated U.S. political morality emptily on "seeking to do God's will" and "has made several attempts to link religious belief to public and private responsibility, most frequently citing the arguments forwarded [offered] by Stephen L. Carter." Pray tell, Mr. President, just what *is* God's will concretely? Does he sanction capital punishment for example?

And is that *your* reason for favoring it? And where does God stand on abortion? Isn't your appeal to God's will just rhetoric?

* * *

Irving Kristol deplores the secularization of American Jewry under the influence of secular humanism, which he tendentiously describes as springing from a "new, emergent religious impulse."[47] As he sees it:

> Because secular humanism has, from the very beginning, incorporated the modern scientific view of the universe, it has always felt itself—and today still feels itself—"liberated" from any kind of religious perspective. But secular humanism is more than science, because it proceeds to make all kinds of inferences about the human condition and human possibilities that are not, in any authentic sense, scientific. Those inferences are metaphysical, and in the end theological.

Kristol muddies the waters: Secular humanists are well aware that scientific knowledge does not suffice to warrant all parts of a moral code. But Kristol darkens counsel by designating the motivation for secular humanism as "religious" and its conception of the human estate as "theological." In so doing, he ignores that the unabridged Webster's Dictionary gives the following *primary* definition of the term "religion": "The service and adoration of God or a god as expressed in forms of worship, in obedience to divine commands, especially as found in accepted sacred writings..."

Although the term "spiritual" has a supernaturalist tinge, Kristol insists that secular humanism springs from a "new philosophical-spiritual impulse":

> What, specifically, were (and are) the teachings of this new philosophical-spiritual impulse? They can be summed up in one phrase: "Man makes himself." That is to say, the universe is bereft of transcendental meaning, it has no inherent teleology, and it is within the power of humanity to comprehend natural phenomena and to control and manipulate them so as to improve the human estate. Creativity, once a divine prerogative, becomes a distinctly human one ....
>
> ... Man's immortal soul has been a victim of progress, replaced by the temporal "self"—which he explores in such sciences as psychology and neurology, as well as in the modern novel, modern poetry, and modern psychology, all of which proceed without benefit of what, in traditional terms, was regarded as a religious dimension.[48]

First, we ought to applaud precisely what Kristol bemoaned when he said: "Creativity, once a divine prerogative, becomes a distinctly human one." The invocation of a divine creator to provide *causal* explanations in cosmology or biology suffers from a fundamental defect vis-à-vis scientific explanations of the effects produced by human agents or by diverse events: As we know from 2,000 years of theology, the hypothesis of divine creation does not even envision, let alone specify, an appropriate *intermediate* causal *process* that would *link* the will of the supposed divine (causal) agency to the effects which are attributed to it. Nor, it seems, is there any prospect at all that the chronic inscrutability of the putative causal linkage will be removed by new theological developments.

In sharp contrast, for instance, the discovery that "an aspirin a day keeps many a heart attack away" has been quickly followed by the quest for a specification of the mode of action that *mediates* the prophylaxis afforded by this drug against coronary infarcts. This is similar for therapeutic benefits from placebos wrought by the mediation of endorphin release in the brain and by the secretions of interferon and of steroids. In physics, there is either an actual specification or at least a quest for the *mediating causal dynamics* linking presumed causes to their effects. In the case of laws of temporal coexistence or simultaneous action at a distance, there is a specification of the concomitant variations of quantified physical attributes by means of functional dependencies.[49]

Indeed, the prominent American Jesuit theologian Michael Buckley makes an important admission as to the hypothesized process of divine creation: "We really do not know how God 'pulls it off.' Catholicism has found no great scandal in this admitted ignorance."[50] But if so, the disbelief in divine creativity, which Kristol bewails, incurs no explanatory loss at all.

Kristol also deplores current disbelief in the immortality of the soul among educated people. Yet, on examination, that tenet is so obscure that it should not be consoling to any reflective person. As Maimonides saw it, the attempt to grasp the nature of eternal bliss in the hereafter while we are alive is akin to the futile effort of a blind person to experience color visually. At any rate, the hypothesis of immortality of the soul collapses in the face of the vast amount of evidence for the dependence of the very existence of consciousness on adequate brain function and, moreover, for the dependence of the *integrity* of our personalities on such function. Witness, for example, the effects of brain tumors, Alzheimer's disease, and various drugs, such as alcohol or mood-altering medications.[51]

But Kristol's principal thesis is that two fundamental flaws undermine the credibility of secular humanism. The first, we learn, lies in its inability to provide a moral code; the second, which is even more fundamental, is that its vision renders our lives meaningless and has become "brain dead."

As to the first, we are told (pp. 24–25):

> We have, in recent years, observed two major events that represent turning points in the history of the 20th century. The first is the death of socialism, both as an ideal and a political program, a death that has been duly recorded in our consciousness. The second is the collapse of secular humanism—the religious basis of socialism—as an ideal, but not yet as an ideological program, a way of life. The emphasis is on "not yet," for as the ideal is withering away, the real will sooner or later follow suit.
>
> … This loss of credibility flows from two fundamental flaws in secular humanism. First, the philosophical rationalism of secular humanism can, at best, provide us with a statement of the necessary assumptions of a moral code, but it cannot deliver any such code itself. Moral codes evolve from the moral experience of communities, and

can claim authority over behavior only to the degree that individuals are reared to look respectfully, even reverentially, on the moral traditions of their forefathers. It is the function of religion to instill such respect and reverence. Morality does not belong to a scientific mode of thought, or to a philosophical mode, or even to a theological mode, but to a practical-juridical mode. One accepts a moral code on faith—not on blind faith but on the faith that one's ancestors, over the generations, were not fools and that we have much to learn from them and their experience. Pure reason can offer a critique of moral beliefs but it cannot engender them.

Elsewhere, Kristol claimed more explicitly: "Secular rationalism has been unable to produce a compelling, self-justifying moral code."[52]

These assertions call for a series of critical comments, showing that fideist theism has hardly succeeded ethically where secular rationalism has failed:

1. What is Kristol's evidence for the purported decline in adherence to secular humanism among educated people who, he tells us, had widely accepted secular humanism as an ideal? Indeed, this alleged collapse, and his prediction of its demise as an ideological program of practical action, is born of wishful thinking. Witness the well-documented massive erosion of religious belief and worship in Western Europe, which is publicly lamented by its religious leaders.

Even in the United States, where avowed religiosity is far greater than in Europe, the Roman Catholic Church faces a crisis in the recruitment of young people for the priesthood. Just this scarcity of recruits has lent urgency to the plea that women be ordained as priests. The widespread disregard for the church's prohibition of "artificial" birth control by American Catholics is likewise well-known. And the pressure to abandon the requirement of celibacy for the priesthood derives practical poignancy from the growing number of lawsuits from practicing Catholics, whose children have been sexually molested by members of the clergy. On the other hand, fundamentalist Protestant evangelism is on the rise and, to the consternation of the Roman Catholic hierarchy, is making considerable inroads in certain segments of its erstwhile faithful. But this headway of fundamentalism is largely confined to the most poorly educated segment of our society. Thus, it is only cold comfort for Kristol.

2. More fundamentally, Kristol erects a straw man when he complains that the philosophical rationalism of secular humanism cannot deliver a moral code. This charge is a red herring for at least two reasons. First, theism as such has turned out to be *morally sterile* no less than atheism or "philosophical rationalism," taken by themselves; in fact, when Kristol urged that "morality does not belong to a scientific mode of thought," he himself conceded that morality *also* does not belong "even to a theological mode, but to a practical-juridical mode." Second, secular humanism can *tack on* moral directives to its atheism on the basis of value judgments made by its adherents, just as, in point of actual fact, theists tack on such directives under the purported aegis of inscrutable divine revelation. Yet, unlike revelationist theists, humanists insist on the liability of their moral convictions to

criticism. Kristol allowed that "pure reason can offer a critique of moral beliefs," but his aim in saying so was not to make a partial concession; instead it was to complete the sentence by saying *one-sidedly*, "but it cannot engender them." Nor, as he fails to see, can theism "produce a compelling, self-justifying moral code."

Kristol draws precisely the wrong lesson from his correct observation that the erosion of belief in theism attenuated the "moral code inherited from the Judaeo-Christian tradition." For, in his view, it tells *against secular humanism* that thereupon "we have found ourselves baffled by the Nietzschian challenge: If God is really dead, by what authority do we say [that] any particular practice is prohibited or permitted?"[53] By now, it should be abundantly clear, however, that in answering the question as to the "authority" for concrete moral yeas and nays, *we are surely no better off if God is alive than if he is dead*! In fact, the threat of moral anarchy or nihilism arises from the erosion of belief in God just because the prevailing moral code had been falsely claimed to derive from Him epistemologically (via revelation), juridically (in the form of divine commandments), and motivationally (from the love or fear of God)!

Evidently, Kristol's echoing of Nietzsche's challenge backfires: The bite of the challenge is injurious to the religious rather than to the secular construal of morality.

3. It is a commonplace that moral codes evolve from the moral experience of communities. But this *genesis* does not warrant Kristol's normative and motivational view that such codes "can claim authority over behavior only to the degree that individuals are reared to look respectfully, even reverentially, on the moral traditions of their forefathers." Surely we ought to winnow the wheat from the chaff in a critical scrutiny of these traditions.

But how, for example, does Kristol's ethical traditionalism enable him to *avoid* asking Jews nowadays to look reverentially at the fact that, at the time of biblical Judaism, women—but not men—were stoned to death for adultery and that the conditions for obtaining a divorce were brutally asymmetrical as between women and men? Political pressure from rabbinical theocrats in Israel has made it impossible for a Jew there nowadays to get a license to marry a Christian.[54] How does Kristol's conservative stance allow him to erect safeguards against such totalitarian tyranny?

Again, are present-day Christians to show respect for the fact that other devout Western Christians performed barbaric clitoridectomies in the nineteenth century to suppress the sexuality of young girls? Or are they to feel pious stirrings on learning that, *with the clergy on his side*, Christopher Columbus could see the holy purpose of initiating slave trading against the people of the Americas, saying: "Let us in the name of the Holy Trinity go on sending all the slaves to Europe that can be sold. The eternal God, our Lord, gives victory to those who follow his way over apparent impossibilities"?[55]

If Kristol were to reply that respect for the repository of ancestral injunctions has to be *selective*, my retort is the one that Sidney Hook gave to Solzhenitsyn:

"What besides the methods of reason and intelligence can enable us to make the proper choice between them?"[56] It seems inescapable that *all* traditional ethical injunctions should be subjected to critical scrutiny and distillation.

Kristol's formula founders on the neglect of the precept afforded by Socrates' insight in the *Euthyphro*: If divinely hallowed injunctions are *deserving* of adoption, then *we* must be the ones—in every epoch anew—to find them worthy. And our only means for doing so are our intelligence and our informed feelings. We have nowhere else to go. Yet Kristol concludes that, since our society no longer defers uncritically or even mindlessly to clerical edicts, contemporary parents are "impotent before such questions" as "what moral instruction should we convey to our children?" (P. 25)

Kristol's application of his traditionalism to contemporary morality features his endorsement, as ancestral divine wisdom, of the inhumane homophobia of biblical Judaism. Referring to the demise of the prohibition of homosexuality as "moral disarray," he says mournfully:

> Reform Judaism has even legitimated homosexuality as "an alternative lifestyle," and some Conservative Jews are trying desperately to figure out why they should not go along. The biblical prohibition, which is unequivocal, is no longer powerful enough to withstand the "why not?" of secular-humanist inquiry. (P. 25)

So much the better for the moral challenge from secular humanism, which produced a humane advance over barbarism and cynical hypocrisy.

But, in Kristol's view, the inability of secular humanism to deliver a "compelling, self-justifying moral code," which he employs as a red herring, is only the first of its "two fundamental flaws." He reserved his supposed *coup de grâce* for the second:

> A second flaw in secular humanism is even more fundamental, since it is the source of a spiritual disarray that is at the root of moral chaos. If there is one indisputable fact about the human condition it is that no community can survive if it is persuaded—or even if it suspects—that its members are leading meaningless lives in a meaningless universe.... Secular humanism is brain dead even as its heart continues to pump energy into all of our institutions. (Ibid.)

But why can secular humanists not lead richly meaningful lives, just because, in their view, the values of life lie *within* human experience itself? How would our lives be more meaningful if we were to suppose that man is the centerpiece of an avowedly inscrutable *overall* divine purpose, which constitutes *the* meaning of our lives but must remain unknown to our finite minds? Being at the focus of elusive cosmic "meaning" is clearly irrelevant to finding value on this earth: Experiencing the embrace of someone we love, the intellectual or artistic life, the fragrance of a rose, the satisfactions of work and friendship, the sounds of music, the panorama of a glorious sunrise or sunset, the biological pleasures of the body, and the delights of wit and humor.

In the movie *Limelight*, Charlie Chaplin put in a nutshell what is wrong with the delusion that there is such a thing as *the* meaning of life: Life, said Chaplin, is not a meaning, but a desire. Yet Václav Havel, who has a penchant for mysticism, lists "the meaning of our being" as a basic human question.[57] And a rabbi demands an "ultimate meaning" in human life: "In the atheistic premise, there is no ultimate meaning to human life. It is just there. Now, no human being behaves as if life had no meaning."[58] But what, pray tell, is *the* meaning of life? Pious cant?

As secular humanists see it, there are as many "meanings" as there are fulfillments of human aspirations. It is sheer fantasy, if not arrogance, on the part of theists to proclaim inveterately that their lives must be more meaningful *to them* than atheists and secular humanists find their own lives to be to themselves. Where is their statistical evidence that despair, depression, suicide, aimlessness, or other dysphoria is more common among unbelievers than among believers? Yet Kristol insists: "It is crucial to the lives of all our citizens, as it is to all human beings at all times, that they encounter a world that possesses a transcendent meaning, a world in which the human experience makes sense."[59] This grandiose assertion is flatly false as a matter of psychological fact.

Regrettably, Kristol did not come to grips with the arguments in Albert Einstein's paper on "Science & Religion" (1941).[60] There Einstein first points out: "Nobody, certainly, will deny that the idea of the existence of an omnipotent, just and omnibeneficent personal God is able to accord men solace, help and guidance; also by virtue of its simplicity the concept is accessible to the most undeveloped mind" (p. 70). But then Einstein issues his cardinal plea, which clashes head-on with Kristol's nostrum: "In their struggle for the ethical good, teachers of religion must have the stature to give up the doctrine of a personal God, that is, give up that source of fear and hope which in the past placed such vast power in the hands of priests" (p. 71).

This rejection of theism as part of Einstein's further explicit denial of supernatural causes impugns Sir Hermann Bondi's reading that Einstein championed a belief in a superintelligence who was the "architect" of the world's complexity.[61] Yet Bondi himself is staunchly antireligious.

It is true, if trite, that if there is deep and widespread demoralization in a community, as well as pervasive disaffection with its institutions, its sociopolitical organization will crumble, and it will become highly vulnerable to its enemies. Kristol transforms this commonplace into an ominous charge against secular humanism. But the supposition that the godless lead meaningless lives is just an ideological phantasm born of moral self-congratulation.

* * *

In an article entitled "Can Atheists Be Good Citizens?"[62] Richard John Neuhaus argues for a *negative* answer to the question posed in its title.

First he tries to cope with the fact that Sidney Hook, a lifelong ardent secular humanist, was a dedicated, fearless critic of totalitarianism for decades who received the Medal of Freedom from the U.S. government. In that attempt,

Neuhaus falls into a confusion between the *semantic content* of a doctrine with the degree of *epistemological confidence* that a given supporter of the doctrine may have in it. The content of theism is the assertion that there is a personal God with specified attributes, while the content of atheism is the denial of that claim. But neither content tells us with what *degree of confidence* a given proponent avows the given tenet.

The Roman Catholic Church claims absolute dogmatic, irrevocable certitude for its theism, while Madelyn Murray O'Hair, for instance, has proclaimed her atheism just as irrevocably. Alternatively, both theism and atheism alike can be espoused with varying lesser degrees of epistemological confidence. Some may regard their belief as a highly probable hypothesis in the light of the evidence, while others may see it more tentatively as the best available working hypothesis.

Theoretical beliefs, however well supported by known evidence, are still fallible or revocable, because of potentially adverse future evidence. It is therefore the better part of wisdom to stop short of espousing one's hypotheses as irrevocably established. Thus, Sidney Hook, Freud, Einstein, and Bertrand Russell, among others, adopted this less-than-dogmatic attitude toward their belief in atheism, but without tampering with its semantic content. Notably, their lack of dogmatism did not, however, constitute a watering down of their atheism into the different doctrine called "agnosticism."

In its technical meaning, agnosticism does not rule out either theism or atheism: It pointedly makes no claim as to the existence of God one way or the other, even tentatively, because it regards the question as unanswerable in principle. Thus, neither theists nor atheists are agnostics. And atheists disavow agnosticism no less than theists do.

This fact was untutoredly overlooked by Robert Bork during his unsuccessful confirmation hearings to become a U.S. Supreme Court Justice. Eyes flashing, Bork told the senators that he is not an agnostic, presumably to convey that he is not irreligious. But Bork's rejection of agnosticism, as such, does not rule out that he could even be an atheist.

Neuhaus (p. 17) denies that Sidney Hook was an atheist, claiming that, instead, Hook was an agnostic. Having wrongly assumed that atheism must be *irrevocably* declared true by its champions, Neuhaus concluded that, since Hook was a *fallibilist*, his rejection of theism must be tantamount to agnosticism after all. But this conclusion is false: Hook's commitment was to atheism, not to agnosticism.

Neuhaus is also led to claim incorrectly (p. 20) that the Enlightenment rationalists were "committed to undoubtable certainty" merely because they were atheists. As a dogmatic theist, opposed to Laplace's statement to Napoleon saying that he sees no need for the "hypothesis" of God, Neuhaus declared sorrowfully: "When God has become a hypothesis, we have traveled a very long way from both the gods of the ancient city and the God of the Bible" (p. 18). But why is that deplorable, if modern knowledge compels the demythologizing of the Bible, as indeed it does?

The principal thesis of Neuhaus's article is that atheists cannot be good citizens. Therefore, Hook's actual atheism commits Neuhaus willy-nilly to the further conclusion that Hook, the recipient of the Medal of Freedom bestowed by the president of the United States, is *philosophically* unfit to be a good citizen. Neuhaus's central argument, no less than Kristol's, turns out to run afoul of the moral sterility of theism. And this ethical infertility undermines his attack on the separation of church and state as well as his irate indictment of those religious people who support that separation.

Yet in his castigation of religious believers who support the organization "Americans United for Separation of Church and State," whom he charges with "political atheism," he abjures even the notion of the "existence" of God as too this-worldly. Indeed, we learn: "The transcendent, the ineffable, the totally other, the God who acts in history was tamed and domesticated in order to meet the philosopher's job description for the post of God" (p. 18). But this jeremiad boomerangs: If God is indeed so totally transcendent as to be ineffable, and if he eludes all intelligibility by being "totally other," how can there possibly be any meaning in the *causal* assertion that He "acts in history"?

Indeed, as I remarked early on, how can we possibly escape the conclusion that talk about such an *avowedly* "totally other" entity is just pretentious babble? Is the insistence on engaging in such discourse not a case of thought pathology, abetted by the penchant to abuse language? If, as we were told in the same vein, Yahweh—the God of Moses—was "above naming and beyond understanding," how can such an entity be *intelligibly* taken on faith even without evidence, let alone be loved or feared?

It is rank political coerciveness for Neuhaus to tell us that unless we are willing to parrot such gibberish, we are poor citizens. In striking contrast, in a letter written in 1790, George Washington explained to a Jewish community leader in Newport, Rhode Island, as reported by Bernard Lewis:

> The citizens of the United States of America ... all possess alike liberty of conscience and immunities of citizenship. It is now no more that toleration is spoken of, as if it was by the indulgence of one class of people that another enjoyed the exercise of their inherent natural rights. For happily the government of the United States, which gives to bigotry no sanction, to persecution no assistance, requires only that they who live under its protection, should demean themselves as good citizens, in giving it on all occasions their effectual support [footnote omitted].[63]

Bernard Lewis articulates George Washington's distinction between mere toleration and genuine coexistence very well:

> In these words, the first president of the United States expressed with striking clarity the real difference between tolerance and coexistence. Tolerance means that a dominant group, whether defined by faith or race or other criteria, allows to members of other groups some—but rarely if ever all—of the rights and privileges enjoyed by its

own members. Coexistence means equality between the different groups composing a political society as an inherent natural right of all of them—to grant it is no merit, to withhold or limit it is an offense.[64]

Yet, significantly, Neuhaus deploys his charge of poor citizenship even against those believers who have felt driven to take intellectual account of post-Enlightenment developments in the modern world. In fact, he levels the charge of deicide against them (p.18).

But the gravamen of Neuhaus's case is yet to come. Having omitted mention of the *fallibilist* kind of atheism held by such secular humanists as Sidney Hook, Neuhaus tendentiously enumerates the doctrines of much less reasonable atheists, and then he asks rhetorically (p. 20):

> Can these atheists be good citizens? It depends, I suppose, on what is meant by good citizenship. We may safely assume that the great majority of these people abide by the laws, pay their taxes, and may even be congenial and helpful neighbors. But can a person who does not acknowledge that he is accountable to a truth higher than the self, external to the self, really be trusted? Locke and Rousseau, among many other worthies, thought not. However confused their theology, they were sure that the social contract was based upon nature, upon the way the world really is. Rousseau's "civil religion" was apparently itself a social construct, but Locke was convinced that the fear of a higher judgement, even an eternal judgement was essential to citizenship.
>
> It follows that an atheist could not be trusted to be a good citizen, and therefore could not be a citizen at all. Locke is rightly celebrated as a champion of religious toleration, but not of irreligion. "Those are not at all to be tolerated who deny the being of a God," he writes in *A Letter Concerning Toleration*. "Promises, covenants, and oaths, which are the bonds of human society, can have no hold upon an atheist. The taking away of God, though but even in thought, dissolves all." The taking away of God dissolves all. Every text becomes pretext, every interpretation misinterpretation, and every oath a deceit.

Neuhaus offers a red herring in his ambiguous rhetorical question: "But can a person who does not acknowledge that he is accountable to a truth higher than the self, external to the self, really be trusted?" A secular humanist's insistence on the indispensability of reliance on the intelligence of the human species patently does not entail, as Neuhaus would have it, that any one of us is morally accountable only to our own self!

Here, he is trading on the vagueness and ambiguity of the phrase "truth higher than the self" to allude to the edicts of purported divine revelation of some sort. Unless he does so, the willingness to acknowledge accountability to a "social contract based on nature—on the way the world really is" obviously does not militate in favor of theism as against secular humanism. Indeed, it is secularism that relies on science to tell us about "the way the world really is."

The statements that Neuhaus then quotes or echoes from John Locke are vitiated by the moral sterility of theism, besides being outrageously false on their face. Indeed, we are being treated to *scurrilous demagogy* when Neuhaus declares that, in the case of an atheist, "every text becomes pretext, every interpretation misinterpretation, and every oath a deceit." This is brazen and insolent defamation!

Ironically, Neuhaus's invocation of Locke boomerangs: According to Locke, citizenship should not be accorded to Roman Catholics either, because these religious believers owe their ultimate allegiance to the foreign Pope rather than to God. Isn't it odd that Neuhaus, the recent convert to Roman Catholicism, made no mention at all of this highly inconvenient fact?

In an important recent article,[65] George Weigel relates and deplores the history of allegations in the United States that "an ascending tyrannical 'Romanism' or 'Papism' poses a threat to the pluralism of American democracy." The burden of his article, however, is a plea against a secularist, antitranscendentalist polity.

Weigel recounts a major episode that, ironically, is a valuable object lesson of the dangers run by adopting politically an "absolute" standard of morality on theological grounds:

> It is of moment … that classic American Protestant anti-Catholicism in the 19th and early 20th centuries simply took it as self-evident that American democracy required a religious foundation: specifically, a Protestant religious foundation. Absent this, it was widely believed there were but two possible outcomes to the American experiment: revival of premodern despotism (linked to Rome), or moral anarchy leading, in short order, to political collapse.

Significantly, Weigel adds that none of the advocates of this Protestant anti-Catholicism "ever dreamed of advocating a secular policy in which religion would be ruled out of the public debate."

Thus, by Weigel's own account, it was not a secularized state that generated the anti-Catholic turn he bewails; it was rather the denominational insistence on an absolute, divinely sanctioned morality amid the conflicting theological revelations. Despite ecumenicism, the strife among the gospels seems ineradicable: Witness the recent breakdown of negotiations between the Vatican and the Anglican Church, which were to yield an ecumenical composition of their theological differences. Or just contemplate the vanishing likelihood that Orthodox Jews will become persuaded of the salvific divinity of Christ![66]

Unaware that his chronicle boomerangs, Weigel concludes by misformulating the clash of ideas between secular humanism and a public policy informed by a religiously transcendent morality. As he would have it, this confrontation (Bismarckian *Kulturkampf*) is "a struggle between those who affirm the classic Jewish and Christian notion of an objective moral order, and those who deny on epistemological grounds that there is any such thing as an 'objective moral norm.'" Having posed the issue in these terms, Weigel speaks conjunctively of "secularism and moral relativism."[67]

But surely the secularist's this-worldly warrant for ethical norms is *neutral* as between an "objectivist" and a "relativist" construal of their epistemological status. To deny that our moral code has a transcendent religious foundation is not to rule out the objectivity of its secular grounds. Nay, ironically, the cacophony of divergent absolutist revelations is effectively tantamount to *moral relativism* as between the rival religious subcultures.

Alas, Neuhaus's and Weigel's gravamen against secular humanism, no less than Kristol's, emerges as a shoddy caricature of the doctrine they attack.

▪ 4. IS THEISTIC MORALITY *MOTIVATIONALLY* SUPERIOR TO SECULAR HUMANISM?

It is well-known that there are theists who were (or are) paragons of morality, such as Francis of Assisi and Mother Teresa, who devoted themselves sacrificially to the poor and to the care of outcasts (e.g., lepers). Yet the great harm done by Mother Teresa's Roman Catholic stance on artificial birth control and her rigid opposition to any and all abortion detract considerably from the moral benefits of her impact on society. Furthermore, the humane services of various religious orders, sects, or denominations in hospitals and in the relief of other suffering (e.g., famine) are legion. Besides, Pope John XXIII, while Archbishop Roncalli of Naples, did his utmost to save the Jews of the Balkans from the Nazis. On the other hand, a Roman Catholic Pope signed concordats with Hitler in addition to Mussolini and Franco.

Incomparably more significantly, and macroculturally, however, the two millennia of Christian history have prompted the German scholar Karlheinz Deschner to characterize much of it as "criminal" in a very widely read multivolume work of documentation: The first three, which are already published (1986, 1988, 1990), are devoted to antiquity; the next three volumes to the Middle Ages; and the last four to modern times (*Kriminalgeschichte des Christentums*, Hamburg: Rowohlt).* Plainly and notoriously, belief in theism is not at all sufficient *motivationally* for the sort of conduct on whose moral worth many theists would agree with secular humanists.

Some Western historians have characterized the Third Reich and the Soviet Union as seats of the two great secular movements of our time. But, even as the theist Seymour Cain concedes that "some egregious horrors connected with traditional Western religions," he opines that "far greater horrors [were] committed by the Third Reich and the Soviet Union," societies that were "anthropocentric without any transcendent norm."[68] It is unclear just how Cain arrives at these comparative measures of evil, but his comparison is, at best, highly and multiply misleading.

In the first place, Cain has to grant that neither of the two societies he names was secular *humanist*; on the contrary, Nazi Germany and Stalinist Russia are anathema

---

* *Editor's Note*: Meanwhile, nine of the ten projected volumes have been published, all with Rowohlt/ Hamburg.

to secular humanists on both moral and scientific grounds. As for the scientific component of secularism, the Nazi racial doctrines were *pseudo*-scientific, as was Stalin's rejection of biological genetics in favor of Lysenkoism—similarly for the governmentally endorsed distortion of scientific theorizing to conform to the prevailing political ideology (e.g., "Nordic" science in Nazi Germany and "proletarian" science in the USSR).

And in regard to the comparative scale of evil, Cain ignores that the technologies of Auschwitz and of the Soviet gulag were simply not available to the Holy office and to Cardinal Torquemada, who had to rely on the thumbscrew and the rack. Nor, to mention only a few examples, were they available to those who fought the Thirty Years' War in Europe for religious stakes, slaughtered the Huguenots, organized the loathsome Crusades, or hanged Quaker women at the stake in Puritan Massachusetts or to the host of others who prompted founding father John Adams to describe the Judeo–Christian tradition as "the most bloody religion that ever existed."[69]

Moreover, even the Stalinists while guilty of their own barbaric persecution of religious believers, did not burn them at the stake, whereas just that was the fate of heretics in Christendom for centuries. And nowadays in Islamic Pakistan, the theocrats are urging that even those who oppose the antiblasphemy laws be put to death.

Indeed, two millennia of doctrinal and often murderous Christian anti-Semitism prepared a climate in Nazi Germany, Vichy France and in Eastern Europe (e.g., in Ustachi Roman Catholic Croatia) that was hospitable to the Holocaust. Even recently, during the Polish election campaign that issued in Lech Walesa's presidency, this devout Roman Catholic demanded that candidates of Jewish origin acknowledge it as a kind of skeleton in their closet, much as those who have a dubious personal past should own up to it. And Reinhard Heydrich, the SS security chief who presided over the genocidal "Final Solution," was a graduate of a Catholic German High School.

Thus, Cain is driven to admit, after all, that *motivationally* theism is not morally superior to secularism. He grants that even the leaders of religious institutions, rather than merely the run-of-the-mill faithful, are no more ethical in practice than are secular leaders. As he acknowledges, both sorts of leaders alike "often put the practical welfare of their institutions above that of higher ethical values" (p. 56).

Thus, it is further grist to my mill, when Cain points out that secularists like Willy Brandt and religionists like Emil Niemoeller and Dietrich Bonhoeffer appealed alike to an authority higher than their secular government in resisting the Nazis.* As Cain notes (p. 56), Willy Brandt's motivation for his anti-Nazi activities:

---

* *Editor's Note*: Willy Brandt (1913–1992) was a German Social Democrat who emigrated to Norway in 1934 and later fled to Sweden. Brandt was chancellor of Germany from 1969 to 1974. Emil Gustav Friedrich Martin Niemöller (1892–1984) was a German Protestant theologian. While at first sympathizing with the Nazis, he later joined the resistance. He was held in the concentration camp

was clearly secular, based on a democratic socialist humanism .... As a member of the Norwegian resistance movement, he became, formally speaking, a traitor to his country, thus challenging the idolatry of the national state so pervasive in modern times. There were many other Germans who resisted Nazi tyranny for purely secular reasons, so far as anything is pure in human existence. Lay religious resisters sometimes found themselves abandoned and disavowed by their church leaders, like the simple Austrian carpenter who inveighed against the Nazi invasion of other countries only to be told by his bishop that he had no business opposing the governing authorities; hence, the church did nothing to prevent his execution. A similar case was that of a young Mormon workingman who engaged in anti-Nazi activities in Germany, only to be excommunicated by his church leaders and executed.

And again (pp. 55–56):

> Take the so-called righteous Gentiles who helped Jews to escape the Nazi murder machine, risking deadly danger for themselves and their families. Some of them may have been acting from a self-sacrificing devotion to values engendered by centuries of Western secular humanism. Others, like the French Huguenots who saved a remarkable number of Jews, were moved by religious motives and identification with the People of the Book. And there may have been French humanistic values mixed in.

Just this record shows that there *can* readily be moral *parity* between secularists and theists rather than the vaunted superiority proclaimed by the theists I challenge.

Furthermore, comparison of the crime statistics in the predominantly theist United States with the largely irreligious countries of Western Europe and Scandinavia resoundingly discredits the recurring claim that the moral conduct of theists is statistically superior to that of secularists, let alone of secular *humanists*. A fortiori, these statistics belie the smug thesis that the fear or love of God is *motivationally necessary*, in point of psychological fact, to assure such adherence to moral standards and good citizenship as there is in society at large.

Thus, the United States has by far the highest percentage of religious worshippers in its population of any Western nation, and presidents from Richard Nixon to Bill Clinton recurrently have given prayer breakfasts. In Great Britain, for example, which has the Anglican state church, only about 3 percent of its citizens attend a place of worship, whereas in the United States the figure is approximately 33 percent, that is, greater by a factor of eleven! In the United States, about 90 percent of the population profess belief in God, whereas in Western Europe and Scandinavia the percentage is considerably below 50 percent. Yet the percentage incidence of

---

Sachsenhausen from 1937 until the end of the war. Dietrich Bonhoeffer (1906–1945) was a Lutheran theologian. From April 1933 on he took a prominent public stand against the Nazis. Around 1938 he joined the resistance movement associated with Wilhelm Franz Canaris. In 1940, he was banned from public speaking and in 1941 from writing and publishing. On April 5, 1943, he was arrested, and two years later (April 9, 1945), on explicit order of Adolf Hitler, he was executed for his purported connection to the Hitler assassination attempt of July 20, 1944.

homicides and other crimes in the God-fearing United States is much higher than in the heavily secularized Western countries. And a corresponding disparity exists between the respective percentages of the prison populations in these societies. But the inveterate clamor for permitting prayer in the public schools of this country invokes the supposed efficacy of such devotionals in fostering "family values."

It emerges thus that theism and atheism as such are not only alike sterile qua theoretical foundation for concrete norms of ethical conduct; motivationally, belief in either of them is far too crude a touchstone to correlate with civilized moral conduct on the personal, social, or national level. If I may use the received androcentric idiom, the brotherhood of man does *not* depend on the fatherhood of God, either normatively or motivationally.

It is time that this major lesson be heeded widely in word and deed, especially by those who are at the levers of power in our polity and vociferously deny it. Thus, Cain (p. 57) was oblivious to the contemporary religio-political climate in the United States, when he wrote:

> I would counsel secular humanists to spend much less of their time accumulating proof texts on the failings and horrors of religion … [they] should stop finding all the good in their own camp and all the evil in that of their adversaries. Bigotry, fanaticism, and the refusal of dialogue are common human failings, affecting secularists as well as religionists. Let's look for the mote in our own eyes.

■ NOTES

I am grateful to my colleague Professor Richard Gale, who made several valuable suggestions in his comments on the first draft of this essay that was originally published as "The Poverty of Theistic Morality," in: K. Gravoglu et al. (eds.), *Science, Mind and Art: Essays on Science and the Humanistic Understanding in Art, Epistemology, Religion and Ethics*, Boston Studies in the Philosophy of Science, 1995, pp. 203–242. The chapter's title is from a shorter version previously published in *Free Inquiry* 12 (1992), 30–39.

1. W. A. Rusher, "Are We Misreading the Signals?" *Las Vegas Review Journal*, May 5, 1992, p. 7B.

2. For example, in the special issue "The Case for Ethical Monotheism," Vol. 7, No. 3, 1991.

3. See A. J. DeBethune, "Catholics in Exile," Letter to the Editor, *New York Times*, October 14, 1993.

4. M F. Perutz, Letter to the Editor, *New York Review of Books*, February 11, 1993, pp. 45–46.

5. See "Anglican Leader Attacks Catholics on Birth Control," *Secular Humanist Bulletin* 8(3) (Fall 1992), p. 9.

6. Sidney Hook, "Solzhenitsyn and Secular Humanism: A Response," *Humanist*, November–December 1978, p. 6.

7. Sigmund Freud, *Standard Edition of the Complete Psychological Works of Sigmund Freud* 21 (1927), p. 32.

8. "Why Are Catholics Afraid To Be Catholics," *New Republic*, February 21, 1994, p. 25.

9. Henry Grunwald, "The Year 2000," *Time*, March 30, 1992, pp. 73–75.

10. Irving Kristol, "Quotable," *Chronicle of Higher Education*, April 22, 1992.

11. *Time* 138(23), December 9, 1991.

12. Stephen L. Carter, *The Culture of Disbelief*, 1993, p. xxx.

13. See Christopher P. Tourney's review of *The Creationist Movement in Modern America* by R. A. Eve and F. B. Harrold, *American Scientist* 80 (May–June 1992), p. 292.

14. Quoted from Martin Buber, "The Dialogue Between Heaven and Earth," in: P. Edwards and A. Pap (eds.), *A Modern Introduction to Philosophy* (3rd ed.), 1973, pp. 394–395.

15. Paul Edwards, "Buber, Fackenheim and the Appeal to Biblical Faith," in: Edwards and Pap, op. cit., p. 395.

16. Buber, *Eclipse of God*, 1952, p. 66; see also pp. 105–106.

17. See Paul Edwards's 1969 Lindley Lecture "Buber & Buberism," p. 34, 1970.

18. Edwards, "Buber, Fackenheim and the Appeal to Biblical Faith," p. 395.

19. William Safire, "God Bless Us," Op-Ed, *New York Times*, August 27, 1992.

20. Emil Fackenheim, "On the Eclipse of God," in: Edwards and Pap, op. cit., Part V, par. 44, pp. 523–533.

21. Ibid., pp. 44–49; see also Edwards, "Buber, Fackenheim and the Appeal to Biblical Faith," pp. 395–398.

22. Quoted in Edwards, "Buber, Fackenheim and the Appeal to Biblical Faith," p. 395.

23. See, for instance, Buber, *Eclipse of God*, pp. 127, 129.

24. Edwards, "Buber & Buberism," p. 34.

25. *London Times*, May 9, 1987.

26. Avishai Margalit, "The Uses of the Holocaust," *New York Review of Books* 41(4), February 17, 1994, p. 7.

27. Cited by Michael J. Prival, Washington Society for Humanistic Judaism, "Hook and Homnick," *Free Inquiry* (Spring 1988), p. 3.

28. Sidney Hook, "A Common Moral Universe?" *Free Inquiry* 7(4) (1987), pp. 29–31.

29. James A. Michener, "God Is Not a Homophobe," *New York Times*, March 30, 1993, p. A15.

30. *New York Times*, February 6, 1994, Book review section, p. 37.

31. Robin L. Fox, *The Unauthorized Version: Truth and Fiction in the Bible*, 1992.

32. Grünbaum, "In Defense of Secular Humanism," *Free Inquiry* 12(4) (1992), pp. 30–39.

33. Seymore Cain, "In Response to Grünbaum's Defense," *Free Inquiry* 14(1) (1993–1994), pp. 55–57.

34. William Safire, "Islam Under Siege," Op-Ed page, *New York Times*, March 18, 1993, p. A15.

35. "The Bomb Threat," Editorial, *New Republic* 208(13), March 29, 1993, p. 9.

36. Yoyssef M. Ibrahim, "Fundamentalists Impose Culture on Egypt," *New York Times*, February 3, 1994, p. A6.

37. R. M. Gale, *On the Nature and Existence of God*, 1991, p. 34.

38. Arthur Schlesinger Jr., "Reinhold Niebuhr's Long Shadow," Op-Ed, *New York Times*, June 22, 1992, p. A13.

39. Sidney Hook, "Solzhenitsyn and Secular Humanism: A Response," *Humanist*, November–December 1978, p. 5.

40. John T. Noonan Jr., "Development in Moral Theology," *Theological Studies* 54(4) (December 1993), pp. 662–677, here: p. 675.

41. Ibid., 667.

42. Cf. Peter Steinfels, "Beliefs," *New York Times*, February 19, 1994, p. 8.

43. Cf. Grünbaum, *Validation in the Clinical Theory of Psychoanalysis*, 1992, ch. 7, "Psychoanalysis and Theism."

44. Stephen L. Carter, *The Culture of Disbelief*, 1993, p. 229.

45. Anna Quindlen, "America's Sleeping Sickness," *New York Times*, October 17, 1993, Section E, p. 17.

46. Gwen Ifill, "Clinton Warns Youths on the Perils of Pregnancy," *New York Times*, February 4, 1994, p. A11.

47. Irving Kristol, "The Future of American Jewry," *Commentary* 92(2) (August 1992), pp. 21–26.

48. Kristol, "The Future of American Jewry," p. 23.

49. See my "Creation as a Pseudo-Explanation in Current Physical Cosmology," *Erkenntnis* 35 (1991), pp. 233–254.

50. Michael Buckley, "Religion and Science: Paul Davies and John Paul II," *Theological Studies* 51 (1990), p. 314.

51. For a fuller discussion, see Paul Edwards, "The Dependence of Consciousness on the Brain," in: Paul Edwards (ed.), *Immortality*, 1992, pp. 292–307.

52. Kristol, "Quotable."

53. Kristol, "The Future of American Jewry," 24.

54. Cf. Ian S. Lastick, *For the Land and The Lord: Jewish Fundamentalism in Israel*, 1988.

55. Quoted in Adolf Grünbaum, "The Place of Secular Humanism in Current American Political Culture," *Vital Speeches of the Day* 54(2), November 1, 1987, p. 43.

56. Hook, "Solzhenitsyn and Secular Humanism," p. 6.

57. Václav Havel, "A Dream for Czechoslovakia," *New York Review of Books*, June 25, 1992, p. 12.

58. Louis Jacobs, *The Book of Jewish Belief*, 1984, p. 10.

59. Kristol, "Quotable."

60. Einstein's paper "Science and Religion" was delivered in 1941 at the Jewish Theological Seminary in New York; reprinted in D. J. Bronstein and H. M. Schulweis (eds.), *Approaches to the Philosophy of Religion*, 1954, pp. 68–72.

61. Herrmann Bondi, "Humanism—The Only Valid Foundation of Ethics," 67th Conway Memorial Lecture, January 24, 1992.

62. R. J. Neuhaus, "Can Atheists Be Good Citizens?" *First Things*, August–September 1991, pp. 17–21.

63. Quoted in Bernard Lewis, "Muslims, Christians, and Jews: The Dream of Coexistence," *New York Review of Books*, March 26, 1992, p. 49.

64. Ibid.

65. George Weigel, "The New Anti-Catholicism," *Commentary*, June 1992, pp. 25–31.

66. Cf. Y. Leibowitz, *Judaism, Human Values, and the Jewish State*, 1992.

67. George Weigel, "The New Anti-Catholicism," p. 30

68. Cain, "In Response to Grünbaum's Defense," p. 55.

69. Cf. Barbara Ehrenreich, *Time*, September 7, 1992, p. 72.

PART III

Philosophy of the 20th-Century Physical Cosmologies and the Critique of Their Theological Interpretations

7   The Poverty of Theistic Cosmology

ABSTRACT

Philosophers have postulated the existence of God to explain (I) why any contingent objects exist at all rather than nothing contingent, and (II) why the fundamental laws of nature and basic facts of the world are exactly what they are. Therefore, we ask: (a) Does (I) pose a well-conceived question that calls for an answer? and (b) Can God's presumed will (or intention) provide a cogent explanation of the basic laws and facts of the world, as claimed by (II)? We shall address both (a) and (b). To the extent that they yield an unfavorable verdict, the afore-stated reasons for postulating the existence of God are undermined.

As for question (I), in 1714, G. W. Leibniz posed the Primordial Existential Question (hereafter "PEQ"): "Why is there something contingent at all, rather than just nothing contingent?" This question has two major presuppositions: (1) A state of affairs in which nothing contingent exists is indeed genuinely possible (the "Null Possibility"), the notion of nothingness being both intelligible and free from contradiction; and (2) *De jure*, there *should* be nothing contingent at all, and indeed there *would* be nothing contingent in the absence of an overriding external cause (or reason), because that state of affairs is "most natural" or "normal." The putative world containing nothing contingent is the so-called Null World.

As for (1), the logical robustness of the Null Possibility of there being nothing contingent needs to be demonstrated. But even if the Null Possibility is demonstrably genuine, there is an issue: Does that possibility require us to explain why it is not actualized by the Null World, which contains nothing contingent? And, as for (2), it originated as a corollary of the distinctly Christian precept (going back to the second century) that the very existence of any and every contingent entity is utterly dependent on God at any and all times. Like (1), (2) calls for scrutiny. Clearly, if either of these presuppositions of Leibniz's PEQ is ill founded or demonstrably false, then PEQ is *aborted* as a *non*starter, because in that case it is posing an ill-conceived question.

In earlier writings,[1] I have introduced the designation "SoN" for the ontological "spontaneity of nothingness" asserted in presupposition (2) of PEQ. Clearly, in response to PEQ, (2) can be challenged by asking the *counter*-question: "But why *should* there be nothing contingent, rather than something contingent?" Leibniz offered an a priori argument for SoN. Yet it will emerge that a priori defenses of it fail and that it has no empirical legitimacy either. Indeed, physical cosmology spells an important relevant moral: As against any a priori dictum on what is the "natural" status of the universe, the verdict on that status depends crucially on empirical evidence. Thus, PEQ turns out to be a nonstarter because its presupposed SoN is ill founded! *Hence, PEQ cannot serve as a springboard for creationist theism.*

Yet Leibniz and the English theist Richard Swinburne offered divine creation ex nihilo as their *answer* to the ill-conceived PEQ. But being predicated on SoN, their cosmological arguments for the existence of God are fundamentally unsuccessful.

The axiomatically topmost laws of nature (the "nomology") in a scientific theory are themselves unexplained explainers and are thus thought to be true as a matter of brute fact. But theists have offered a theological explanation of the specifics of these laws as having been willed or intended by God in the mode of agent causation to be exactly what they are.

A whole array of considerations is offered in Section 2 to show that the proposed theistic explanation of the nomology fails multiply to transform scientific brute facts into specifically explained regularities.

■ 1. WHY IS THERE SOMETHING RATHER THAN NOTHING?

1.1 Refined Statement of Leibniz's Primordial Existential Question

Leibniz's 1714 essay "Principles of Nature and of Grace Founded on Reason"[2] is the locus classicus of the question: "Why is there something rather than nothing?" In the German translation of his French original, this question is, "warum es eher Etwas als Nichts gibt?"[3]

We shall speak of this query as the Primordial Existential Question and will use the acronym PEQ to denote it for brevity. But we must refine the statement of PEQ to preclude a *trivialization*, which Leibniz certainly did not intend when he asked this question. As we shall see, he believed that God is "a necessary being, bearing the reason of its existence within itself" in order to provide a "sufficient reason" for "the existence of the [*contingent*] universe."[4] But if there is a *necessary* being, there can be no question whether *it* exists, rather than not, because such a being could not possibly *fail* to exist. Therefore, it would clearly *trivialize* Leibniz's cardinal PEQ, if it were asked concerning a "something" comprising one or more entities whose existence is logically or metaphysically *necessary*.

Hence, the scope of the term "something" in his PEQ must obviously be *restricted* to entities whose existence is logically *contingent*: entities whose *non*existence is *logically possible.* And similarly for the scope of the term "nothing." Accordingly, we can formulate Leibniz's nontrivial construal of PEQ as follows: "Why is there something contingent at all rather than just nothing contingent?" Philip Quinn has usefully characterized that articulation of PEQ as an "explanation-seeking contrastive why-question."[5] He calls it "contrastive" because it features the contrastive locution "rather than."

William Craig is oblivious to the aforementioned *non*trivial construal of PEQ. Thus, in a paper published in the *British Journal for the Philosophy of Science* and entitled "Professor Grünbaum on the 'Normalcy of Nothingness" in the Leibnizian and Kalam Cosmological Arguments,"[6] which is directed against my earlier essay in the same journal,[7] Craig obfuscates and eviscerates Leibniz's primordial question, which drives him to an exegetical falsehood as follows: "It must be kept in mind that for Leibniz (in contrast to Swinburne) … a state of nothingness is logically impossible" (p. 377). But Craig's assertion here is a red herring precisely because, for both Leibniz and Swinburne, a state of affairs in which there is nothing *contingent* is indeed logically possible. If Craig is to be believed and Leibniz had regarded a state of nothingness to be logically impossible, then his PEQ would have been tantamount to asking fatuously: Why is there something rather than a specified logically *impossible* state of affairs? This alone, it appears, is a reductio ad absurdum of Craig's exegesis of Leibniz.

As we shall see in some detail in Section 1.9, Swinburne[8] deems the existence of God to be logically *contingent*, and therefore he *excludes* God from a state of affairs in which there is nothing contingent. But Swinburne as well as Leibniz was all too aware that, if there are entities that exist *necessarily*, then even a state in which there is nothing contingent cannot exclude such entities. Moreover, Leibniz had inferred that God exists necessarily qua sufficient reason for the "existence of the [contingent] universe."[9] Hence, Leibniz deemed the existence of God to be compossible with a state featuring nothing contingent, whereas Swinburne denied that compossibility, having concluded that God exists only contingently.

It is crucial to note at the outset that PEQ rests on important presuppositions. If one or more of these presuppositions is either ill founded or demonstrably false, then PEQ is aborted as a *non*starter, because it would be posing a *non*issue (pseudo-problem). And, in that case, the very existence of something contingent, instead of nothing contingent, does *not* require explanation. In earlier writings,[10] I have used the rather pejorative term "pseudo-problem"—"Scheinproblem" in German—to reject "a question that rests on an ill-founded or demonstrably false presupposition."[11] But since the term "pseudo-problem" was given currency by the Vienna Circle, I immediately issued the caveat that, in my own use of it, "I definitely do not intend to hark back to early positivist indictments of 'meaninglessness.'"[12] Terminology aside, PEQ will indeed turn out to be a nonstarter, because one of its crucial presuppositions is demonstrably ill founded. As we shall see, that presupposition is a corollary of a distinctly Christian doctrine, which originated in the second century C.E.

What are the most important presuppositions of PEQ? Clearly, one of them is that the notion of a state of affairs in which absolutely nothing *contingent* exists is both *intelligible* (meaningful) and *free from contradiction*. Let us call such a putative state of affairs the "Null Possibility," as the English philosopher Derek Parfit does.[13] And let us speak of a supposed world in which there is nothing contingent as the "Null World."

Yet it is vital to recognize that the Null Possibility is *not* shown to be logically genuine by the premise that each contingent entity, taken individually, might possibly not exist. After all, this premise is entirely compatible with the *denial* of the Null Possibility. Indeed, the familiar fallacy of composition is being committed if one infers that all entities, taken *collectively*, might possibly fail to exist merely because each contingent entity, taken *individually*, might possibly fail to exist.

In just this way, both Derek Parfit[14] and the English theist Richard Swinburne[15] seem to have fallaciously inferred the logical robustness of the Null Possibility after enumerating a finite number of actual entities, *each* of which individually may possibly fail to exist. And their commission of the fallacy of composition then blinds them to their obligation to justify the Null Possibility as logically sound before posing PEQ. Alternatively, they may just have taken the Null Possibility for granted peremptorily. Thus, Parfit gave the following version of PEQ:

> Why is there a Universe at all? It might have been true that nothing [contingent] ever existed: no living beings, no stars, no atoms, not even space or time. When we think about this ["Null"] possibility, it can seem astonishing that anything [contingent] exists.[16]

In this statement, Parfit presumably construed the term "nothing" to mean "nothing *contingent*," as Leibniz did. Evidently, Parfit inferred the Null Possibility without ado, declaring: "It might have been true that nothing ever existed." But he gave no cogent justification for avowing this logical possibility to be genuine: he just assumed peremptorily that the *nihilistic proposition* "There is nothing," or "The Null World obtains," is both intelligible and free from contradiction. Instead of providing a conceptual *explication* of the Null Possibility, Parfit has evidently offered a mere *open-ended enumeration* of the absence of familiar ontological furniture from the Null World: "no living beings, no stars, no atoms, not even space or time." Thereupon, he enthrones PEQ on a pedestal: "No question is more sublime than why there is a Universe [i.e., some world or other]: why there is anything rather than nothing."[17] Besides presupposing that the Null Possibility is logically robust, Parfit's motivation for PEQ tacitly *pivots* on the supposition that, de jure, there should be nothing contingent.

1.2 Is It Imperative to Explain Why There *Isn't* Just Nothing Contingent?

Parfit told us that "when we think about this ['null'] possibility, it can seem astonishing that anything exists." And assuming such an astonished response, he feels entitled to ask why the Null Possibility does *not* obtain; that is, why there is something after all rather than just nothing. But I must ask: Why should the *mere contemplation* of the Null Possibility reasonably make it "seem astonishing that anything exists"?

If some of us were to consider the logical possibility that a person might conceivably metamorphose spontaneously into an elephant, for example, I doubt strongly that we would feel even the slightest temptation to ask why that *mere logical possibility is not realized.* But what if someone were to reply that, in such a case, we are not puzzled because, as we know empirically with near certainty, people just don't ever turn into elephants? Then I would retort: Indeed, and what could possibly be more commonplace empirically than that something or other does exist? On the other hand, consider, as just a thought experiment, that *per impossibile*, a person actually metamorphoses into an elephant. If we were suddenly to witness such a spontaneous transformation, we would all be aghast, and we would ask urgently: Why, oh why, did this monstrous transformation occur?

Why then, I put it to Parfit, should anyone reasonably feel astonished at all that the Null Possibility, if genuine, has remained a mere logical possibility and that something does exist *instead*? In short, why *should* there be just nothing, merely because it is logically possible? This *mere* logical possibility, I claim, does *not suffice* to legitimate Parfit's demand for an explanation of why the Null World does *not* obtain, an explanation he seeks as a philosophical anodyne for his misguided ontological astonishment.

1.3 Must We Explain Why Any and Every *De Facto Unrealized* Logical Possibility Is Not Actualized?

To justify a negative answer to this question, let us inquire quite generally: For *any* and *every* de facto unrealized logical possibility, is it well conceived to demand an explanation of the fact that it is *not* actualized? As we know, Leibniz's Principle of Sufficient Reason (PSR) has been used to answer affirmatively that every fact has an explanation. Yet, as we shall see in section 1.7.1, Leibniz himself did not regard that principle as itself an adequate justification for his PEQ, because he also relied on its presupposition SoN to convey that the existence of something contingent *is not to be expected at all* and therefore calls for explanation. But even his PSR is demonstrably unsound.

To appraise his Principle of Sufficient Reason, consider within our universe the grounds for the demise of Laplacean determinism in quantum theory. This *empirically* well-founded theory features irreducibly stochastic probability distributions governing such phenomena as the spontaneous radioactive disintegration of atomic nuclei, yielding emissions of alpha or beta particles and/or gamma rays. In this domain of phenomena, there are not only logically but also nomologically (i.e., law-based) possible particular events that *could* but do *not* actually occur under specified initial conditions. Yet it is impermissibly legislative ontologically to insist that merely because these events are thus possible, there *must* be an explanation entailing their specific *non*occurrence and, similarly, of course, for stochastically governed, actually occurring events. This lesson was not heeded by Swinburne, who avowed entitlement to *pan*-explainability, declaring: "We expect all things to have explanations."[18] In our exegesis of Leibniz in section 1.7.1, we shall deal further with his PSR.

The case of quantum theory shows that an empirically well-grounded theory can warrantedly discredit the tenacious demand for the satisfaction of a previously held ideal of explanation, such as Leibniz's Principle of Sufficient Reason. To discover that the universe does not accommodate rigid prescriptions for explanatory understanding is not tantamount to scientific failure; instead, it is to discover positive reasons for identifying certain coveted explanations as phantom. And to reject the demand for them is legitimate in the face of Charles Sanders Peirce's heuristic injunction not to block the road to inquiry, for this rejection does not abjure the search for a new, better theory in which the original explanatory quest may appear in a new light.

The demise of the PSR at the hands of micro-physics spells a moral for Parfit's question why the Null Possibility does *not* obtain: the *mere* logical possibility of the Null World—assuming it to be genuine—does *not suffice* to legitimate Parfit's demand for an explanation of why the Null Possibility does *not* obtain, rather than something contingent.

Nonetheless, Richard Swinburne declared: "It remains to me, as to so many who have thought about the matter, a source of extreme puzzlement that there should exist anything at all."[19] And, more recently, he opined: "It is extraordinary that there should exist anything at all. Surely the most natural state of affairs is simply nothing: no universe, no God, nothing."[20] It is here, incidentally, that Swinburne apparently commits the fallacy of composition, as Parfit did, in trying to vouchsafe the Null Possibility by an enumeration of contingent entities, each of which, taken individually, may possibly fail to exist.

1.4 Is a World Not Containing Anything Contingent Logically Possible?

We need to be mindful of a further imperative to demonstrate that the Null Possibility hypothesized by PEQ is logically authentic, if indeed it is: some philosophers have explicitly denied the intelligibility of a kindred possibility. Thus, Henri Bergson has argued relatedly against nothingness: "The idea of absolute nothingness has not one jot more meaning," he tells us, "than a square circle."[21] True enough, Richard Gale has given a number of detailed reasons for rejecting Bergson's claim of unintelligibility.[22] Yet Gale's own proposed explication of the hypothetical claim that "Nothing exists" is itself so qualified as to drive him to the following unfavorable conclusion: "it is not [logically] possible for there to be [absolutely] Nothing."[23]

To state the nub of his reasoning, let me again use the locution "Null World" to speak of a putative world in which the Null Possibility in fact obtains. Then we can say that the Null World is devoid of space–time, no less than of all other contingent objects. But according to Gale's account,[24] the Receptacle of space and time (extension and duration), along with the "positive" properties or "forms," "are the ontological grounds for the possibility of there being Nothing." Hence,

Gale contends that "there is no possibility of *their* not existing. Put differently, it is not possible for there to be [absolutely] NOTHING, for there must at least be the [spatio-temporal] Receptacle and the forms."[25] Thus, Gale diverges from Parfit's view that space and time exist only contingently: For Gale, they exist necessarily and hence exist even in the Null World, but for Parfit, they are excluded from the Null World, qua existing only contingently. Therefore, it is puzzling that, in the face of this exclusion, Parfit used the seemingly *temporal* term "ever" when he told us that "it might have been true that nothing ever existed."

But as Edward Zalta has pointed out to me (private communication), it is unclear how Gale's avowal of space and time as existing *necessarily* and Bergson's indictment of meaninglessness are relevant to the issue of the intelligibility of the Null Possibility. That possibility pertains to contingent existents, not to necessary ones. After all, as we saw, Leibniz's Null World contains necessarily existing entities like his God, while being devoid of all contingent ones. Hence Gale's argument has not gainsaid the pertinent sort of Null Possibility. And, as for Bergson, he is addressing the hypothesis that "absolutely nothing exists" rather than the hypothesis that nothing *contingent* exists. And the latter may be meaningful, even if the former is not.

But are there positive arguments that establish the meaningfulness of the Null Possibility? The reader is referred to philosophical misgivings or challenges issued by Edward Zalta of Stanford University, which place the burden of proof on those who deem PEQ to be well conceived. In any case, it should be borne well in mind that the provision of a viable explication of the Null Possibility is surely not *my* philosophical responsibility but rather belongs to the protagonists of PEQ, who bear the onus of legitimating their question. In the absence of assurance that the Null Possibility is logically authentic, PEQ might well be aborted as a nonstarter for that reason alone.

How, then, are we to understand more deeply the *tenacity* with which PEQ has been asked not only by some philosophers but also even in our culture at large? An illuminating set of answers is afforded, it seems, by delving critically into three kinds of impetus for this ontological question, as follows: (1) historically based assumptions going back to the second century of the Christian era, which served to inspire PEQ; (2) explicitly a priori logical justifications of PEQ put forward by Leibniz, Parfit, Swinburne, and Robert Nozick; and (3) hypothesized emotional sources articulated by Arthur Schopenhauer in his magnum opus *The World as Will and Representation (Die Welt als Wille und Vorstellung)*.

Let us consider these three sorts of impetus for PEQ seriatim.

1.5 Christian Doctrine as an Inspiration of PEQ

On Maimonides' reading of the opening passage of the Book of Genesis, the Mosaic God created the world *out of nothing*. Yet there is recent biblical exegesis contending that this doctrine of creation ex nihilo was not avowed in the Book of Genesis. Though the doctrine may have had a prehistory, it was first widely held by Christian theologians, beginning in the second century C.E., as a *distinctly Christian* precept.[26] Thus, in an exegetical essay on "Genesis's account of creation" in the Old Testament, the Jewish scholar Norbert Samuelson wrote four years ago: "This [Hebraic] cosmology presupposes that initially God is not alone. Prior to God's act of creation ... the earth [and] water are the stuff from which God creates."[27] But Christian writers regard their specific conception of divine creation ex nihilo as a philosophical advance over the account in the Book of Genesis, if only because they held that an omnipotent God had no need for preexisting materials to create the universe. Thus, as one such writer noted rather patronizingly: "The abstract notion of nothing does not seem to have been reached by the Israelite mind at that time."[28] And, evidently, the notion of nothingness was essential to generate PEQ.

According to traditional Christian ontological doctrine, the *very existence* of any and every contingent entity other than God himself is utterly dependent on God at any and all times. Let us denote this fundamental Christian axiom of total ontological dependence on God by "DA," for "Dependency Axiom." Clearly, DA entails the following cardinal maxim: "Without God's [constant creative] support [or perpetual creation]," the world "would instantly collapse into nothingness."[29] This assumption played a crucial role in subsequent philosophical history. Thus, in later centuries, precisely this hypothesis DA was avowed, as we shall see, by such philosophers as Thomas Aquinas and Descartes, among a host of others.

Evidently, DA in turn entails that, in the absence of an external cause, the spontaneous, natural, or normal state of affairs is one in which nothing contingent exists at all. As will be recalled, I have denoted the assertion of this ontological spontaneity of nothingness by "SoN." As before, we shall usually speak of the putative state of affairs in which no contingent objects exist at all as "the Null World," a locution that is preferable to the term "nothingness." In that parlance, SoN asserts the ontological spontaneity of the Null World.

As we see, the fundamental Christian ontological axiom DA of total existential dependence on God *entails* SoN. In other words, logically the truth of SoN is a *necessary condition* for the truth of the fundamental ontological tenet of Christian theism. In this clear sense, SoN is a *presupposition* of DA, which will turn out to be a heavy doctrinal burden indeed. SoN is "a heavy doctrinal burden," because, as we shall see, it is *completely baseless*.

According to SoN, the actual existence of something contingent or other—qua deviation from the supposedly spontaneous and natural state of nothingness—automatically requires a *creative external cause* ex nihilo, a so-called ratio essendi. And such a supposed creative cause must be distinguished, as Aquinas emphasized, from a merely transformative cause: Transformative causes produce changes of state in contingent things that *already exist in some form*, or the transformative causes generate new entities from previously existing objects, such as in the building of a house from raw materials.

Furthermore, in accord with the traditional Christian commitment to SoN, creation ex nihilo is required at every instant at which the world exists in some state or other, whether it began to exist at some moment having no temporal predecessor in the finite past or has existed forever. More precisely, having presupposed SoN, traditional Christian theism makes the following major claim: In the case of any contingent entity E other than God himself, if E exists, or begins to exist *without* having a *transformative* cause, then its existence must have a creative cause ex nihilo rather than being externally *uncaused*.

Yet, as some scholars have pointed out, "to the ancient Indian and Greek thinker the notion of *creation* [ex nihilo] is unthinkable."[30] Indeed, as John Leslie has pointed out informatively: "To the general run of Greek thinkers *the mere existence of things* [or of the world] *was nothing remarkable*. Only their *changing* patterns provoked [causal] inquisitiveness" (italics added). And he mentions Aristotle's views as countenancing the acceptance of "reasonless existence."[31]

It is a sobering fact that, before Christianity molded the philosophical intuitions of our culture, those of the Greeks and of many other world cultures[32] were basically different ontologically. No wonder that Aristotle regarded the material universe as uncreated and eternal. In striking contrast, SoN is deeply ingrained in the traditional Christian heritage, even among a good many of those who reject Christianity in other respects. And the Christian climate lends poignancy to Leslie's conjecture that "when modern Westerners have a tendency to ask why there is anything at all, rather than nothing, possibly this is *only* because they are heirs to centuries of Judaeo-Christian thought" (italics added).[33] So much for the Christian historical contribution to PEQ via its SoN doctrine.

1.6 Henri Bergson

Early in the twentieth century, Henri Bergson was alert to the often beguiling, if not insidious, role of SoN in metaphysics, and he aptly articulated that assumption as inherent in PEQ. In 1935, speaking of occidental philosophy, Bergson lucidly wrote disapprovingly concerning PEQ as follows:

> Part of metaphysics moves, consciously or not, around the question of knowing why anything exists—why matter, or spirit, or God, rather than nothing at all? But the question presupposes that reality fills a void, that underneath Being lies nothingness, that *de jure* there should be nothing, that we must therefore explain why there is de facto something.[34]

Bergson's concise formulation of SoN as a presupposition of the Primordial Existential Question is that "de jure there should be nothing." But as a rendition of this cardinal presupposition of PEQ, his formulation that de jure there *should* be nothing is significantly incomplete: It needs to be amplified by the further claim that there indeed *would* be nothing in the absence of an overriding external cause or reason! Thus, let us bear in mind hereafter that SoN makes the following very strong ontological assertion: De jure, there *should* be nothing contingent at all rather than something contingent, and indeed, there *would* be just nothing contingent in the absence of an overriding external cause (reason).

In a chapter devoted to PEQ, Robert Nozick notes, as Bergson had, that this inveterate question is predicated on SoN, a presupposition avowing, in his words, "a presumption in favour of nothingness." As he puts his view there: "To ask 'why is there something rather than nothing?' assumes that nothing(ness) is the natural state that does not need to be explained [causally], while deviations or divergences from nothingness have to be explained by the introduction of special causal factors."[35]

Importantly, SoN can be challenged by the counter-question: "But why *should* there be nothing contingent, rather than something contingent?" And, indeed, why *would* there be nothing contingent in the absence of an overriding external cause (reason)? In effect, Leibniz[36] endeavored to disarm this challenge, as we are about to see, when he tried to legitimate SoN—albeit unsuccessfully—as part of a *twofold* a priori justification of PEQ.

Since PEQ is predicated on SoN, PEQ will be undermined in due course by the failure of a priori defenses of SoN and by the unavailability of any empirical support for it. Hence, I am not persuaded by Nicholas Rescher's claim that PEQ—which he calls "the riddle of existence"—"does not seem to rest in any obvious way on any particularly problematic presupposition."[37] Although the defects of SoN are indeed not obvious, SoN will, in fact, turn out to be a "particularly problematic presupposition" of PEQ.

1.7 A Priori Justifications of PEQ by Leibniz, Parfit, Swinburne, and Nozick

A number of writers have used ideas such as simplicity, nonarbitrariness, naturalness, and probability in an attempt to justify PEQ a priori. In each case, the argument seems to be that a state of affairs in which nothing contingent exists has a crucial property (e.g., simplicity, nonarbitrariness, naturalness, high probability) which we would a priori expect the world to have. This a priori expectation is then presumed to validate the second presupposition of PEQ, which entails that the Null World would be actual in the absence of an external cause. Typically, the property in question (e.g., simplicity) has not even been sufficiently articulated, but even if its character is taken for granted the governing principle (e.g., the world should be simple) has not been justified, as we shall see now.

1.7.1 Leibniz

In stark contrast to Bergson, both Leibniz[38] and Swinburne[39] maintained that SoN is a priori *true*. And their reason was that the Null World is simpler, both ontologically and conceptually, than a world containing something contingent or other. This very ambitious assertion poses two immediate questions: (a) Is the Null World really a priori simpler and indeed the *simplest* world ontologically as well as conceptually? And (b) even assuming that the Null World *is* thus simpler, does its supposed maximum dual simplicity *mandate ontologically* that there *should* be just nothing de jure and that, furthermore, there *would* be just nothing in the absence of an overriding cause (reason), as claimed by SoN?

As for question (a), of maximum twofold simplicity, the Swedish philosophers Erik Carlson and Erik Olsson speak of the Null World as "the intrinsically simplest of all possibilities," and they add that they "have not seen it questioned."[40] But as for the question of *conceptual* simplicity, there is one caveat, which needs to be heeded.

To see why, note first that Leibniz couched his original 1697 statement of PEQ in terms of "worlds" when he demanded "a full reason why there should be any world rather than none."[41] This formulation suggests that, conceptually, the very notion of the Null World may well range—by *negation* or *exclusion*—over all of the possible contingent worlds or objects other than itself which are *not* being actualized in it. But this collection of unrealized, noninstantiated contingent worlds is *super-denumerably infinite* and is of such staggering complexity that it boggles the mind!

As we have remarked, the champions of the maximum simplicity of the Null World have not given us a demonstrably viable explication of the notion of the Null World as being logically authentic or robust. Therefore, they cannot claim to have *ruled out* that, conceptually, this notion is highly complex, instead of being the simplest. So much for the caveat pertaining to the purported maximum *conceptual* simplicity of the Null World.

Beyond this caveat, we do not need to address the ramified issues raised by (a) in dealing with Leibniz's defense of SoN except to say that, to my knowledge, the purported conceptual and *ontological* maximum simplicity of the Null World has not been demonstrated by its proponents.

But let us assume, just for the sake of argument, that Leibniz and Swinburne could warrant a priori the maximum conceptual and ontological simplicity of the Null World, as avowed by Leibniz, when he declared: "'nothingness' is simpler and easier than 'something.'"[42] It is of *decisive importance*, I contend, that, even if the supposed maximum ontological simplicity of the Null World were warranted a priori, that presumed simplicity would *not* mandate the claim of SoN that *de jure the thus simplest world must be spontaneously realized ontologically* in the absence of an overriding cause. Yet, to my knowledge, neither Leibniz nor Swinburne nor any other author has offered any cogent reason at all to posit such an ontological imperative.

Let us quote and then comment on the context in which Leibniz formulates his PEQ and then seeks to *justify* it at once by relying carefully on both of the following two premises: (1) his Principle of Sufficient Reason (PSR); and (2) an a priori argument from simplicity for the presupposition SoN inherent in PEQ:

Sec. 7. Up till now we have spoken as *physicists* merely; now we must rise to *metaphysics*, making use of the *great principle*, commonly but little employed, which holds that *nothing takes place without sufficient reason*, that is to say that nothing happens without its being possible for one who has enough knowledge of things to give a reason sufficient to determine why it is thus and not otherwise. This principle having been laid down, the first question we are entitled to ask will be: *Why is there something rather than nothing?* For "nothing" [the Null World] is simpler and easier than "something." Further supposing that things must exist, it must be possible to give a reason *why they must exist just as they do* and not otherwise.

Sec. 8. Now this sufficient reason of the existence of the universe cannot be found in the series of contingent things, that is to say, of bodies and of their representations in souls.... Thus the sufficient reason, which needs no further reason, must be outside this series of contingent things, and must lie in a substance which is the cause of this series, or which is a necessary being, bearing the reason of its existence within itself; otherwise we should still not have a sufficient reason, with which we could stop. And this final reason of things is called *God*.[43]

These two major passages in Leibniz's sections 7 and 8 invite an array of comments:

(1) Right after enunciating his PSR, Leibniz poses PEQ, "Why is there something rather than nothing?" as "the first question we are entitled to ask." And immediately after raising this question, he relies on simplicity *to justify* its presupposition SoN that, de jure, there should be nothing contingent at all, rather than something contingent: "For 'nothing' [the Null World] is simpler and easier than 'something.'" But, in the class of logically contingent entities to which this claim of greater simplicity pertains, "nothing" (the Null World) and "something" (a world featuring something) are mutually exclusive and jointly exhaustive. Thus, Leibniz is telling us here, in effect, that the Null World is the a priori *simplest* of all, besides being "the easiest." But, alas, he does not tell us here in just what sense the Null World is "the easiest."

This thesis of the intrinsically greatest a priori ontological and conceptual simplicity of the Null World has been a veritable mantra in the literature, which is why Carlson and Olsson wrote that they "have not seen it questioned."

(2) It is vital to appreciate that Leibniz explicitly went beyond his PSR to justify his PEQ on the heels of enunciating PSR and posing PEQ: fully aware that PEQ presupposes SoN, he clearly did *not* regard PEQ to be justified by PSR alone, since he explicitly offered a *simplicity argument* to justify the presupposed SoN immediately after posing PEQ. Most significantly, he is *not* content to rely on PSR to ask *just* the truncated question: "Why is there something contingent?" Instead he uses SoN in his PEQ to convey his dual thesis that (i) the existence of something contingent *is not to be expected at all*, and (ii) *its actual existence therefore cries out for explanation in terms of the special sort of noncontingent causal sufficient reason he then promptly articulated in his Section 8.*

Thus, the soundness of Leibniz's justification of his PEQ evidently turns on the cogency of his PSR as well as of his a priori argument for SoN. As for the correctness of his PSR, recall our preliminary objections to it from Section 1.3, which were prompted there by Parfit's erotetic musings. The modern history of physics teaches that PSR, which Leibniz avowedly saw as metaphysical, cannot be warranted a priori and indeed is untenable on empirical grounds. The principle asserts that every event—in Leibniz's parlance, anything that "takes place" or "happens" ["geschieht," i.e., "sich ereignet" in the German translation of his original French text]—has an *explanatory* "reason [cause] sufficient to determine *why it is thus and not otherwise*" (italics added). Leibniz's inclusion of the locution "and not otherwise" is presumably intended to emphasize an important point: his PSR guarantees the existence of a sufficient reason not only for the actual occurrence of a given specific event **E** but also for the actual *nonoc-currence* of any and every *specific* event that is *different* from **E** in some respect or other. Quite reasonably, therefore, PSR has been taken to avow the existence of a reason sufficient to explain any and every *fact* pertaining to an individual event.

In sum, PSR is untenable: Irreducibly stochastic laws in quantum physics tell us that some events have no individual explanations but occur as a matter of brute fact. And, assuming with Leibniz that there is no infinite regress of explanations, the history of science strongly supports the view that, no matter how we axiomatize our body of knowledge, every such axiomatization will feature some contingent fundamental laws or other that are unexplained explainers in that axiomatization and codify brute facts. And, as for Leibniz's a priori argument from simplicity for SoN, we saw earlier in this Section 1.7.1 that it does not pass muster.

(3) To set the stage for a further instructive commentary on the subtle deficits of PSR, recall from Section 1.3 Swinburne's own formulation of SoN, which reads in part: "Surely *the most natural* state of affairs is simply nothing."[44] As will be shown in section 1.7.3, this formulation of SoN entails the following consequence: it would be natural—though *not* "most natural"—for *our* world or universe U_o *not* to exist rather than to exist. Let us denote that corollary of SoN by "SoN(U_o)." Now, though Leibniz's PSR turned out to be untenable, suppose just for argument's sake that we were to grant him his PSR for now. Then someone may be tempted to believe that its explanatory demand could suffice after all, *without* a separate additional argument for SoN's corollary SoN(U_o), to legitimate the question "Why does our universe U_o exist, rather than not?," a question which presupposes that corollary. In short, the issue is whether PSR can *single-handedly* license the injunction to explain the existence of our universe in particular.

To evaluate the claim that it can, note that Leibniz called for an explanatory "sufficient condition for the existence of the universe [U_o]." But, importantly, this explanatory demand generates at least three *distinct* questions, which differ both from each other *and* from PEQ:

(Q_1): Why does U_0 exist, rather than not?

(Q_2): Why does U_0 exist, rather than just nothing contingent?

(Q_3): Why does U_0 exist, featuring certain laws L_0, rather than some different sort of universe U_n, featuring logically possible different laws of nature L_1?

Moreover, note that each of these questions, no less than PEQ, is predicated on a presupposition of its own, which is asserted by the *alternative* stated in its "rather than" clause. The relevant presupposition is that, *de jure*, the corresponding alternative *should* obtain and indeed might well or would obtain in the absence of an overriding cause (reason). And the question then calls for an explanation of the *deviation* from the supposedly *de jure* alternative.

In question Q_1, the alternative A_1 is that U_0 should not exist. Hence, Q_1 is predicated on SoN(U_0). But its alternative, A_1, is noncommittal as to whether *other* universes likewise should *not* exist. Thus, A_1 differs from the alternative A_2 in Q_2 that nothing contingent at all should exist, which asserts SoN. A_2, in turn, differs from the alternative A_3 in Q_3 that something else exists in lieu of U_0. Thus Q_1, Q_2, and Q_3 are different questions whose answers therefore may well be different. Furthermore, observe that Q_2, besides differing from Q_1 and Q_3, also differs from PEQ: Whereas Q_2 asks specifically why our U_0 exists, rather than nothing contingent at all, PEQ asks why *something or other* exists, as against nothing at all.

In effect, Q_3 demands an explanation of why *our* world U_0 exists, as contrasted with a logically possible different sort of universe featuring different laws of nature. Interestingly, Leibniz called for an answer to *that* question only after having assumed that: "Further, supposing that things must exist, it must be possible to give a reason *why they must exist just as they do* and not otherwise." By way of anticipation, note that, as Section 2 will show, the volitional theological answer to question Q_3 by such theists as Leibniz, Swinburne, and Quinn completely fails to deliver on their explanatory promises.

We can now deal specifically with the aforestated issue: if Leibniz's PSR were granted, could it *single-handedly* legitimate question Q_1 ("Why does U_0 exist, rather than not?") *without* having to justify Q_1's presupposition SoN(U_0) by an additional argument? This question will now enable us to demonstrate the inability of PSR to serve solo as a warrant for Q_1, just as it turned out previously to be incapable of licensing PEQ without a further argument justifying the latter's presupposition SoN. As will be recalled, Leibniz himself recognized that limitation of PSR by his recourse to the supposed greater simplicity of the Null World as the ontological underwriter of the presupposition SoN of his PEQ.

In what sense, if any, might his PSR underwrite his demand for a "sufficient reason for the existence of the universe [U_0]"? This injunction is clearly more specific than the imperative to supply a sufficient reason for the existence of something contingent or other. In his statement of PSR, Leibniz asserts the existence of a sufficient reason for what "*takes place*" or "happens." And, very importantly, this reason for the occurrence of events, he tells us, is "sufficient to determine *why it is thus and not otherwise."* Evidently, his contrasting alternative to "what *takes place"* is "*otherwise."* But, crucially, that alternative demonstrably fails, however, to be *univocal!*

An example that, alas, has become commonplace since the two attacks on the World Trade Center in New York will illustrate the considerable ambiguity of Leibniz's notion of "otherwise." When we explain ordinary sorts of events, we typically know instances of their occurrence as well as instances of their nonoccurrence. Furthermore, we have evidence concerning the conditions relevant to both kinds of instances and often have information as to their relative frequency. That information often tells us which of these sorts of events, if either, is to be expected routinely.

Indeed, when we explain the occurrence of a given event, the *contrasting non-* occurrence of that event can *take different forms:* We may wish to explain why the event occurred rather than some *specified* other sort of event that might have occurred *instead*, or we may just ask why the given event occurred rather than not.

Thus, when a presumably well-built skyscraper collapsed, we may ask why it did rather than withstand an assault on it, though not without considerable damage. Or we may ask why a presumably well-built skyscraper collapsed rather than just staying intact. The latter contrasting alternative of staying intact constitutes the "natural" career of well-built skyscrapers in the absence of an overriding external cause.

In the context of Leibniz's inquiry into "the sufficient reason for the existence of the universe $[U_o]$," the relevant event or happening is the existence of the universe through time. Hence, we must ask in that context: What becomes of his call for a sufficient reason that determines "why it [an event] is thus and not otherwise"?

The ambiguity of this request for a sufficient reason is shown by the different contrasting alternatives in questions Q_1, Q_2, and Q_3, questions which pertain to the existence of U_o. One such alternative to the existence of U_o is simply that it does not exist, another that nothing contingent exists, yet another alternative is that some other sort of universe U_n exists *instead* of U_o. Yet the demand to explain the existence of U_o, as issued by PSR, does *not* dictate *which* one of these alternatives is presupposed by it as "otherwise"! Thus, PSR fails to license the question Q_1 single-handedly by failing to single out its presupposed $SoN(U_o)$ as against the very different presuppositions of questions Q_2 and Q_3.

For precisely this portentous reason, it turns out that to ask "Why does U_o exist?" is *to ask an incomplete question!* Hence, Leibniz's PSR is incapable of showing that if the existence of U_o is to be explained, it must be explained qua deviation from its nonexistence, as against qua deviation from some other alternative. In short, PSR itself does not license Q_1 as against Q_2 or Q_3. The tempting belief that it does so single-handedly is a will-o'-the-wisp.

The failure of his PSR to underwrite the particular question Q_1 "Why does U_o exist, rather than not?" also emerges from one of the reasons for rejecting $SoN(U_o)$ as ill founded. As we saw, when we call for an explanation of such events as the collapse of a skyscraper, we do so against the background of having observed specified instances of their *non*occurrence. Indeed, these nonoccurrences may well be so very frequent that we are warranted in taking them to be "natural." But obviously, yet very importantly, we have never ever observed an event constituted by the *non*-existence of our U_o, let alone found evidence that its nonexistence would be "natural." Nor does the empirically supported Big Bang cosmology feature such a temporal event.[45] Accordingly, it is ill founded to regard the *non*existence of our U_o to be "natural," such that PSR could then warrant the question why U_o exists, rather than not.

(4) As we saw in Section 1.1 on refining Leibniz's PEQ to preclude its unintended trivialization, its articulated version states: "Why is there something *contingent* at all rather than just nothing *contingent?*" Alas, this rather straightforward construal of PEQ was lost on Craig. As we saw in Section 1.1, his bizarre, misguided reading of Leibniz turns PEQ into the ludicrously fatuous question: Why is there something rather than a specific logically impossible state of affairs? And it will be recalled that, contrary to Craig, for Leibniz no less than for Swinburne, the Null World—being devoid of all *contingent* existents, though clearly not of any necessary existents—is indeed logically possible. Yet Craig denied this exegetical fact and offers a non sequitur:

> It must be kept in mind that for Leibniz (in contrast to Swinburne) God's existence is logically necessary, so that a state of nothingness is logically impossible. Hence, Leibniz *cannot* be assuming that a state of absolute nothingness is the natural or normal state of affairs, as Swinburne does.[46]

But here Craig has irrelevantly fabricated for himself a notion of "a state of (absolute) nothingness" that is indeed logically impossible, because he himself incoherently banished all *necessary* existents from it rather than only all contingent existents, as in Leibniz's Null World. Thus, Craig clearly offers a non sequitur in claiming that "for Leibniz (in contrast to Swinburne) God's existence is logically necessary, so that a state of nothingness is logically impossible," a baseless conclusion. Moreover, once Craig had concocted his own incoherent state of "absolute nothingness," which is logically impossible by excluding necessary existents, he manufactured an incoherent *pseudo*-version of SoN which features such phantom nothingness. Thereupon, he informs us irrelevantly and misleadingly that Leibniz "cannot" be claiming such absolute nothingness to be "the natural or normal state of affairs, as Swinburne [allegedly] does."

But surely nobody of just ordinary intelligence, let alone Swinburne, would avow Craig's incoherently devised pseudo-version of SoN! Swinburne's Null World excludes only all *contingent* existents, while including, of course, any necessary existents. Thus, his conception of SoN plainly does not pertain to Craig's inane artifact of "absolute nothingness," with which Craig saddled him.

Unfortunately, Craig used an ignoratio elenchi and an exegetical jumble to reject my attribution of SoN to Leibniz, although SoN is a patent presupposition of Leibniz's PEQ, as Bergson and Nozick explicitly recognized.

(5) In Section 1.3, and under item (3) of the current Section 1.7.1, there is a preliminary mention of Swinburne's alternative formulation of SoN, which reads in part: "Surely *the most natural* state of affairs is simply nothing,"[47] a state devoid of all and only *contingent* existents. And as we shall see in Section 1.7.3, this formulation of SoN entails the following consequence: It would be natural—though *not* "most natural"—for our world or universe U_o of inanimate matter, biological organisms, and Homo sapiens *not* to exist, rather than to exist, a corollary of SoN that we have denoted by "SoN(U_o)." Since Leibniz evidently presupposed SoN in his PEQ, he is likewise committed to its corollary SoN(U_o).

But the existence of our U_o is a *deviation* from its purportedly "natural" *non*-existence, as avowed by SoN(U_o). And any deviation from naturalness calls for *explanation* in terms of a suitable cause or reason. No wonder that, in the opening sentence of Leibniz's aforecited Section 8, he demanded that the answer to his PEQ provide a "sufficient reason" for "the existence of the universe [U_o]."

Yet, true to form, Craig tells us blithely:

Even with respect to the physical universe, moreover, Leibniz did not hold that the natural or normal state of affairs is the non-existence of the physical universe, for he held (notoriously) that God's creation of the world is, like God Himself, necessary.[48]

But, as Nicholas Rescher has argued in a thorough chapter significantly entitled "Contingentia Mundi: Leibniz on the World's Contingency":

From the earliest days of his philosophizing Leibniz insisted upon the contingency of the world. It was always one of his paramount aims to avert a Spinozistic necessitarianism, and he regarded the contingency of the world's constituents and processes as an indispensable requisite towards this end, one in whose absence the idea of divine benevolence would be inapplicable.[49]

To the further detriment of Craig's exegesis, Rescher explains:

Leibniz distinguishes between two different modes of necessity. The one is the *metaphysical* necessity of that whose opposite is logico-conceptually impossible. And the other is the *moral* necessity of that whose opposite is ethically unacceptable. Only the former is absolute and categorical, the latter is standard-relative and dependent upon ultimately evaluative ethical considerations.[50]

And Rescher elaborates crucially:

God's choice of the best available alternative for actualization [i.e., "God's creation of the world," in Craig's words.], while indeed a certain fact regarding God, is only *morally* and not metaphysically necessary—*and thereby contingent* [footnote omitted].

And while God's moral perfection as creator of this best of worlds is itself a morally necessary truth, it is emphatically not metaphysically necessary.[51]

As shown by Rescher's account of Leibniz's views, Craig used an exegetically false premise to deny incorrectly that Leibniz was committed to SoN(U_o). It would seem that the remainder of Craig's misguided gloss on Leibniz does not merit discussion.

1.7.2 Derek Parfit

Parfit has gone beyond Leibniz and Swinburne in laying down alleged a priori ontological imperatives. Claiming that the Null World is not only the a priori simplest but also the "least arbitrary," Parfit believes that SoN is warranted all the more qua presupposition of PEQ. And hence he thinks that the supposed minimum arbitrariness of the Null World confers even greater urgency on PEQ. Thus, he asks, "Why is there a Universe at all? Why doesn't reality take its simplest and least arbitrary form: that in which nothing ever exists?"[52]

But why, one must ask, *should* ontological reality spontaneously be "least arbitrary," even assuming that the Null World is a priori "least arbitrary" in Parfit's intended sense? He develops his reasoning in his essay "The Puzzle of Reality: Why Does the Universe Exist?" First he declares quite generally: "If some possibility would be less puzzling, or easier to explain, we have more reason to think that it obtains."[53] And then, echoing Leibniz's undemonstrated belief that the Null World is the a priori simplest both conceptually and ontologically, Parfit offers that world as his paradigm example (p. 421): "Is there some global possibility whose obtaining would be in no way puzzling? That might be claimed of the Null Possibility .... Perhaps, of all the global possibilities, this would have needed the least explanation. It is much the simplest and it seems the easiest to understand."

Parfit's view that the Null Possibility "seems easiest to understand" may well have been suggested by Leibniz's dictum that "'nothing' is simpler and easier than 'something,'" a claim that Leibniz offered unsuccessfully to justify SoN qua presupposition of his PEQ. But the Null Possibility may well not be "easiest to understand": As we saw in section 1.7.1, conceptually the very notion of the Null World may well range—by *negation* and *exclusion*—over a mind-boggling nondenumerable infinitude of possible contingent worlds. This complexity seems to have been tacitly discerned by Parfit in his implicitly *open-ended* enumeration of objects that are *excluded* from the Null World (p. 420).

In any case, Parfit has to retract his view that the Null Possibility is the least problematic (p. 420): "Even if this possibility would have been the easiest to explain, it does not obtain. Reality does not take its simplest and least puzzling form."

Why then did Parfit claim, in the first place, that we have more reason to think that the allegedly simplest and "least arbitrary" Null World would obtain than that some other unremarkable possible world would be actualized? Without ado, he appeals to some concept of "coincidence," saying:

> Coincidences can occur. But it seems hard to believe. We can reasonably assume that, if all possible worlds exist, that is *because* that makes reality as full as it could be.
>
> Similar remarks apply to the Null Possibility. If there had never been anything, would that have been a coincidence? Would it have merely happened that, of all the possibilities, what obtained was the *only* possibility [i.e., the *one* possibility] in which nothing exists? That is also hard to believe. Rather, if this possibility had obtained, that would have been because it had that feature. (P. 424)

Yet now Parfit no longer claims that some remarkable global feature **F** of some possible world **W** (e.g., maximum simplicity) warrants *the presumption that **W** would obtain* rather than some unremarkable world. Instead, now we learn that *if* **W** does obtain and features some remarkable global property **F**, then we can infer explanatorily that **W** was ontologically mandated by the fact that **F** was ontologically *self-realizing*! Relying on such purported mandatoriness, Parfit says: "Our world may seem to have some feature that would be unlikely to be a coincidence. We might reasonably suspect that our world exists, not as a brute fact but because it has this feature" (pp. 426–427).

But how can Parfit tell whether a given global feature **F** of our world or of some other possible world in fact is "unlikely to be a coincidence"? He told us that, if every conceivable world were actualized in a so-called plenary universe, it would be very unlikely to be a coincidence. And if the Null Possibility were to obtain, that too would be very unlikely to be coincidental. On the other hand, we learn (p. 424), if a universe of 57 worlds were to exist, its cardinality of 57 "could hardly" be self-actualizing and hence *would be coincidental.* In the case of the plenary universe and of the Null World, Parfit (ibid.) is struck alike by the fact that "of all the global possibilities, the one that obtains" is just that particular one. But, clearly, in the 57-worlds-universe as well, *only* one of the "countless global possibilities" is instantiated.

Why then, one must ask, would the obtaining of the *extremities* of the full range of possibilities *not* be coincidental and therefore cry out for being explained, whereas the actualization of one of the possibilities *in between* would be just coincidental? Why, indeed, isn't the 57-member world-ensemble just as a priori *improbable* as either the plenary universe or the Null World?

Apparently, Parfit determines which instantiations are not coincidental, and which ones are, by *tacitly* appealing to some metric of a priori probabilities for the actualization of possible worlds that he has not even begun to articulate, let alone to justify. A fortiori, he has not shown that the then a priori improbable actualizations call for *ontological* explanation just because they are a priori improbable. For these reasons alone, it seems, his bizarre ontology of potentially self-actualizing explanatory global properties does not enlarge our philosophical horizons.

Furthermore, Swinburne objects cogently to Parfit's self-actualizing scenario for our world:

> Parfit's suggestion that there might be some non-causal explanation of the existence of the Universe involves his claiming that there is some kind of principle at work in producing the Universe, which is never operative in producing more limited effects within the Universe. But then we have absolutely no reason for supposing that that kind of principle is ever at work, or that such a principle explains anything at all.[54]

Note, incidentally, that when Parfit purports to explain that our universe U_o exists by recourse to its possession of some remarkable self-actualizing property **P**, what he is, in effect, claiming to explain is why U_o is actualized, *rather than* some other universe U_n that *lacks* **P**.

Parfit even generalizes his notion of self-actualizing remarkable global features into a comprehensive doctrine of cosmological explanation:

> If some possibility obtains because it has some feature, that feature selects what reality is like. Let us call it the *Selector*. A feature is a *plausible* Selector if we can reasonably believe that, were reality to have that feature, that would not merely happen to be true.
>
> There are countless features which are not plausible Selectors. Suppose that fifty-seven worlds exist. Like all numbers, 57 has some special features. For example, it is the smallest number that is the sum of seven primes. But that could hardly be *why* that number of worlds exist.
>
> I have mentioned certain plausible Selectors. A possibility might obtain because it is the best, or the simplest, or the least arbitrary, or because it makes reality as full as it could be, or because its fundamental laws are as elegant as they could be. There are, I assume, other such features, some of which we have yet to discover.
>
> For each of these features, there is the *explanatory* possibility that this feature *is* the Selector. That feature then explains why reality is as it is. (P. 424)

Among the a priori selectors, which Parfit countenanced as "plausible," he included the minimization of a priori arbitrariness: "A possibility might obtain because it is ... the simplest, or the least arbitrary" (p. 424). Thus, when he inquired why the Null Possibility does *not* obtain, he asked why reality does *not* "take its simplest and least arbitrary form."[55]

Parfit's a priori view that the minimization of arbitrariness is *ontologically legislative* had been championed years earlier by Peter Unger, who took a leaf out of Robert Nozick's[56] fanciful "fecundity principle," which richly populated the universe with an infinite "plenitude" of isolated worlds on avowedly altogether a priori grounds.[57] Though Unger admits to being "uncertain" as to what he means by the conception of a "highly unarbitrary" world, he reaches a very gloomy *empirical* verdict concerning the realization of *non*arbitrary features in our world:

> Well, then, what *is* the available empirical evidence, and what *does* it indicate about the actual world? As empirical science presents it to us, is the world we live in, the world of which we are a part, a world notable for its lack of natural arbitrariness? Far from it, the actual world, our evidence seems to indicate, is full of all sorts of fundamental arbitrary features, quirks that seem both universal for the world and absolutely brute....
>
> According to available evidence, and to such a theory of our actual world as the evidence encourages, the actual world has nowhere near the lack of arbitrariness that rationalist intuitions find most tolerable. To satisfy the rationalist approach, our evidence tells us, we must look beyond the reaches of our actual space and time, beyond our actual causal network. For there to be a minimum of arbitrariness in the universe entire, indeed anything anywhere near a minimum, we might best understand the universe as including, not only the actual world, but infinitely many other concrete worlds as well.[58]

Thus, neither Parfit nor Unger has supplied any cogent reason for believing that a priori nonarbitrariness—insofar as that notion is clear at all—is *ontologically legislative* in the sense of mandating what sort of world is actualized.

As an ontological injunction, the minimization of arbitrariness seems to be just a case of apriorism run amok.

1.7.3 Richard Swinburne and Thomas Aquinas Vis-à-Vis SoN

Interestingly, Swinburne's alternative formulation of SoN, already mentioned in section 1.7.1, can serve as a point of departure for arguing, as a lesson from the history of science, that SoN stands or falls *on empirical but not on* a priori grounds. Swinburne's alternative formulation of SoN reads in part: "Surely the most natural state of affairs is simply nothing."[59] Let us develop the implications to which Swinburne commits himself by his thesis that the Null World is "surely the most natural state of affairs." The articulation of the corollaries of his version of SoN will then enable us to make an *empirical* judgment of its validity.

Assume, for argument's sake, that the Null World was actualized. And consider the set **W** of logically possible contingent worlds (or objects) that *fail to be instantiated* or realized in the Null World. Then Swinburne's version of SoN makes the following claim: it is *most natural* that *none* of the possible member-worlds or objects in **W** are actualized. But if this collective failure of actualization is "most natural," what is the ontological status, in regard to "naturalness," of the individual member-worlds of **W**, taken singly or distributively? As we are about to see, Swinburne's SoN entails that, for each individual member of **W**, taken singly, it is *natural*—though *not* "most natural"—that it *not* be actualized rather than that it exist de facto.

To see that this conclusion follows from his version SoN, note that if it *were* natural for *some one* member world of **W** to *exist actually rather than not*, then the state of affairs in which *no* member is actualized (i.e., the Null World) could *not* be the *most* natural of all. Notice, incidentally, that the stated inference from SoN does *not* commit the fallacy of division, since it reasons from the superlative attribute "most natural" of the collection **W** to the merely positive attribute "natural" of its individual members.

Thus, it is a corollary of Swinburne's version of SoN that it is natural for our world or universe U_0 of inanimate matter, biological organisms, and Homo sapiens not to exist, rather than to exist. As before, let us denote this particular corollary by "SoN(U_0)." A like corollary pertains to each individual member-world of a putative mega-universe which features universes in addition to ours in a world ensemble. But the actual existence of our U_0 is a deviation from its allegedly de jure "natural" nonexistence, as avowed by SoN (U_0). Hence its existence calls for explanation in terms of a suitable external cause or reason. Indeed, precisely because Swinburne claims that there should be nothing instead of our universe U_0, he is driven to ask *why* our cosmic abode U_0 does exist, even though, *naturalistically*, it supposedly should not.

By the same token, he, like other theists before him, appeals to SoN(U_0) to demand an explanation of how our world is *kept in existence* instead of reverting to its supposedly natural state of not existing.[60] And, again in concert with other theists, he therefore holds the view that "God keeps the universe in being, whether he has been doing so forever or only for a finite time."[61] This is the Christian doctrine of perpetual divine creation, which is labeled creatio continuans.

Thomas Aquinas is one of the early major theists who simply assumes SoN *tout court* in his metaphysics of essence and existence.[62] Logically contingent existing entities, he holds, are "composed" of both essence and existence. The ontologically *spontaneous* state of affairs, in his view, is *à la* SoN that *no* logically contingent objects exist, but that their essences constitute *potentialities* for actualization. Yet since logically contingent objects do exist, Aquinas's assumption of SoN prompts him to conclude that there must be a ratio essendi, a creative "act of being" which actualizes their essences. And that act, we have learned, is performed by God. Moreover, the divine bestowal of being has to take place at every instant of their existence. Thus, he concluded: "The being of every creature [i.e., every logically contingent thing] depends on God, so that not for a moment could it subsist, but would fall into nothingness were it not kept in being by the operation of Divine power."[63]

Aquinas's peremptory assumption of SoN as a presupposition of his argument above for perpetual divine creation is further evident from his view that God could *annihilate* objects by *merely ceasing to conserve them*, without any destructive act, since they would then *spontaneously* lapse into nonexistence.[64]

But, now suppose, for the sake of argument, that one conceptualizes any logically contingent object with Aquinas as being "composed" of its essence and existence, which is quite dubious anyway. Even then, it is baseless to assume that the essence of the object must be *ontologically prior*, as it were, to its existence *such that SoN is true*. Evidently, Aquinas does *not* justify SoN but begs the question by taking it for granted in the service of his creationist theological agenda.

William Craig concedes that "Grünbaum is correct in seeing the spontaneity of nothingness … to lie at the heart of the Thomist cosmological argument [for the existence of God]." But then he objects: "It is up to Grünbaum to explain why he rejects the Thomistic metaphysics that underlies the argument, which he has not even begun to do."[65] Although Craig himself recognized that Aquinas's cosmological argument is predicated on SoN, he did not see that this argument is a petitio principii, since Aquinas begs the question with respect to SoN. Thus, contrary to Craig, I am entitled to reject Aquinas's cosmological argument as ill founded after having impugned his baseless contention that essence is ontologically prior to existence *such that SoN is true*.

Alas, the sixteenth-century Jewish Kabbalist Meir ben Gabbai encumbers his affirmations of SoN with primitive word magic. Yet without even a hint of intellectual disapproval, Nozick reports that, according to ben Gabbai, "only God's continuing production of the written and oral Torah maintains things in existence."[66] Speaking of the alleged dire ontological consequences of any interruption at all of God's continuing production of the written and oral Torah, that Kabbalist wrote: "were it [i.e., this divine production of Hebrew words] to be interrupted for even a moment, all creatures would sink back into their non-being." Thus, ben Gabbai relies on divine word magic to replace the ontological role that Aquinas assigned to divine creation ex nihilo. But plainly, the Torah-scribes are the ones who keep producing the written Torah. And it is ongoing human verbal communication among the faithful which preserves the Torah orally. Hence, if ben Gabbai is to be believed, human language users are ontologically necessary at all times to prevent the mighty cosmos from lapsing into nothingness in accord with SoN, a patent absurdity that no one should ever have taken seriously.

But, as we are about to appreciate, SoN is altogether ill founded *empirically*, so that *any* cosmological argument or doctrine that is predicated on it is likewise empirically unwarranted. And since a priori defenses of SoN are seen to have failed, it will then emerge as an induction from various episodes in the history of science that SoN stands or falls on *empirical grounds*.

1.7.4 *The "Natural" Status of the World as an Empirical Question*

Consider the corollary of SoN pertaining to our own world U_0, that is, $SoN(U_0)$, in its own right. As we recall, the latter corollary asserts that it is *natural* for our world U_0 *not* to exist, rather than to exist. As against any a priori dictum on what is the "natural" status of our world, the verdict on that status will now be turning out to depend crucially on empirical evidence. Two cosmological examples will spell this empirical moral:

(a) The *natural evolution* of one of the Big Bang models of the universe countenanced by general relativistic cosmology is a clear cosmological case in point. This model, the so-called Friedmann universe, is named after the Russian mathematician Alexander Friedmann. In the context of the general theory of relativity (GTR), the Big Bang dust-filled "Friedmann universe" has the following features:[67]

(i) It is a spatially closed spherical universe (a "3-sphere"), which expands from a Big Bang to a maximum finite size and then contracts into a crunch.

(ii) It exists altogether for only a finite span of time, such that no instants of time existed prior to its finite duration or exist afterward.[68]

(iii) As a matter of natural law, its total rest mass is conserved for the entire time period of its existence so that, during that time, *there is no need for a supernatural agency to prevent it from lapsing into nothingness à la* Aquinas, or as in René Descartes' scenario in his Third Meditation.[69]

Evidently, the "natural" dynamical evolution of the Friedmann Big Bang universe *as a whole* is specified by the *empirically supported* general relativistic theory of cosmology. And if there is a world ensemble of Big Bang worlds, the "natural" evolution of the members of this mega-universe would likewise be based on such physical laws as are hypothesized to prevail in them. But the epistemic warrant for these presumed laws likewise cannot dispense with empirical evidence, if they are to become warrantedly known to us. Thus, the "natural" behavior of Big Bang worlds is not vouchsafed a priori.

(b) The same epistemic moral concerning the empirical status of cosmological naturalness is spelled by the illuminating case of the now largely defunct Bondi and Gold steady-state cosmology of 1948, if only because its account of the hypothesized steady state of the expanding universe as being natural owes its demise to contrary empirical evidence.

The 1948 steady-state theory features a spatially and temporally infinite universe in which the following steady-state cosmological principle holds: As a matter of natural law, there is large-scale conservation of matter *density*.[70] Note that this conservation is *not of matter* but of the *density* of matter over time. The conjunction of this constancy of the density with the Hubble expansion of the universe then entails a rather shocking consequence: Throughout space–time, and without any matter-generating agency, matter (in the form of hydrogen) pops into existence completely *naturally* in violation of matter–energy conservation.[71] Hence the steady-state world features the accretion or formation of *new* matter as its *natural*, normal, spontaneous behavior. And although this accretive formation is ex nihilo, it is clearly *not* "creation" by an external agency. Apparently, if the steady-state world was actual, it would impugn the ontology of the Medieval Latin epigram "Ex nihilo, nihil fit."

Its spontaneous matter accretion occurs at the rate required by the constancy of the matter density amid the mutual galactic recession, and it populates the spaces vacated by the mutual galactic recession. Thus, in the hypothesized Bondi and Gold world, the spontaneous accretion of matter would be explained deductively as *entirely natural* by the conjunction of two of its fundamental physical postulates. But the rate of this spontaneous cosmic debut of new matter is small enough to leave the received matter–energy conservation law essentially intact locally (terrestrially).

Precisely because the new matter is held to originate *spontaneously* in the steady-state world, it is salutary to use the *agency-free* term "matter *accretion*" to describe this hypothesized process. And we must shun the use of the misleading *agency-loaded* term "matter *creation*," because the noun "creation" denotes an act of causing something to exist by an agency *external* to its object. Thus the notion of creation calls for a creator.

It was therefore quite misleading that the cosmologist Hermann Bondi, who is a dedicated secular humanist, equated the problem of the origin of the universe with the alleged "problem of creation" and declared that the steady-state theory solves this problem of creation *scientifically*, whereas "other theories," such as the Big Bang theory, do not, but hand it over to metaphysics.[72] But since SoN is turning out to be groundless, the purported problem of creation emerges to be a nonissue.

The steady-state theory owes its demise to the failure of its predictions and retrodictions to pass observational muster in its competition with the Big Bang cosmology. This episode again teaches us that *empirically based scientific theories are our sole epistemic avenue to the "natural" behavior of the universe at large*, though only fallibly so, of course, since such theories are liable to be replaced by others in the light of further empirical findings.

In earlier writings,[73] I have given other examples from the history of physics (Aristotelian and Newtonian mechanics) and from the history of biology (spontaneous generation of life from nonliving substances), showing how evolving empirically based theories in these domains provided *changing* conceptions of the "natural," spontaneous behavior of *subsystems* of the universe. By the same token, these episodes illustrate *how misguided it is to persist in asking for an external cause of the deviations of such systems from the pattern that an empirically discredited theory erroneously affirms to be "natural."*

Thus, as I emphasized furthermore in previous writings,[74] the history of empirical science has legitimated the *theory-relative rejection* of certain why questions. Bas van Fraassen referred to my legitimation of such rejections and has aptly encapsulated their upshot: "The important fact for the theory of explanation is that not everything in a theory's domain is a legitimate topic for why-questions; and that *what is* [legitimate], *is not determinable a priori.*"[75]

Thus, it is fitting that we should ask: What is the empirical verdict on SoN(U_0), a corollary of Swinburne's SoN, which asserts that "it is natural for our universe not to exist, rather than to exist"? Its proponents have offered no empirical evidence for it from cosmology, let alone for SoN itself, believing mistakenly, as we saw, that it can be vouchsafed a priori *à la* Leibniz.

But what if they were to counter the injunction to supply empirical evidence, objecting that it demands the impossible. Impossible, it might be said, because, in principle, there just can be no evidence from within our actual world which might show that it is the "natural" state of our cosmos *not* to exist in the first place. To this, the retort is twofold: It is not obvious that this epistemic predicament is genuine, but even if it were it would redound only to the baselessness of SoN(U_0) rather than tell against the legitimacy of demanding evidence for any and every averral of cosmic naturalness, including SoN(U_0).

Philip Quinn agrees that, as he put it, "current scientific theories and the empirical evidence on which they rest provide little or no support for SoN." But he then chides me, declaring that, "the de facto history of science falls far short of establishing the strong modal conclusion that this issue [the credentials of SoN] *cannot* be settled a priori because only empirical evidence could have a bearing on it." And he opines that it is "scientistic"—in von Hayek's[76] pejorative sense of being philosophically imperialistic—to maintain, as I do, that "only empirical evidence of the sort that supports scientific theories could have a bearing on the acceptability of SoN."[77]

Quinn's complaint of scientism prompts several responses:

(a) There is an important *asymmetry*, I submit, between an empirical and an aprioristic adjudication of the truth of SoN, because of their *very different track records* in making determinations of naturalness. A priori defenses of SoN seem to have failed, as we have already seen. But the *scientific* record of determining *what transpires naturally* is *brilliant*, not only in physics but also in biology. Witness the history of the theories as to the *spontaneous generation of life* from inorganic materials to which we alluded before, starting with Louis Pasteur but including Alexander Oparin and Harold Urey in the twentieth century. Furthermore, the rich history of the *disintegration of Kantian apriorism* in regard to the external world spells a strong caveat against the expectation of an a priori *vindication* of SoN.

(b) As Quinn noted, I appeal inductively to the substantial evidence from the history of science to infer that settling the merits of SoN a priori is pie-in-the-sky. But I allow, of course, that this induction is *fallible*. Hence I would surely be open to correction, if someone were unexpectedly to come up with a cogent a priori argument for SoN. But I trust I can be forgiven if I do not expect that to happen at all. And I do not see that I am being rash in my epistemological attitude.

Indeed, I deem it very important that Quinn makes both of the following concessions, the first of which bears repetition from before:

(i) "It seems to me that current scientific theories and the empirical evidence on which they rest provide little or no support for SoN."

(ii) "If only empirical evidence can settle the issue of whether SoN is true [as claimed by Grünbaum] and it [SoN] is not well supported by empirical evidence, then SoN—[qua presupposition of PEQ]—is, indeed, ill founded and the *pseudo-problem charge* [against PEQ] *has been established* [as claimed by Grünbaum]."[78]

(c) I attach great importance to pointing out why one should generally be left rather unmoved by the charge of scientism in metaphysics and epistemology, not only in this context but also in others. It is easy enough to raise the red flag of scientism, whenever scientific reasoning impugns hypotheses or presumed ways of knowing that someone wishes to immunize against scientific doubts. But, as I see it, he who would level the charge of scientism *responsibly* incurs a major obligation: to come up with a *positive vindication* of the principles or methods that this critic wishes to *contrapose* to the scientific ones. In the present case, I would ask for *positive* reasons to expect that a priori methods can settle the merits of SoN. But neither Quinn nor, to my knowledge, anyone else has given any such reason.

As an example of an irresponsible use of the charge of scientism, I mention the hypothetical case in which scientific findings are adduced against the demonic possession theory of insanity, only to be rejected as "scientistic." The charge would be that the scientific ontology is too impoverished to have room for Satan or other demons, who spring the confines of its allegedly narrow horizons.

1.7.5 Robert Nozick

Robert Nozick has offered some distinctive views on SoN and PEQ in his *Philosophical Explanations*. He notes correctly that SoN is presupposed by the question "Why is there anything at all, rather than nothing?" a presupposition he deems to be "a very strong assumption" in the following sense: "To ask this question [i.e., PEQ] is to presume a great deal, namely, that nothingness is a natural state requiring no explanation, while all deviations from nothingness are in need of explanation."[79] But the explanation of *deviations from the* natural state does not preclude that the natural state *also* may have an explanation of its own. Indeed, as shown by our scientific cosmological illustrations, Nozick was not entitled to suppose that the hypothesized "natural" state would *itself* simply require "no explanation," if it were to obtain.

Strangely, Nozick goes on to shroud the natural state of the world in blanket agnosticism, declaring *tout court* (ibid., p. 126): "The first thing to admit is that we do not know what the natural state is." But surely the *fallibility* of the evolving verdicts of our empirically supported scientific theories as to the natural behavior of the universe is not tantamount to our wholesale ignorance of the natural state of affairs.

Indeed, if Nozick were right that the natural behavior of the world is unknown to us *tout court*, then none of those who endorse PEQ as an authentic question, himself included, could even entertain the claim of its presupposition SoN that the *Null World* is the most natural cosmic state. And that loss would then abort PEQ even before it is posed.

Yet despite his agnostic disclaimer concerning SoN, Nozick is undaunted in tackling PEQ a priori after first noting that this question is deeply, if not uniquely, problematic for the following reason:

> The question [i.e., PEQ] appears impossible to answer [footnote omitted]. Any factor introduced to explain why there is something will itself be part of the something to be explained, so it (or anything utilizing it) could not explain all of the something—it could not explain why there is *anything* at all. Explanation proceeds by explaining some things in terms of others, but this question seems to preclude introducing anything else, any explanatory factors. (Ibid., p. 115)

In effect, Leibniz[80] had anticipated Nozick's objection here by arguing that if PEQ is to have an answer, the sufficient reason for the existence of something could not be provided by a series of *contingent* somethings, because they would form an *infinite* explanatory regress; instead, he contended, the required sufficient reason (cause) terminates the regress by existing *necessarily*.

Though Nozick has posed a forbidding difficulty for PEQ, he insists that PEQ is "not to be rejected" and writes:

> This chapter [i.e., his chap. 2, on PEQ] considers several possible answers to the question [PEQ]. My aim is not to assert one of these answers as correct (if I had great confidence in any one, I wouldn't feel the special need to devise and present several); the aim, rather, is to loosen our feeling of being trapped by a question with no possible answer—one impossible to answer yet inescapable.... The question cuts so deep, however, that any approach that stands a chance of yielding an answer will look extremely

weird. Someone who proposes a non-strange answer shows he didn't understand this question. Since the question is not to be rejected, though, we must be prepared to accept strangeness or apparent craziness in a theory that answers it.

Still, I do not endorse here any one of the discussed possible answers as correct. It is too early for that. Yet it is late enough in the question's history to stop merely asking it insistently, and to begin proposing possible answers. Thereby, we at least show how it is possible to explain why there is something rather than nothing, how it is possible for the question to have an answer. (Ibid., p. 116)

Alas, the hospitality then displayed by Nozick to avowed "extreme weirdness" and "apparent craziness" does not stop short of countenancing explanations vitiated by gross logical improprieties or crude abuses of language. And he is plainly not offering them tongue-in-cheek. Let us cite one of his proposed answers to PEQ, although it will turn out to be a mere farce.

Recall that the Null World, which is assumed to obtain de jure by PEQ, also excludes the existence of time. Yet Nozick will now offer us a *temporal* scenario a priori from which, he claims, one could conclude that "there is something rather than nothing because the nothingness there once was nothinged itself, thereby producing something [thereafter]" (ibid., p. 123). Nozick then depicts the grotesque scenario as follows:

> Is it possible to imagine nothingness being a natural state which itself contains the force whereby something is produced? One might hold that nothingness as a natural state is derivative from a very powerful force toward nothingness, one any other forces have to overcome. Imagine this force as a vacuum force, sucking things into nonexistence or keeping them there. If this force acts upon itself, it sucks nothingness into nothingness, producing something or, perhaps, everything, every possibility. If we introduced the verb "to nothing" to denote what this nothingness force does to things as it makes or keeps them nonexistent, then (we would say) the nothingness nothings itself. (See how Heideggerian the seas of language run here!) Nothingness, hoisted by its own powerful petard, produces something.

When Nozick speaks of "the nothingness there once was," he means, I take it, that at one time, the Null World obtained. He envisions further that the Null World itself contains "a very powerful force toward nothingness." Even this much already seems incoherent, since the Null World is presumably *devoid* of all forces, physical fields, or forms of energy. If there were such a force, we learn, it would annihilate (destroy) any preexisting things permanently, without residue.

Nozick describes this putative action of the force metaphorically and misleadingly by speaking of the force "sucking things into nonexistence and keeping them there." But there are no things to be destroyed in the Null World. Hence, if *per impossibile*, a thing-consuming "vacuum force" were operative in the Null World after all, on what does Nozick think it can act? Its function, he tells us, would be to act "upon itself," presumably to suspend or annihilate itself as an agency of potential destruction. But clearly, his putative "force toward nothingness" is not *itself* identical with nothing or "nothingness." Hence even if the annihilating force can intelligibly act on itself, it would *not* be annihilating "nothing" or the Null World as the object of its self-destruction. Besides, "nothing" (the Null World) is not a thing-like substance such as a fluid or a gas that could be "sucked out" (evacuated), let alone *itself* be annihilated, leaving "something" in its wake!

Thus, Nozick is misformulating his own scenario by beguilingly saying that, when acting on itself, the force "sucks nothingness into nothingness, producing something or, perhaps, everything, every possibility." Needless to say, Nozick's metaphysical Potemkin village is impervious to appraisal by empirical evidence. Alas, in his attempt to propose this answer to PEQ, Nozick's imagination seems to have gone berserk, leaving only bewildered indignation in its wake.

Nozick also takes logical liberties in another tack he explores toward dealing with PEQ. But now he is considering *undermining* the question, instead of proposing an answer to it (ibid., p. 130): "Why is there something rather than nothing? There isn't. There's both." Here he invokes a so-called "fecundity assumption," which asserts (p. 128) that "all possibilities are realized" in the following sense: "All the possibilities exist [are realized] in independent noninteracting realms, in 'parallel universes'" (p. 129). But, as shown by our discussion of the Null Possibility, the obtaining of the Null World, which Nozick declares to be *compossible* with the existence of a super-abundance of different actualized universes, does logically *exclude* the realization (actualization) of any and all logically possible contingent worlds *other than itself.* Thus, the Null World cannot be one of the "noninteracting realms" alongside parallel universes that constitute other actualized possibilities. Therefore, Nozick's fecundity principle cannot serve to undermine PEQ, although that question has turned out to be a nonstarter because it presupposes the truth of the baseless SoN.

Yet the Nobel laureate physicist Steven Weinberg[81] entertains Nozick's "fecundity principle" even to the extent of declaring: "If this principle is true, then our own quantum-mechanical world exists but so does the Newtonian world of particles orbiting endlessly and so do worlds that contain nothing at all." So, Weinberg countenances a *plurality* of "non-interacting" *Null* Worlds! That seems unintelligible.

So much for Nozick's a priori treatment of PEQ.

■ 1.8 HYPOTHESIZED PSYCHOLOGICAL SOURCES OF PEQ

It would be appropriate to consider possible emotional inspirations of PEQ, if we are to understand the tenacity with which it has been asked.

As Charles Larmore has emphasized, the unflinchingly pessimistic Arthur Schopenhauer held, contrary to Kant, that it is not Reason as such which drives us to pose questions such as PEQ.[82] In a chapter on "Man's Need for Metaphysics," Schopenhauer wrote:[83]

> Undoubtedly it is the knowledge of death, and therewith the consideration of the suffering and misery of life, that give the strongest impulse to philosophical reflection and metaphysical explanations of the world. *If our life were without end and free from pain, it would possibly not occur to anyone to ask why the world exists, and why it does so in precisely this way, but everything would be taken purely as a matter of course.*

Elaborating further on "Man's Need for Metaphysics," Schopenhauer declared:

> In fact, the balance wheel, which maintains in motion the watch of metaphysics that never runs down, is the clear knowledge that this world's non-existence is just as possible as its existence .... What is more, in fact, we very soon look upon the world as something whose non-existence is not only conceivable, but even preferable to its existence .... Accordingly, philosophical astonishment is at bottom one that is dismayed and distressed. (Ibid., p. 171)

But to the detriment of Schopenhauer's diagnosis of the emotional inspiration of PEQ, he leaves fundamentally *unexplained* why that question has apparently been posed *only*—or at least principally—by the heirs of the distinctly Christian doctrine SoN. After all, the thinkers in other cultures who did *not* raise it were just as conscious of death and the miseries of life as the legatees of traditional Christian doctrine.

Yet it would be interesting to investigate empirically the motives of philosophers who embrace PEQ as an *authentic* question, so as to learn to what extent, if any, such philosophers are driven by emotions of the sort conjectured by Schopenhauer. It is perhaps not implausible that our deeply instilled fear of death has prompted some of us to wonder why we exist so precariously. And some of us may then have extrapolated this precariousness, more or less unconsciously, to the existence of the universe as a whole. But whatever the emotional inspiration of PEQ, no such motivation can vindicate it as a *philosophically* viable question, since its presupposed SoN is baseless.

Disappointingly, after declaring PEQ to be the most fundamental question of metaphysics, Heidegger[84] *psychologized* it away as inspired by existential anxiety, thereby essentially echoing Schopenhauer's ideas on the psychology of PEQ.

But PEQ dies hard. In 1999, it was the focus of a massive book of over 750 pages by the Swiss philosopher Ludger Lütkehaus. Published in German, its title in English becomes *Nothing: Farewell to Being, End of Anxiety*. Let Lütkehaus speak for himself in stating the aim of his opus (1999, p. 29) in his German original, to which I append my English translation in the endnote:

> Es [dieses Buch] versucht, die Präokkupationen eines Denkens zu revidieren, das seinsfixiert, "ontozentrisch" in seinen Werthierarchien, "ontomorph" in seinen Begriffen und Vorstellungen und bedingungslos "ontophil" in seinen Antrieben ist.

Den symptomatischsten Ausdruck hat dieses Denken in seinem paranoiden, "nihilo-phoben" Verhältnis zum Nichtsein, zum "Nichts" gefunden. Nicht die "Seinsvergessen-heit," wie es die Todtnauberger Schule beklagt—die *Nichtsvergessenheit* bezeichnet das wahre "Schwarze Loch" seiner ontologischen Amnesie. Nichtsvergessenheit, Nichtsangst und Seinsgier bilden sein "ontopsychologisches," "ontopathologisches" Syndrom. Und gerade damit arbeitet dieses Denken der Vernichtung und Selbstvernichtung zu. Das ist die—vielleicht tragische—Ironie der so seinsfixierten westlichen Seinsgeschichte.[85]

After this avalanche of words, I have no idea just what Lütkehaus would have each of us do to overcome our alleged "ontopathological syndrome." Should we forsake all joie de vivre?

▪ 1.9 PEQ AS A *FAILED* SPRINGBOARD FOR CREATIONIST THEISM: THE COLLAPSE OF LEIBNIZ'S AND SWINBURNE'S THEISTIC COSMOLOGICAL ARGUMENTS

We are now ready to appraise the theistic creationist answers given to PEQ by Leibniz and Swinburne respectively as part of a *cosmological argument* for the exis-tence of God.

Swinburne has argued cogently against Leibniz that, if there is a God, his exis-tence is logically contingent no less than that of the universe. As he reasoned carefully:

> It seems coherent to suppose that there exists a complex physical universe but no God, from which it follows that it is coherent to suppose that there exists no God, from which in turn it follows that God is not a logically necessary being. If there is a logically nec-essary being, it is not God [footnote omitted].[86]

And having deemed the existence of God to be logically contingent, the sweep of Swinburne's version of SoN *excludes* God along with our contingent universe from the Null World. Thus, as we recall, Swinburne formulated SoN as follows: "Surely the most natural state of affairs is simply nothing: no universe, no God, nothing."[87] But, on the basis of his SoN, he had demanded, in response to PEQ, a suitably potent external divine cause to explain the existence of the universe qua *deviation from nothingness*. And he issued this explanatory demand for a creator ex nihilo as a challenge to atheists.[88]

But clearly, what is sauce for the goose is sauce for the gander: If God does exist contingently, as Swinburne claims, then the contingent existence of the deity also constitutes a *deviation* from the allegedly most natural state of nothingness! Thus, on Swinburne's version of SoN, the existence of God requires causal explanation *in answer to PEQ* no less than the existence of the universe does. Hence, Swinburne is not entitled to take the existence of God for granted, as he does, to explain the existence of the universe *in answer to PEQ*. To point out against Swinburne that, on his version of SoN, God and the universe *alike* require causal explanation is *not* to saddle him with an infinite regress of explanations.

How, then, does he deal with the following inescapable challenge from his version of SoN? If he is going to give an answer to PEQ, as he does, he needs to explain why God exists, *rather than just nothing contingent*, fully as much as he needs to explain why our universe exists, rather than just nothing contingent. Yet Swinburne is *oblivious* to this major challenge as emanating from his SoN!

Thus, unencumbered by this explanatory debacle, Swinburne opines one-sidedly: "the view that there is a God ... explains the fact that there is a universe at all."[89] And, again in accord with his SoN, he claims furthermore that God also keeps "the many bits of the universe" in existence.[90]

But in regard to the imperative to explain why God exists rather than just nothing, Swinburne is driven to concede that "inevitably we cannot explain" it.[91] Thus he claims that "the choice is between the universe as [explanatory] stopping point and God as stopping point."[92] Yet Swinburne defaulted on his explanatory debt when he conceded that the existence of God "inevitably" defies explanation. He had assumed that debt by embracing his version of SoN, which excludes God from the Null World and turns the existence of the deity into a deviation from the "most natural" state of nothingness. Hence, contrary to Swinburne, on his own premises, God does not qualify as an explanatory "stopping point" after all.

Thus Swinburne has indeed deservedly incurred a jibe akin to the one Schopenhauer famously issued against those who demand a creative cause of the existence of the universe but then suspend a like demand to explain the existence of God: Swinburne has treated SoN like a hired cab that he dismissed, just when it reached his intended theological destination.

But let there be no misunderstanding of my use of Schopenhauer's simile of the hired cab against Swinburne. I am emphatically *not* maintaining generally that a theological hypothesis **T** can be explanatory only if **T** itself can, in turn, be explained; instead, I am contending that Swinburne hoists himself with his own petard: in answer to PEQ, his recourse to SoN to call for a theistic explanation of the very existence of the universe *boomerangs*, because his SoN likewise requires a causal explanation of the existence of God, which, he tells us explicitly, is not to be had.

Leibniz (ibid., sections 7 and 8) and Swinburne[93] have offered the prima facie most persuasive of the traditional "first cause" cosmological arguments for the existence of God as creator of the universe ex nihilo. Though they differ as to whether God exists necessarily (Leibniz) or contingently (Swinburne), the common core of their cosmological arguments can be encapsulated as follows: (i) The Null World, which is devoid of all contingent existents, is the simplest (ontologically); (ii) SoN is true: De jure, the Null World should obtain qua being the simplest, and indeed it would obtain as the most "natural" or normal state of affairs in the absence of an external cause (or "reason"); (iii) But the de facto existence of our universe of contingent objects is a massive *deviation* from the Null World mandated by SoN;

(iv) This colossal existential deviation from ontological "normalcy" cries out for explanation by a suitably potent cosmic cause, making an answer to PEQ imperative. The required cause is a creator ex nihilo. Hence, the God of theism exists.

Thus, it is clear that the theistic creationist answers given to PEQ by both Leibniz and Swinburne are each predicated on a version of SoN in the face of contingently existing things. Their versions differ somewhat: Leibniz tells us that, in the absence of an overriding reason (cause), the nihilistic state of affairs is ontologically imperative, because it is "simpler and easier" than the state of something contingent, whereas Swinburne claims that the Null World is "the most natural state of affairs." But I have been at pains to argue that both of these versions of SoN are baseless, each for reasons of its own. Yet they both avow the central claim of SoN that de jure there *should* be nothing contingent and that there would indeed be nothing contingent in the absence of an overriding external divine creative cause.

But the ill-founded SoN is clearly a presupposition of PEQ. Therefore, the cosmic ontological question PEQ is a nonstarter by posing a pseudo-issue. Yet the purported imperative to answer precisely this global question is the basis for Leibniz's and Swinburne's cosmological arguments for the existence of God. Thus, *PEQ cannot serve as a springboard for creationist theism.* Hence, Leibniz's and Swinburne's cosmological arguments are fundamentally unsuccessful.

■ 2. DO THE MOST FUNDAMENTAL LAWS OF NATURE REQUIRE A THEISTIC EXPLANATION?

We now consider theistic arguments that have been offered, not about the existence of contingent objects but about the explanation of the natural laws that are exhibited by their behavior, laws which are sometimes called the "nomological structure" of the world, the "nomic structure" or, briefly, its "nomology." Theists have claimed to explain the nomology as having been willed or intended by God in the mode of agent causation to be exactly what it is. We shall speak of this supposed theistic explanation as "the theological volitional explanation of the nomology." And the principal issue in this Section 2 will be whether creationist theism succeeds in explaining the specific content of the nomological structure, as claimed by its advocates.

2.1 The Ontological Inseparability of the Laws of Nature from the Furniture of the Universe

Unless there is an infinite regress of explanations, every explanatory theory will feature some set of *unexplained explainers*. Yet any theory can be axiomatized in *alternative* ways. For example, in Euclid's synthetic plane geometry, the famous Parallel Postulate (Number 5) can be interchanged with the theorem that the sum of the interior angles of a rectilinear triangle is 180 degrees as follows: this angle sum theorem now becomes the 5th Postulate, while the previous Parallel Postulate now becomes a theorem. Similarly, Newtonian dynamics has been alternatively axiomatized by means of the so-called calculus of variations. Thus, the fundamentality of a postulate is *not* absolute but depends on the axiomatization. Hereafter, when we speak of "the most fundamental laws" of the nomology, this characterization is to be understood within just such axiomatic relativity.

In a scientific theory pertaining to the laws of nature and featuring unexplained explainers, the most fundamental of these laws (in a given axiomatization) will hold as a matter of brute fact. But, as we just saw in Section 1.9, in a theistic system, the existence of God is its avowed unexplained explainer and thus is *its* brute fact. Hence, as we noted, Swinburne declared: "The choice is between the universe as [explanatory] stopping point and God as stopping point."[94]

The nomology consists of the law-like regularities exhibited by the physical, biological, and biopsychological constituents of the universe. It is of cardinal importance here that these nomic patterns *inhere* in the behavior of the world's furniture and do not exist independently alongside it. Theists and atheists can agree that the laws do not hover over the universe, as it were, in some separate realm. Swinburne rightly rejected this hypostatization: "Talk about laws of nature is really only talk about the power and liabilities of bodies."[95]

In short, the laws are inextricably intertwined with the material *content* of the universe. Hence, we can speak of their intrinsic entanglement as "the *ontological inseparability*" of the nomology from the world's furniture.

But just this inextricability has a very important corollary pertaining to *a posteriori* teleological arguments for the existence of a designer God, a corollary that seems to have been overlooked heretofore. If the theistic God is to endow the laws of nature with teleological features—such as permitting the formation of intelligent life—then he must do so precisely *by means of* creating the material content of the world ex nihilo as his handle on the laws. Thus, the designer role which the theist attributes to the deity cannot be fulfilled by God without his being the creator ex nihilo. Yet the a posteriori argument for a designer God cannot *also* shoulder the probative burden of warranting divine creation ex nihilo, an onus which was borne by the received cosmological argument. Indeed, if the teleological argument had the probative resources to argue not only for a *designer God* but also for a *creator God* ex nihilo—the latter being the conclusion of the cosmological argument— then the teleological argument would make the cosmological one superfluous! But it does not. Instead, the two arguments complement one another as follows: The proponent of the design argument for the existence of God is engaged in *adding* to the cosmological conclusion, which is that God the creator exists, the *further* a posteriori conclusion that a *designer* God exists, who is also a creator God. Hence, an argument for God the *designer* that uses as a premise the existence of God the *creator* does not "beg the question" as to the existence of a designer God.

Accordingly, the assertion of divine creation ex nihilo, which is the *conclusion* of the failed cosmological argument for the existence of God, is now seen to be a tacit *premise* of the traditional teleological argument that seemingly *goal-directed* features of the world call a posteriori for a cosmic *designer*. In other words, if God is to implement his inferred role as cosmic designer of the nomology, he must be assumed to be the creator ex nihilo of the substantive fabric and texture of the universe.

But absent a successful cosmological argument for the occurrence of divine creation ex nihilo, the teleologist must bear the enormous additional probative burden of somehow warranting the very framework of creation ex nihilo in which the teleological arguments are inevitably anchored. In short, the teleologist is in dire need of some kind or other viable, cogent *substitute* for the received cosmological argument, an argument whose most persuasive versions (by Leibniz and Swinburne) turned out to be genuinely flawed in Section **1.9**. Yet no such substitute is extant.

2.2 The Probative Burden of the Theological Explanation of the World's Nomology

The theological volitional explanation of the nomology, which we shall develop in the next section, has to shoulder a multiple heavy probative burden as follows:
(a) Since the theist purports to explain the laws of nature as the product of divine intention, the ontological inherence of the laws of nature in the cosmic furniture commits him or her to the claim that God brought the nomic structure into existence *by means of creating* ex nihilo *the world's furniture* from which that structure is inseparable; thus, just like the theistic argument for a *designer* God, the theological volitional explanation of the nomic structure is in dire need of a successful substitute for the failed cosmological argument for divine creation ex nihilo.

(b) According to theism, God is the creator ex nihilo of all logically contingent existing entities, whenever they exist, though of course he does not create himself. If that claim were true, then God would *automatically also* be the simultaneous creator ex nihilo of such laws of nature **L** as govern the content of the universe, precisely because the nomic structure **L** is *intrinsic* to the furniture of the universe.

(c) Yet we must heed a caveat: It would *not* follow from the reliance of the theistic volitional explanation of the nomology on creation ex nihilo that if the universe is *not* the product of divine creation ex nihilo, then no sort of supernatural agency—such as the phantasmic demiurge in Plato's *Timaeus*—might have been the craftsman of the laws **L** by *transforming prior chaos* into a cosmos. But the traditional theist is unwilling to countenance a divine cosmic craftsman who merely transforms a preexisting chaotic world into a nomic universe, holding that an omnipotent God had no need for preexisting substances to create a universe. Hence the notion of a mere cosmic transformative craftsman is unavailable to the theist, and hence would not enable the theistic explanation of the nomology to dispense with the *equivalent* of a cosmological argument for creation ex nihilo.

But, as we saw in Section 1.9, the received cosmological argument for divine creation ex nihilo is erected in response to PEQ on the quicksand of SoN. And since this cosmological argument is thus *ill founded*, neither the divine volitional explanation of the nomic structure nor the aforementioned teleological argument for the existence of God can build on it *cumulatively*; instead, they must then bear the enormous *additional* probative burden of somehow warranting the very *framework* of creation ex nihilo from which the theistic volitional explanation of the world's nomology and boundary conditions is inseparable. But no such warrant is in sight.

Philip Quinn[96] has tried to parry my claim that the commitment to a divine volitional explanation of the nomic structure confronts the theist with the probative burden of providing the equivalent of a successful cosmological argument for creation ex nihilo. Quinn denies that burden and chides me for the "error" of "underestimating the number of sources from which justification for the existence of the God of theism can be derived." And he explains that the contributions from these various sources "can combine to yield a cumulative case argument."

Yet the issue before us is specifically the warrant, if any, for the theist's purported explanation of the world's nomology as having been intended by God in the mode of agent-causation to be exactly what it is. And how does Quinn think his envisioned cumulative argument can *dispense* with the *specific* demonstration that the theistic volitional explanation of the nomology must show the creation of the nomic structure to be part and parcel of the divine creation ex nihilo of the material content of the world? Any such specific demonstration, it seems, does bear the same heavy probative burden as the received cosmological argument, which failed.

But, serious though it is, the need for a successful substitute for the failed cosmological argument is *merely one of a whole array of defects of the theistic explanation of the nomology*. We shall turn to these other failings after first articulating the proposed volitional theistic explanation of the nomology as developed by Swinburne and Quinn.

2.3 The Theistic Explanation of the Cosmic Nomology

In a 1993 *Festschrift* for me, Philip Quinn set the stage for advocating a theological explanation of the nomology, which purportedly transforms scientific *brute* facts into specifically explained regularities. Quinn says:

> The conservation law for matter-energy is logically contingent. So if it is true, the question of why it holds rather than not doing so arises. If it is a fundamental law and only scientific explanation is allowed, the fact that matter-energy is conserved is an inexplicable brute fact.... If there is a[ny] deepest law, it will be logically contingent, and so the fact that it holds rather than not doing so will be a brute fact.[97]

Quinn now proceeds to draw two inferences from the *scientific* brute fact status of the most fundamental laws of nature, assuming that there are such "ultimate" laws. He writes:

> There are, then, genuine explanatory problems too big, so to speak, for science to solve. If the theistic doctrine of creation and conservation is true, these problems have solutions in terms of *agent-causation. The reason why there is a certain amount of matter-energy and not some other amount or none at all is that God so wills it, and the explanation of why matter-energy is conserved is that God* [creatively] *conserves it* [as required by SoN]. (Ibid., italics added)

In the same vein as Quinn, Swinburne characterizes explanation in terms of agent causation as "intentional" or "personal."[98] And speaking of the laws of nature **L**, Swinburne endeavors to prepare the ground for that sort of theistic explanation of facts of nature:

> Why does the world contain just that amount of energy, no more, no less? [The laws] **L** would explain why whatever energy there is remains the same; but what **L** does not explain is why there is just this amount of energy.[99]

Evidently, Quinn and Swinburne presume to *quantify* the "amount of matter-energy" univocally. But even in elementary Newtonian mechanics, after integration of the equation of motion to derive its law of conservation of dynamical energy, the numerical value of the total energy is dependent on the arbitrarily (i.e., humanly) chosen *zeros* of the component potential and kinetic energies. How then is divine volition to explain that "there is just this [numerical] amount of energy"? Does God create to within the zeros?

But let us consider more generally the context of the volitional theological explanations offered by Swinburne and Quinn in answer to their question of why the actual world's nomology is what it is. This question singles out the presumed ultimate laws and facts of nature for explanation. And our two theists demote science for its inability to answer their question. As they make clear,[100] they consider their theistic volitional explanation of the ultimate nomology to be a major explanatory advance over scientific brute fact. Yet neither Swinburne nor Quinn *spelled out* their very ambitious deductive theistic explanation of the nomic structure.

Therefore, I now offer a reconstruction of essentially the deductive explanatory reasoning that, I have good reason to believe, they originally had in mind. Quinn (private communication) authenticated my reconstruction in regard to the view he held before 2003. After codifying Swinburne's and Quinn's original versions of the purported theistic explanation of the basic laws of nature, we shall address their more recent accounts.

As for the earlier versions of Swinburne's and Quinn's explanation of the nomology qua product of divine agency, let me significantly refine my earlier formulation of it.[101] To determine whether their explanation redeems the very ambitious claims they made for it, let us have in mind, for the sake of concreteness, Swinburne's own example, originally endorsed by Quinn, of explaining theologically the supposed specific amount of total energy in the universe, which they depict as a *scientific brute fact*. Or, just for argument's sake, suppose that the nonlinear partial differential equations codifying Einstein's theory of the gravitational-cum-metric field *were* ultimate laws of nature. How, then, did Swinburne and Quinn envision that explanatory recourse to divine agency would *transform* these and all other specific putative scientific *brute* facts into volitionally *explained* facts?

I have schematized their originally presumed theistic explanations in a deductive argument, using the familiar term "explanandum" to denote *what is to be explained*, which is asserted in the conclusion of the deductive argument below. But in this schematic reconstruction, the purportedly explained actual specifics of the most basic laws and facts are patently *not* stated in either the Conclusion or in the Premises, if only because they are not known; instead each of the Premises and the Conclusion speak of the unspecified laws of nature by means of *place-holders*. Yet whatever these specifics actually are, this explanatory schema is presumably the theistic solution—in Quinn's words—to "genuine explanatory problems too big, so to speak, for science to solve."[102] With these understandings, the supposed volitional explanation becomes schematically:

Deductive Theistic Volitional Explanation of the Presumed Ultimate Laws and Facts of Nature:

Premise 1. God freely chose (intended or *willed*) that the contents of our world exist and that they exhibit the laws which inhere in them.

Premise 2. Being omnipotent, God was, and is, perpetually able to cause directly (i.e., creatively bring about ex nihilo) the existence of the world's contents, so that they exhibit the laws which inhere in them.

Premise 3. If God chooses that p, and is able to cause it to be the case that p, then p.

Conclusion/Explanandum: The contents of our world exist and exhibit the laws which inhere in them.

This deductive argument invites some elaboration:

Premise 1 is to be understood more explicitly as entailing that God chose or intended or willed the realization of the possible world which is in fact actual so as to be nomologically precisely what it is, rather than the actualization of another possible world featuring alternative fundamental laws or facts, such as a different value of the presumed numerical total energy.[103]

Swinburne uses the lowercase letter e to denote the *explanandum*, which states the facts to be explained by the explanatory argument. And he articulated the substance of Premises 2 and 3 in the following two statements:

(a) "Clearly whatever [the explanandum] e is, God, being omnipotent, has the power to bring about e. He will do so, *if he chooses to do so*." And the e that God chooses to bring about will be *compatible* with the assumed omnibenevolence of his aims.[104]

(b) "God, being omnipotent, cannot rely on [mediating] causal processes *outside his control* to bring about effects, so his range of easy control must *coincide* with his range of *direct control* and include all states of affairs which it is *logically possible* for him to bring about."[105]

Claiming that the nomology inherent in the world's content is explained as the product of divine intention, the theist's explanation requires that God brought the nomology into being *by means of creating* ex nihilo *the cosmic furniture* in which it inheres: Evidently, in the absence of a *cogent*, viable *substitute* for the failed received cosmological argument for divine creation ex nihilo, the theistic volitional explanation of the nomology relies crucially on the *conclusion* of the cosmological argument as its underwriter. Thus, the theistic volitional scenario inherits the epistemic liabilities of that argument, set forth in Section 1, much as does the a posteriori argument for a *designer* God, as we saw in Section 2.1.

Yet Swinburne claims to offer *a scientized* epistemology for creation ex nihilo:

> *The very same criteria which scientists use to reach their own theories* leads us to move beyond these theories to a creator God who sustains everything in existence.

Moreover, Swinburne asserts theistic *pan*-explainability, declaring:

> Using those same [scientific] criteria, we find that the view that there is a God explains *everything* we observe, not just some narrow range of data [italics in original]. It explains the fact that there is a universe at all [*via* SoN], [and] that scientific laws operate within it.[106]

But, as we shall see further on, in a paper of 2003 which was cited in Section 1.7.4 a propos of the epistemic status of SoN, Quinn parted company with Swinburne and modified his earlier version of the theistic explanation of the nomology. It will turn out that, in this latest version, Quinn distanced himself, in effect, from Swinburne's aforecited 1996, purportedly *scientized* epistemology of theistic pan-explainability.

Yet in Swinburne's Reply to my lengthy essay of the same year,[107] he claimed to offer a clarification of his account of explaining the nomic structure theologically. Alas, on the contrary, this supposed clarification will be seen to muddy the waters. As will emerge under "Objection 4" which follows in section 2.4, it features a *conflation* of the Bayesian *confirmation* of the hypothesis that God exists, on one hand, with the volitional theistic *explanation* of the specific content of the basic nomic structure, on the other. Hence, as in the case of Quinn, we shall discuss Swinburne's views in *two* stages, deferring scrutiny of his supposed clarification.

As against the thesis that theism solves "genuine explanatory problems too big ... for science to solve," I now offer a series of further cardinal objections to the purported divine volitional explanation of the nomology. Even if that theistic explanation did not depend on demonstrated creation ex nihilo, these impending additional major objections thoroughly undermine it. In developing these animadversions, let us be mindful of Swinburne's aforecited claims that theism is of a piece, epistemologically, with scientific theorizing while transcending science by offering pan-explainability and transforming scientific brute facts into specifically explained states of affairs.

2.4 Further Major Defects of the Theological Explanation of the Fundamental Laws of Nature

OBJECTION 1. How does Swinburne reason *epistemically* that God *actually chose* to bring about the specific de facto *e* of the explanandum? He surely needs to validate this premise in order to attribute the prevailing *e* causally to divine creative volition. Obviously, that premise is not vouchsafed at all by Swinburne's *conditional* assurance that God will bring about *e* "if he chooses to do so." Equally patently, it would beg the question, if one were to answer our question here by claiming that God must have chosen to produce *e*, since *e* is actually the case! In sum, although the premise that God actually chose to produce *e* is explanatorily crucial for Swinburne and Quinn, no independent evidential support for it is in sight.

Indeed, just this epistemic gaping hole *alone fundamentally undermines* Swinburne's purported theistic volitional explanations of the *specific* content of the world's *ultimate* nomic structure and of such presumed basic facts as the envisioned specific amount of total cosmic energy.

OBJECTION 2. To the detriment of Quinn and Swinburne, the volitional theological explanation of the nomology features *a built-in sort of* ex post facto *defect* which prevents the evidence in the explanandum *e* from providing a check on the validity of the explanatory theistic premises! And this liability is not only anathema in the epistemology of scientific theories *but is unacceptable in any sort of explanation based on evidence.*

Thus, let us consider a hypothetical situation in which the steady-state world of the 1948 Bondi and Gold theory was actual, a world which we had occasion to discuss in Section 1.7.4. If that world were actual, the theistic explanatory premises would be that omnipotent God willed the law of the constancy of mass–*density* as well as the Hubble expansion of the galaxies. Alternatively, suppose that the actual world were one exhibiting mass–energy conservation as well as Swinburne's and Quinn's envisioned specific amount of total energy. In the latter event, our theists would explain this *different* state of affairs equally confidently by telling us that the deity intended and chose to implement mass–energy conservation, rather than the *density* conservation of the steady-state world.

Thus, whichever of the two cosmologies actually materializes, *the evidence in the explanandum e provides no check on the validity of the explanatory premises!* And the crux of this *immunity* from evidential check is *achieved formally* by the following bizarre device: Whatever the content of the explanandum in the conclusion, that same explanandum is *identically built* into the premises!

In this way, the theistic explanation of the nomology is purchased effortlessly in advance on the cheap. But no building of the explanandum identically into the premises is found in the respected sciences, as illustrated by the following explanations in physics and biology:

(1)   The Newtonian gravitational explanation of the orbit of the moon.

(2)   The deductive-nomological explanations of optical phenomena furnished by Maxwell's equations, which govern the electromagnetic field, or, in a statistical context.

(3)   The genetic explanations of hereditary phenotypic human family resemblances.

In short, the range of the explanatory latitude of the theistic volitional explanation is prohibitively permissive, in clear contravention of Swinburne's aforecited declaration: "*The very same criteria which scientists use to reach their own theories* lead us to move beyond these theories to a creator God who sustains everything in existence."[108] Moreover, the building of the explanandum identically into the premises is unacceptable *in any sort of explanation based on evidence*. Nor can it be made acceptable by abjuring "scientism"!

OBJECTION 3. As we recall, Swinburne wrote:

> Why does the world contain just that amount of energy, no more, no less? **L** [the basic laws of nature] would explain why whatever energy there is remains the same; but what **L** does not explain is why there is just this amount of energy.[109]

Let us denote by E_o the putative specific amount of total energy in the universe, which Swinburne and Quinn[110] each consider well defined, and which they characterized as a *scientific brute fact*. Their point in so doing is to claim that explanatory recourse to divine agency would *transform* this specific unexplained fact, as well as the specific content of scientifically ultimate laws of nature, into volitionally *explained* items. Thus, we recall, Swinburne wrote grandiosely: "The view that there is a God explains *everything* we observe, not just some narrow range of data."[111]

To make good on his thesis of such explanatory *specificity*, Swinburne would need to be able to justify the following contention: given the hypothesis h that God exists in conjunction with assumed relevant background knowledge k, the specific pertinent e is a *deductive consequence* of the conjunction of h and k; that is, the probability of the *explanandum e* on this conjunction is *1*. Presumably, Swinburne's example of explaining the specific total amount of energy E_o theistically is intended to make the general point of such deductive explainability far beyond E_o: it is to illustrate his global contention that the theistic hypothesis h "explains *everything* we observe" (italics in original).

But how does Swinburne see himself as vindicating this mind-bogglingly ambitious claim? Surely one is entitled to have expected him *to spell out the details* of the explanatory argument for at least one major case. Very disappointingly, that very reasonable expectation is dashed. Instead, immediately after having avowed theistic volitional *pan*-explainability, he greatly weakens his explanatory thesis. Speaking of the universe, he now maintains just that the existence of God explains *generically* that *there are* laws of nature, and we learn that *h* explains much more modestly "that there is a universe at all [via SoN], [and] that scientific laws operate within it." And in his earlier opus, *The Existence of God*, in a summary of several of its chapters, he wrote, again generically: "What science cannot explain [but theism can] is why the laws of nature are of the character they are."[112]

Yet it simply won't do to offer a theistic argument for the likelihood of a generic nomic structure, even if successful, as a substitute for redeeming the vaunted theistic pan-explainability of the *specific content* of the fundamental scientific laws and facts, items which science avowedly leaves unexplained as brute facts. Indeed, it is regrettably misleading philosophically to offer demonstrably hollow pan-explainability as an improvement upon the scientific explanatory enterprise.

OBJECTION 4. For the sake of the discussion, suppose that Swinburne had articulated a formally valid deductive theistic volitional argument for the basic laws and facts *e* of the universe. Importantly, even the provision of such an argument would not suffice to qualify the deduction as *explanatory*.

A deliciously hilarious example cited by Wesley Salmon over thirty years ago makes this point tellingly by featuring a *pseudo-explanation* of why John Jones did not become pregnant during the past year.[113] The purported cause is that he took birth control pills all year, and the causal hypothesis is that no man who takes such pills ever becomes pregnant. It then follows impeccably that Jones did not become pregnant last year. But plainly, the birth control pills are causally irrelevant here.

An elementary classroom example of a causal pseudo-explanation is that, other things being equal, victims of the common cold who are coffee drinkers recover from it within one month because drinking coffee is therapeutic for the common cold. Note that although this pseudo-explanation is stupendously predictive, it is nonetheless unacceptable causally: As we know, the afflicted cold sufferers recover equally well if they do *not* ingest any caffeine at all.

Quite generally, ever since Francis Bacon taught, it has been known that, at least in the case of *causal* hypotheses, the *mere* deducibility of some data from some such hypotheses (together with known initial conditions) does *not* suffice to qualify the hypotheses as explanatory; nor does it qualify the data as supporting evidence for the hypotheses. To believe that it does is to indulge in dubious *hypothetico-deductive pseudo-confirmation*. What is being overlooked by such a belief is that, although the causal hypotheses (in conjunction with the known initial conditions) entail the particular data, the hypothesized causal factors are often actually *causally irrelevant* to the data that are to be explained.

If the causal hypotheses are to be explanatory, they need to meet further well-known epistemic requirements, such as furnishing suitable "controls" instantiating actual causal relevance. Thus, in the present case, the theist's claim that God is the creative cause of the existence of the world and thereby the architect of its laws of nature should offer evidence *against* the rival *null* hypothesis that *no* external creative cause ex nihilo is required. If SoN *were* at all evidentially warranted, it could serve to *rule out* that rival null hypothesis. But, as shown in Section 1.7, SoN is baseless and hence unavailable to rule out the rival hypothesis that no creative cause ex nihilo is needed at all.

OBJECTION 5. The premises in the theistic volitional explanation yield that a divine volitional state, though itself uncaused, issued in God's creatively causing ex nihilo the existence of our nomological world. Yet, again, given the demise of SoN, *transformative* causation is the only kind of causation for which we have evidence—be it agent causation or event causation—rather than creative causation ex nihilo.

Thus, as emerges from the preceding considerations in Objections 1 through 5, Swinburne did not score a point against atheism when he wrote:

> The only plausible alternative to theism is the supposition that the world with all the characteristics I have described just is, has no explanation. That however is not a very probable alternative. We expect all things to have explanations.[114]

But this assertion does not even cohere with Swinburne's claim[115] that the existence of God has no explanation, as we saw in Section 1.9.

OBJECTION 6. As mentioned in Section 2.3, more recently Swinburne offered a reply to my prior objections of the same year to his account of theistically explaining the world's ultimate nomic structure and other scientific brute facts. Recall his showcase paradigm example of the putative total cosmic energy E_o[116] and, more generally, his claim of theistic volitional pan-explainability of the *specifics* of "*everything* we observe."[117] Astonishingly, in his Reply to me, he *sabotages* his erstwhile grandiose vision of a theistic explanatory edifice as follows:

> The hypothesis *h* which I consider to explain the data *e* is not, "there is a God and he causes *e*" (which is what Grünbaum may be supposing on his p. 20), but "there is a God" (as he explicitly recognizes on p. 36). Given *h*, it follows that God can bring about *e*, but how probable it is that he will, depends on whether (in virtue of his perfect goodness) he has good reason to do so. (God's perfect goodness, I claim, follows from his omniscience and his perfect freedom, that is his freedom from influences other than rational considerations.) Quite a bit of my writing is devoted to showing that he does have such good reason—e.g. that simple regularities in nature give to finite beings the power to grow in power and knowledge, etc., and that that is a good thing.[118]

As we know, Bayes's theorem in the calculus of probability, if used to probabilify *hypotheses*, is a device for updating the *evidential* appraisal of a hypothesis on the basis of new, or previously unavailable, or unconsidered evidence. Thus, as Wesley Salmon has emphasized, "Bayes' theorem belongs to the context of confirmation, not to the context of explanation."[119] And this important distinction is, of course, *not* lessened at all by the fact that, once a hypothesis is sufficiently confirmed, it can qualify *epistemically* to serve as a premise in an explanation.

In my earlier critique of Swinburne, I cited Salmon's reiteration of Carl Hempel's caveat that: "Explanation-seeking why-questions solicit answers to questions about why something occurred, or why something is the case. Confirmation-seeking why-questions solicit answers to questions about why *we believe* that something occurred or something is the case."[120] And, being mindful that Bayes's theorem belongs to the context of *confirmation*, I wrote:

> Swinburne ... muddies the waters. He tries to use Bayes' theorem both to probabilify (i.e., to increase the confirmation of) the [hypothesis of the] existence of God, on the one hand, and, on the other, to show that theism offers the best [simplest] explanation of the known facts, assuming that God exists. And his [Swinburne's] account of the notation he uses in his statement of the theorem reveals his failure to heed the Hempel-Salmon distinction.[121]

In his Reply to me, Swinburne turned a deaf ear to the relevance of the Hempel-Salmon distinctions.[122] And it was thus lost on him that *I* was explicitly speaking of the theistic hypothesis which he was trying to *confirm* (incrementally) *à la* Bayes, when I went on to say:

> It is vital to be clear on what Swinburne takes to be the hypothesis *h* in his Bayesian plaidoyer for the existence of the God of theism. He tells us explicitly: "Now let *h* be our hypothesis—'God exists' (Swinburne [1991], p. 16)."[123]

But, contrary to Swinburne's Reply to me,[124] I absolutely *never, ever* "explicitly recognized" that the hypothesis *h* which he took to be *sufficient* to *explain* the data *e* was just the *parsimonious* one "there is a God" or "God exists." This *confinement* of the explanatory premises to *h* never even occurred to me, because such a parsimonious hypothesis obviously could not possibly redeem Swinburne's mantra that theism explains the *specifics* of "*everything* we observe."[125]

After all, as I pointed out emphatically under my OBJECTION 1, to make good on that omnivorous explainability, it is hopelessly insufficient to declare with Swinburne that "given *h*, it follows that God *can* bring about *e*, but how probable it is that he will, depends on whether (in virtue of his perfect goodness) he has good reason to do so."[126] Nor does it help rescue Swinburne's forlorn all-encompassing explanatory pretensions to point out, as he does, that "he [God] does have such good reason" as, for example, "that simple regularities in nature give to finite beings the power to grow in power and knowledge."

To have even a hope of redeeming his explanatory mantra, Swinburne does indeed require at least the following *conjunctive* theistic hypothesis, which he mentions but rejects: "God exists *and* he chose to cause *e* ex nihilo"[127]—a *stronger* hypothesis which I articulated in the deductive volitional explanation I have already set forth. In short, in effect Swinburne has now *repudiated* his erstwhile signature doctrine of all-encompassing theistic explainability rather than having offered a relevant cogent rebuttal to me.

For his part, Quinn[128] has come to appreciate these serious defects in Swinburne's views so that, by 2003, he developed a quite different conception of the theistic explanation of the ultimate nomic structure and basic facts of the world. Now Quinn mentions three positive answers to the question, "Why does the possible world that is in fact actual obtain, rather than another?" and he suggests that, presented with the three answers to it, the majority of contemporary theists would prefer the explanation that "God had a sufficient reason to actualize it [i.e., the de facto existing world], *but this reason is utterly beyond our ken.*"[129] Yet this sort of surrogate explanation belongs to fideist rather than natural theology! Therefore, I cannot see why a theist would expect anyone who does not *antecedently* believe in God to embrace theism as *explanatory*, if it features, as this forlorn surrogate explanation does, resort to the old chestnut that God's sufficient reason passes all human understanding.

To be sure, the intellectual humility expressed by it is ingratiating. But that explanation forsakes any conjecture as to God's specific reason for choosing the actual nomic structure, *as against an alternative one.* Yet Quinn's erstwhile *plaidoyer* for a theistic explanation was precisely, like Swinburne's, that it transforms scientific brute facts into specifically *explained* states of affairs. As I have argued, it does nothing of the kind: Neither Swinburne nor Quinn has redeemed at all their vaunted promise to explain theologically what science leaves unexplained.

▨ 3. CONCLUSION

In parts 1 and 2 of this essay, I have argued for "the poverty of theistic cosmology" in the following *two* respects: Neither the theistic answer to the question, "Why is there something contingent rather than nothing contingent?" nor the theological explanation of the ultimate nomological architecture of the world withstands evidential scrutiny.

▨ ADDENDUM*

This chapter is reprinted from my article "The Poverty of Theistic Cosmology" in the *British Journal for the Philosophy of Science* 55 (2004), pp. 561–614. I have published a much shorter sequel to this chapter under the title "Why Is There a Universe at All, Rather Than Just Nothing?" That sequel is the text of my Presidential Address, August 9, 2007, at the 13th quadrennial International Congress of the Division of Logic, Methodology, and Philosophy of Science, which is one of the two divisions of the International Union of History and Philosophy of Science. My Presidential Address appeared in the volume of the proceedings of the 13th Congress, which is titled *Logic, Methodology and Philosophy of Science* and is edited by Clark Glymour of the United States, Wei Wang of China, and Dag Westerståhl of Sweden. These proceedings were published by King's College Publications, London, UK, in 2009.

---

* *Editor's Note*: This Addendum was added by Professor Grünbaum for this volume.

In advance of that publication, the journal *Free Inquiry* also published a variant of its text in two installments: Part 1 appeared in the June–July 2008 issue (vol. 28, no. 4, pp. 32–35); Part 2 in the August–September 2008 issue (vol. 28, no. 5, pp. 37–41).

As will be recalled from section 1.7.1 on Leibniz, I contended that the hypothesized a priori maximum ontological simplicity of the Null World does *not* mandate the claim of SoN that, de jure, the thus simplest world must be spontaneously realized ontologically in the absence of an overriding cause, because a priori simplicity is *not* ontologically legislative.

But Swinburne has given a special theological twist to simplicity in the 2004, second edition of his book *The Existence of God* (Oxford University Press). There, he tells us (p. 336): "If there is to exist something, it seems impossible to conceive of anything simpler (and therefore a priori more probable) than the existence of God." And in his 1997 monograph *Simplicity as Evidence of Truth*, as extended in his 2001 Oxford book *Epistemic Justification*, he has stated empirical conditions under which, he claims, the simpler of two rival hypotheses is most probably true. Might Swinburne then be able to claim that the theistic existential hypothesis is inductively more likely to be true than its atheistic competitor? Elsewhere, I have thoroughly justified a decidedly negative answer to this question: See my 2008 "Is Simplicity Evidence of Truth?" especially the section "A Coda on Atheism Versus Theism."

■ NOTES

I greatly benefited from the scholarship, advice, and comments of several valued colleagues as well as of three graduate research assistants. I wish to express my warm gratitude to them.

Philip Quinn, who was my doctoral dissertation student at the University of Pittsburgh nearly four decades ago, was a steadfast and generous interlocutor on the entire spectrum of the issues I treat, as can also be gleaned from within my text.

Teddy Seidenfeld was a great and indefatigable resource in specifically appraising Swinburne's *Bayesian* argument for the existence of God. My Pittsburgh colleagues Richard Gale, Gerald Massey, Nicholas Rescher, and the late Wesley Salmon served helpfully as sounding boards on one or another subtopic.

My graduate assistants Emily Aiken, Alan Love, and James Tabery ably processed and/ or summarized some relevant literature for me. Jim Tabery also went over the drafts of this essay with a fine-toothed comb, making good suggestions to improve my exposition.

1. Grünbaum, "A New Critique of Theological Interpretations of Physical Cosmology," *British Journal for the Philosophy of Science* 51 (2000), p. 5; included in the present volume as chapter 9.

2. Leibniz, "Principles of Nature and of Grace Founded on Reason" (1714), in: G. H. R. Parkinson (ed.), *Leibniz: Philosophical Writings*, 1973, sec. 7, p. 199.

3. Leibniz, *Vernunftprinzipien der Natur und der Gnade; Monadologie = Principes de la Nature et de la Grace fondés en Raison; Monadologie*, 1714/1956, p. 13.

4. Leibniz, "Principles of Nature," sec. 8, p. 199.

5. Quinn, "Cosmological Contingency and Theistic Explanation," *Faith and Philosophy* 20 (2003), Special Issue no. 5, pp. 583–584.

6. Craig, "Professor Grünbaum on the 'Normalcy of Nothingness' in the Leibnizian and Kalam Cosmological Arguments," *British Journal for the Philosophy of Science* 52 (2001), sec. 2, pp. 375–78.

7. Grünbaum, "New Critique."

8. Swinburne, *The Existence of God*, rev. ed., 1991, p. 128.

9. Leibniz, "Principles of Nature," sec. 8, p. 199.

10. Grünbaum, "Theological Misinterpretations of Current Physical Cosmology," *Philo* 1 (1998), p. 16 (see chapter 8 in the present volume); "New Critique," pp. 5, 19.

11. Grünbaum, "New Critique," p. 19.

12. Ibid.

13. Parfit, "The Puzzle of Reality: Why Does the Universe Exist?" in: P. van Inwagen and D. Zimmerman (eds.), *Metaphysics: The Big Questions*, 1998, p. 420.

14. Parfit, "Why Anything? Why This?" Pt. 1, *London Review of Books* 20(2) (January 22, 1998), p. 24.

15. Swinburne, *Is There a God?* 1996, p. 48.

16. Parfit, "Why Anything?" p. 24.

17. Ibid., column 1.

18. Swinburne, *Existence of God*, p. 287.

19. Ibid., p. 283.

20. Swinburne, *Is There a God?* 1996, p. 48.

21. Bergson, *The Two Sources of Morality and Religion*, trans. R. A. Audra and C. Brereton, 1974, p. 240.

22. Gale, *Negation and Non-Being*, 1976, pp. 106–13.

23. Ibid., p. 116.

24. Ibid., pp. 115–16.

25. Ibid., p. 116.

26. See May, *Schöpfung aus dem Nichts: Die Entstehung der Lehre von der Creatio Ex Nihilo*, 1978.

27. Samuelson, "Judaic Theories of Cosmology," in: J. Neusner, A. Avery-Peck, and W. Green (eds.), *The Encyclopedia of Judaism*, Vol. 1., 2000, p. 128.

28. Loveley, "Creation: 1. In the Bible," in: *New Catholic Encyclopedia*, Vol. 4, 1967, p. 419.

29. Hasker, "Religious Doctrine of Creation and Conservation," in: E. Craig (ed.), *Routledge Encyclopedia of Philosophy*, Vol. 2., 1998, p. 695; cf. Edwards, "Atheism," in: P. Edwards (ed.), *The Encyclopedia of Philosophy*, Vol. 1, 1967, p. 176.

30. Bertocci, "Creation in Religion," in: P. Wiener (ed.), *Dictionary of the History of Ideas: Studies of Selected Pivotal Ideas*, Vol. 1, 1968/1973, p. 571.

31. Leslie, "Efforts to Explain All Existence," *Mind* 87 (1978), p. 185.

32. See Eliade, *Essential Sacred Writings From Around the World*, 1992.

33. Leslie, "Efforts to Explain All Existence," p. 185.

34. Bergson, *Two Sources of Morality and Religion*, pp. 239–40.

35. Nozick, *Philosophical Explanations*, 1981, p. 122.

36. Leibniz, "Principles of Nature," sec. 7, p. 199.

37. Rescher, "On Explaining Existence," in: Rescher, *The Riddle of Existence*, 1984, p. 19.

38. Leibniz, "Principles of Nature," sec. 7, p. 199.

39. Swinburne, *Existence of God*, pp. 283–284.

40. Carlson and Olsson, "The Presumption of Nothingness," *Ratio* 14 (2001), p. 205.

41. Leibniz, "On the Ultimate Origination of Things" (1697), in: G. H. R. Parkinson (ed.), 1973, *Leibniz: Philosophical Writings*, 136.

42. Leibniz, "Principles of Nature," sec. 7, p. 199.

43. Ibid., secs. 7 and 8, p. 199. It merits mention that, in the German translation and French original version of his 1714 essay (*Vernunftprinzipien der Natur und der Gnade/Principes de la Nature et de la Grace fondés en Raison*, 1956, pp. 13 and 12, respectively), the word "nothing" in the sentence "For 'nothing' is simpler and easier than 'something'" in sec. 7 is rendered by the respective *nouns* "das Nichts" and "le rien." Note also that although the English translation by Parkinson and Morris of the first sentence of Leibniz's Section 8 speaks of the sufficient reason "of" the existence of the universe, I shall hereafter replace their "of" by "for," since the German translation of Leibniz's text uses the term "für."

44. Swinburne, *Is There a God?*, p. 48 (italics added).

45. Grünbaum, "Theological Misinterpretations," sec. 5, pp. 25–26.

46. Craig, "Professor Grünbaum on the 'Normalcy of Nothingness,'" p. 377.

47. Swinburne, *Is There a God?*, p. 48 (italics added).

48. Craig, "Professor Grünbaum on the 'Normalcy of Nothingness,'" p. 378.

49. Rescher, "Contingentia Mundi: Leibniz on the World's Contingency," in: *On Leibniz*, 2003, p. 45.

50. Ibid., 46.

51. Ibid., p. 49 (italics added)

52. Parfit, "Why Anything?" p. 25.

53. Parfit, "The Puzzle of Reality: Why Does the Universe Exist?" in: P. van Inwagen and D. Zimmerman (eds.), *Metaphysics: The Big Questions*, 1998, p. 420.

54. Swinburne, "Response to Derek Parfit," in P. Van Inwagen and D. W. Zimmerman (eds.), *Metaphysics: The Big Questions*, 1998, p. 428 (footnote omitted).

55. Parfit, "Why Anything?" p. 25.

56. Nozick, *Philosophical Explanations*, 128.

57. Unger, "Minimizing Arbitrariness: Toward a Metaphysics of Infinitely Many Isolated Concrete Worlds," *Midwest Studies in Philosophy* 9 (1984), pp. 45–49.

58. Ibid., pp. 48–49.

59. Swinburne, *Is There a God?*, p. 48.

60. Ibid., p. 49.

61. Quoted in Quinn, "Creation, Conservation, and the Big Bang," in: J. Earman et al. (eds.), *Philosophical Problems of the Internal and External Worlds*, 1993, p. 593.

62. *Summa Theologiae* Ia, 50, Art. 2, ad. 3.

63. Quoted in Quinn, "Creation, Conservation, and the Big Bang," p. 593.

64. See ibid., pp. 593–594.

65. Craig, "Professor Grünbaum on the Normalcy of Nothingness," p. 383, and n. 6 there.

66. Nozick, *Philosophical Explanations*, p. 122, note*, second paragraph.

67. Wald, *General Relativity*, 1984, pp. 100–1.

68. See Grünbaum, "Theological Misinterpretations," pp. 25–26.

69. Descartes, "Meditation III. Of God: That He Exists," in: *The Philosophical Works of Descartes*, vol. 1, 1967, p. 168.

70. Bondi, *Cosmology*, 2nd ed., 1960.

71. Ibid., pp. 73–74, 140, 152.

72. Ibid., p. 140.

73. Grünbaum, "Theological Misinterpretations," sec. 3, pp. 22–23; "A New Critique of Theological Interpretations of Physical Cosmology," *British Journal for the Philosophy of Science* 51 (2000), pp. 5–7.

74. Grünbaum, *Philosophical Problems of Space and Time*, 2nd ed., 1973, pp. 406–07; "Theological Misinterpretations," secs. 2–4.

75. van Fraassen, *The Scientific Image*, 1980, pp. 111–112 (italics added).

76. von Hayek, *The Counter-Revolution of Science: Studies on the Abuse of Reason*, 1952.

77. Quinn, "Cosmological Contingency and Theistic Explanation," pp. 583–584.

78. Ibid. (italics added).

79. Nozick, *Philosophical Explanations*, 1981, p. 126.

80. Leibniz, "On the Ultimate Origination of Things," pp. 136–137; "Principles of Nature," sec. 8, p. 199.

81. Weinberg, *Dreams of a Final Theory*, 1993, p. 238.

82. Schopenhauer, *The World as Will and Representation*, vol. I, Appendix.

83. Schopenhauer, *The World as Will and Representation*, vol. II, ch. 17, p. 161 (italics added).

84. Heidegger, *Einführung in die Metaphysik*, 1953, p. 1.

85. Lütkehaus, *Nichts: Abschied vom Sein, Ende der Angst*, 1999, p. 29. In my English translation, it reads:

This book is an attempt to revise the preoccupation of a mode of thinking that is fixated on being, "ontocentric" in its hierarchy of values, "ontomorphic" in its concepts and ideas, and unconditionally "ontophilic" in its motivations. This mode of thought has found its most symptomatic expression in its paranoid, "nihilophobic" relation to non-being, to nothingness. Not the forgetting of being, as deplored by the Todtnauberg School, but rather the forgetting of nothingness is the genuine "black hole" of its ontological amnesia. The forgetting of nothingness, fear of nothingness and ontological greed constitute its "ontopsychological, 'ontopathological' syndrome. And precisely thereby this way of thinking conduces to annihilation and self-annihilation. Thus fixated on being, this is the perhaps tragic irony of the Western history of being.

86. Swinburne, *Existence of God*, 1991, chap. 7, p. 128.

87. Swinburne, *Is There a God?* 1996, p. 48.

88. Ibid., pp. 48–49; p. 2.

89. Ibid., p. 2.

90. Ibid., p. 49.

91. Ibid.

92. Swinburne, *Existence of God*, p. 127.

93. Ibid., pp. 121–130.

94. Ibid., p. 127.

95. Ibid., p. 43.

96. Quinn, "Creation, Conservation, and the Big Bang," p. 592.

97. Ibid., p. 607.

98. Swinburne, *Is There a God?* pp. 21–22.

99. Swinburne, *Existence of God*, p. 125.

100. Ibid., p. 125; Quinn, "Creation, Conservation, and the Big Bang," p. 607.

101. Grünbaum, "New Critique," p. 20.

102. Quinn, "Creation, Conservation, and the Big Bang," p. 607.

103. Swinburne, *Existence of God,* p. 125; Quinn, "Creation, Conservation, and the Big Bang," p. 607.

104. Swinburne, *Existence of God,* p. 109 (italics added).

105. Ibid., p. 295 (italics added).

106. Swinburne, *Is There a God?* p. 2 (italics added); cf. also Swinburne, *Existence of God,* ch. 4 on "Complete Explanation."

107. Swinburne, "Reply to Grünbaum's 'A New Critique of Theological Interpretations of Physical Cosmology,'" *British Journal for the Philosophy of Science* 51 (2000), pp. 481–85.

108. Swinburne, *Is There a God?* p. 2 (italics added).

109. Swinburne, *Existence of God,* p. 125.

110. Quinn, "Creation, Conservation, and the Big Bang," p. 607.

111. Swinburne, *Is There a God?* p. 2.

112. Swinburne, *Existence of God,* p. 287.

113. Salmon, "Statistical Explanation," in: Salmon, *Statistical Explanation and Statistical Relevance,* 1971, p. 34.

114. Swinburne, *Existence of God,* p. 287.

115. Swinburne, *Is There a God?* p. 49.

116. Swinburne, *Existence of God,* p. 125.

117. Swinburne, *Is There a God?* p. 2.

118. Swinburne, "Reply to Grünbaum," p. 482.

119. Salmon, "Explanation and Confirmation: A Bayesian Critique of Inference to the Best Explanation," in: G. Hon and S. S. Rackover (eds.), *Explanation: Theoretical Approaches and Applications,* 2001, p. 79.

120. Grünbaum, "New Critique," p. 35. Citation is from Salmon, "Explanation and Confirmation," p. 79.

121. Grünbaum, "New Critique," p. 35.

122. Swinburne, "Reply to Grünbaum," p. 482.

123. Grünbaum, "New Critique," p. 36.

124. Swinburne, "Reply to Grünbaum," p. 482.

125. Swinburne, *Is There a God?* p. 2.

126. Swinburne, "Reply to Grünbaum," p. 482 (italics added).

127. Ibid.

128. Quinn, "Cosmological Contingency and Theistic Explanation," p. 593.

129. Ibid. (italics added).

8 Theological Misinterpretations of Current Physical Cosmology

■ ABSTRACT

In earlier writings, I argued that *neither* of the two major physical cosmologies of the twentieth century *supports* divine creation, so that atheism has nothing to fear from the explanations required by these cosmologies. Yet theists ranging from Augustine, Aquinas, Descartes, and Leibniz to Richard Swinburne and Philip Quinn have maintained that, at *every* instant anew, the existence of the world *requires* divine creation ex nihilo as its cause. Indeed, according to some such theists, for any given moment *t,* God's volition that the world should exist at *t* supposedly *brings about* its actual existence at *t.*

In an effort to establish the current viability of this doctrine of perpetual divine conservation, Philip Quinn[1] argued that it is entirely compatible with *physical* energy conservation in the Big Bang cosmology as well as with the physics of the steady-state theories.

But I now contend that instead there is a logical *incompatibility* on both counts. Besides, the stated tenet of divine conservation has an additional defect: It speciously purchases plausibility by trading on the multiply disanalogous volitional explanations of human actions.

■ 1. INTRODUCTION

It has been claimed that the Big Bang cosmogony—and also the now largely unpopular steady-state cosmology—pose a *scientifically insoluble* problem of matter–energy creation and fail to explain why the world does not lapse into nonbeing at any given moment. We are told that this alleged conundrum is solved by postulating divine intervention as an external cause. If there is a first moment at which the universe begins to exist, we learn, then this creative supernatural intervention occurs at that moment and ever after. In any case, divine creative intervention is allegedly required *throughout all existing time,* no matter whether the universe has a temporal beginning or not.

In the case of the Big Bang theory, the champions of this thesis have ranged from Pope Pius XII in 1951, as he told the Pontifical Academy of Sciences, to the British astronomer Bernard Lovell, the American astronomer Robert Jastrow, and the theistic philosophers Richard Swinburne at Oxford and Philip Quinn at Notre Dame University in the United States. Lovell had made the same claim à propos of the steady-state cosmology.

In my earlier papers of 1989 through 1991[2] I disputed this theological twist. And I maintained more generally that *atheism has nothing to fear at all* from these two major twentieth-century physical cosmologies, because neither of them supports the idea of God the creator.[3] But, I shall now argue further that, conversely, perpetual divine creationism actually has a great deal to fear from both of these cosmologies.

The familiar meaning of the word "creation" lends itself to the insinuation of a creative role of a supernatural agency *without argument*. As Webster's Dictionary tells us, in its primary use, the term "creation" means: "act of causing to exist, or fact of being brought into existence by divine power or its equivalent; especially the act of bringing the universe or this world into existence out of nothing." Evidently, the transitive verb "to create" calls for a *subject* as well as an object. And in a *cosmological* context, the verb is laden with the notion of a divine *agency* or cause *external* to the entire world.

In a 1989 paper, which was reprinted in John Leslie's 1990 volume *Physical Cosmology and Philosophy*,[4] I argued that the question of whether the universe had a temporal origin had been *fallaciously transmuted* into the *pseudo-problem* of the creation of the world with its matter–energy by a cause *external* to the universe.

In a 1991 paper in *Erkenntnis*,[5] I extended my arguments so as to include a critique of the thesis of the English physicist C. J. Isham. According to Isham, the Hartle and Hawking version of quantum cosmology lends itself to supporting Augustinian creation ex nihilo. Writing in a 1988 Vatican Observatory volume, Isham extolled as "profound" Augustine's doctrine that God created both time itself and matter.[6] Yet, as I shall explain at the end of Section 5, I contend that Augustine's view is fundamentally unsound.

My 1989 paper provoked three responses, only one of which will concern me here because it pertains to the most influential of the creationist scenarios: *perpetual* divine creation.

The theist Philip L. Quinn of Notre Dame University has recently offered a cosmological defense of divine creation and conservation,[7] which I shall challenge here.

In the 1989 paper, I had not confined myself to the minimalist doctrine that God created the world all at once. Instead, I had also taken explicit issue with Descartes' thesis of *perpetual divine conservation* of matter vis-à-vis Lavoisier's hypothesis of *natural* spontaneous matter conservation through time. The Cartesian doctrine asserts that the preservation of matter in existence requires divine *repetition* of an act of creation at every moment. That thesis of creatio continuans was espoused by a historically long succession of theists.[8] I shall argue, however, that it fails altogether for an array of reasons.

The upshot of this article will strengthen considerably, I trust, my earlier objections to theological creationism. As already noted, previously I had argued mainly that atheism has nothing to fear from the physical cosmologies of the past half-century because they provide no evidential support for divine creation. Philip

Quinn's challenge, among others, now prompts me to offer the following *stronger* indictment of creationist natural theology: The Big Bang model of general relativity theory as well as the steady-state theory are *each logically incompatible* with the theological doctrine that divine *creatio continuans* is *required* in both of their worlds. Moreover, that doctrine is vitiated by major epistemological and conceptual difficulties, as I shall try to show.

The well-known American Roman Catholic Jesuit theologian Michael Buckley at Notre Dame University, in a critique of Paul Davies's "wishy-washy" theology, comes fairly close to conceding unintentionally my impending thesis that divine volitional creation offers a *pseudo*-explanation, when Buckley makes a major concession concerning the hypothesized process of divine creation. As he admits: "We really do not know how God 'pulls it off.' Catholicism has found no great scandal in this admitted ignorance."[9] But if theology is thus admittedly ignorant, then the theological hypothesis of creation ex nihilo adds no articulated *causal* understanding of the existence of matter–energy to *any* physical model of cosmogony! We have no evidence at all for effective volitional actions that are *causally unmediated* by a nervous system and yet conform to the practical syllogism. As we know, in that syllogism an action is explained by a desire-cum-belief set. And it would, of course, be entirely illicit for the theist to trade tacitly on the picture of *transformative* causation to defend creation ex nihilo. Yet Pope Pius XII[10] and many, many others have told us that science is explanatorily defective in a basic way without the hypothesis of divine creation ex nihilo.

In rejecting creationist theological appropriations of the steady-state and Big Bang cosmologies alike, I need not make any claims concerning their respective technical scientific merits, which currently vary somewhat in the case of the Big Bang model, although the major features of the model continue to command much loyalty from cosmologists. For example, until 1990, when NASA's satellite COBE found wrinkles in the previously thought uniform density of the cosmic microwave radiation, the Big Bang model conspicuously lacked an explanation of the genesis of the galaxies! But when the Berkeley physicist George Smoot announced the detection of the density fluctuations in April 1991 to the American Physical Society, he electrified the newspapers and some of the faithful by declaring: "If you are religious, this is like looking at God."[11]

Though probably unintended, the journalistic moral seems to be that you have a better chance to behold the Almighty with a differential microwave radiometer than by praying! Yet the evidential fortunes of the *Big Bang theory* are not entirely secure, although it has been almost universally victorious among cosmologists so far over the rival steady-state theories. Now it confronts the embarrassing discrepancy between the age of the oldest stars and the newly calculated lesser age of the universe since the Big Bang. Yet it would seem most recently that the cosmic expansion will continue forever instead of being followed by a cosmic collapse and annihilating crunch. Thus, our own galaxy will be ever more "alone" in the cosmos, a prospect that some people may find depressing but which I myself view with complete equanimity.

The steady-state cosmology now has few adherents among physical scientists, with such notable exceptions as Fred Hoyle and Jayant V. Narlikar and perhaps others. Previously, I criticized the specific theistic reading that the English radio astronomer Sir Bernard Lovell gave of the steady-state world. It is philosophically instructive, however, I believe, that, despite the serious empirical difficulties of the steady-state theory, I examine further critically its theological creationist interpretation, as articulated in 1993 by Quinn. And it will be expeditious to discuss it before I deal with the Big Bang theory.

■ 2. THE STEADY-STATE COSMOLOGY

The steady-state theories were pioneered in the late 1940s by Fred Hoyle and by Hermann Bondi and Thomas Gold. Very recently, Hoyle published a modification of his 1948 theory in the journal *Astrophysics and Space Science*.[12] But for my philosophical purposes here, which pertain to attempted *theological appropriations* of physical cosmology, I need to focus on the simplest of the 1948 versions. That original form of the theory features a *violation* of matter–energy conservation by the formation of new matter without any transformative causation, that is, "out of nothing," whereas the modification of the theory in the 1980s and since no longer features such a violation. At the hands of such astronomers as Lovell,[13] divine intervention was claimed to be required by the *non*conservative formation of the new matter that had been deduced in the original 1948 theory. (But in the modified recent version, the positive energy of the new matter is balanced by the negative energy of the so-called C-field.)

The steady-state theory postulated originally *as a matter of natural law* that, while the galaxies are receding from each other everywhere in the universe, the matter *density* nonetheless ubiquitously *remains constant* through time. This constancy is enunciated by the so-called perfect cosmological principle—hence, the name "steady-state" for this cosmic scenario of eternal constancy of density. But, if there is such constancy of density alongside the galactic recession, then completely *new* matter must pop into existence out of nowhere in violation of matter conservation, such that it fills, at the requisite rate, the spaces vacated by the galactic recession. Yet the ensuing rate at which the presumed new matter would make its cosmic debut is so small as to presumably elude detection in the laboratory, at least foreseeably.

Lovell[14] asked, in effect: What is the *external cause* of the coming into existence of the new hydrogen atoms in the Bondi and Gold universe, which come into being in violation of matter–energy conservation? Thereupon he complains that the "steady-state theory has no solution to the problem of creation of [new] matter." Note that Lovell uses the theologically tinged causal term "creation," instead of the neutral descriptive term "accretion."

Now observe that Lovell's demand for an *external cause* of the new matter is unfortunately loaded with tacitly taking the law of energy conservation for granted, as is clear from his complaint that the steady-state theory makes no provision for "the *energy input* which gave rise to the created [hydrogen] atom."[15] But the steady-state theory explicitly *denies* that energy-conservation law. Thus, Lovell's conservationist assumption of the need for other energy as the source of the matter "input" *contradicts* the steady-state theory! After all, the steady-state theory had deduced an *altogether natural* violation of energy conservation from its postulate of density constancy in an expanding universe. Hence, Lovell simply begged the question when he asked for the energy source or transformative cause of the new hydrogen.

It is granted, of course, that the postulate of density constancy may be questioned as long as there is insufficient evidence for it. Thus, Lovell and everyone else are entitled to ask for the *observational credentials* of the steady-state theory. But, as we saw, *that* was *not* his question, since he did not challenge that theory epistemologically but only ontologically. I am happy to report that, at a 1986 meeting in Locarno, Lovell conceded my point, and he said so in the published proceedings.[16]

In my 1989 article, I had drawn the following conclusion: "In the steady-state theory, … *non*-conservative matter-accretion [or popping into existence ex nihilo] is claimed to transpire *without any kind of external* [or supernatural] *cause*, because it is held to be cosmically the spontaneous, natural, unperturbed behavior of the physical world!"[17] Quinn objects: "But neither does the steady-state theory *rule out* a [required] divine cause for the [eternal] coming to be of its new hydrogen."[18] Yet I shall now argue here against Quinn that his claim of such a required divine creative role is indeed *ruled out* as definitely *inconsistent* with the steady-state cosmology.

As Quinn emphasizes, several contemporary theists besides himself echo the doctrine of *creatio continuans* championed by Aquinas, Descartes, Berkeley, Leibniz, Locke, Jonathan Edwards, et al. Thus, Quinn maintains explicitly that perpetual divine creative activity is crucial for such mere physical energy or matter conservation as holds in a Big Bang universe, no less than for the hypothesized coming into existence of new matter in the steady-state world. And, as Quinn tells us, Richard Swinburne attributes to theists the view that "God keeps the universe in being, whether he has been doing so for ever or only for a finite time."[19] Indeed, the British physicist-theologian John Polkinghorne sees just the doctrine of *perpetual* rather than initial creation as the essence of the Christian scenario, although his views should not be equated with Quinn's in other respects. In short, as Descartes claimed in the Third Meditation, creation and conservation require the same divine power and action. And, as Berkeley explained, divine conservation is simply continued and repeated creation.

Thus, in the traditional theistic account, it is held that "all contingent things are continuously dependent upon God for their existence."[20] I shall challenge that claim as *ill conceived* from the outset. On Quinn's view, "God not only creates all contingent things but also conserves them in existence, moment by moment, in a way that is tantamount to continuously creating or recreating them."[21]

According to Quinn, the relevant "relation of metaphysical dependence or causation" is a primitive relation rendered by the following locution: "God willing that x-exists-at-t brings about x-existing-at-t."[22] I disregard here my multiple malaises with this sort of notion of divine *volitional* causation but just recall my brief objections to it à propos of the Jesuit Buckley's agnostic disclaimer as to the mediating causal process. Yet, as every paralytic and paraplegic knows all too well, a mediating causal process involving the adequate functioning of the nervous system needs to be specified when we explain *in the context of existing evidence,* say, a particular outcome as the product of human volitional action. If, for example, Jones wants an electric light bulb to be turned on, it won't do to explain the lit state of the bulb by *merely* saying in the manner of the Book of Genesis: "Jones willed: Let there be light!" We have no evidence at all for this kind of unmediated causation, which is reminiscent of *word magic.*

Quinn emphasizes that his relation of divine bringing about volitionally "must have the following marks in order to serve its theological purposes":[23] (a) "What does the bringing about [i.e., divine volition] is the *total* cause of what is brought about; nothing else is required by way of causal contribution in order for the effect to obtain," because the divine will is causally *sufficient;* and (b) "the bringing about is the *sole* cause of what is brought about; causal *overdetermination* is ruled out,"[24] since it allows more than one sufficient cause.

Quinn is concerned to rule out sufficient causes other than divine volition in order to claim that God's creative and conservative actions are *necessary* for the existence of the physical entities. In short, as Quinn has it, God is the *total* and *only* cause of the existence of things. And the crucial underlying assumption is that this very existence *must have a cause at all,* a posit that I shall discredit in Section 4.

Now note that the cardinal postulate of the theories of Hoyle and of Bondi and Gold is the so-called perfect cosmological principle. Rightly or wrongly, it asserts, as a matter of natural law, that there is conservation of matter *density.* But it is of *decisive* importance that, in conjunction with that law of density conservation, the so-called expansion of the universe or mutual galactic recession is *causally sufficient* for the *completely natural* coming into existence ex nihilo of new matter! Equally crucial is the fact that, *without* this cosmic expansion, density conservation *alone* would *not* issue in matter accretion.

Thus, Leibniz could get his coveted *sufficient* reason for the existence of the new matter *from the physics itself without God,* if he could have known the content of the 1948 steady-state theory of Bondi and Gold. Indeed, this natural *physical* causal sufficiency is decisive, because it obviously rules out the theistic claim, made by Quinn and Lovell, that external creative divine intervention in the universe is *required* for such formation of new matter.

It has been wrongly claimed that the Bondi and Gold explanation of the rate of the formation of new matter is suspect as being *teleological,* since it is *seemingly* dictated by the *outcome* state of *density* conservation during the expansion. But this objection is without merit. Density conservation is no more teleological than energy conservation or charge conservation. The outcome states result from the prior state in accord with the pertinent laws. One might object, equally fallaciously, that neutrino production during radioactive decay, as postulated by Wolfgang Pauli and Enrico Fermi, is teleological, because it is governed by the outcome state of *energy* conservation, given that the fragments of the radioactive decay have a smaller total mass–energy than their undecayed ancestor. Relatedly, the claim that teleology dictates the formation of new matter in the steady-state world cannot sustain the theistic creationist interpretation of the nonconservative formation of new matter in the steady-state theory.

Thus, contrary to Quinn, the steady-state cosmology is indeed *logically incompatible* with his and Lovell's claim that divine creative intervention is causally necessary for the *non*conservative popping into existence of new matter in the steady-state universe.

But that is not all. In Quinn's theistic scenario, we recall, the divine creative will is both the *total* and the *sole* cause of the matter accretion. This alleged totality and exclusivity of *God's* causal role in the existence of the new matter entails the *bizarre conclusion* that the physics of the steady-state universe makes no causal contribution at all to the popping into existence of the new matter.

Let me emphatically reject as completely futile and evasive the reply that, at any moment in the steady-state world, it is within God's power to *suspend* its density-conservation principle, much as a government can revoke the normativity of its statutory laws. Note at once the dubious analogy between revoking a statutory, *normative* law, which does not describe actual behavior, and "suspending" a *descriptive* law. But suppose that someone would try to disarm the physical causal sufficiency for the genesis of new matter which I have demonstrated, declaring: God does his creative job *indirectly* by keeping the law of density conservation in place during the cosmic expansion. In this way, it might be thought, the doctrine of required indirect divine creation might be made compatible with the physics after all. But such an attempt to neutralize my critique simply fails.

In the first place, Quinn asserted the logical compatibility of the required theistic creationist scenario with the *assumed truth* of the steady-state cosmology. But that cosmology categorically features *as given* the eternal temporal invariance of density conservation in an expanding universe. Second, but no less importantly, Quinn, citing Leibniz and a 1988 work by David Braine, told us explicitly that divine creative causation is direct in the form of *unmediated* bringing about the existence of matter, rather than only indirect, such as via the density-conservation law.[25] As Quinn explained: In characterizing the causal relation in his account of creation and conservation, he had "specified that what does the bringing about causes what is brought about *immediately* rather than remotely by means of instruments such as secondary physical causes."[26]

Thus, it would completely beg the question in this context to seek refuge in the deus ex machina of the alleged divine ability to suspend the density-conservation law, as it were, or to stop the expansion of the universe. Theists are free to take that supposed divine ability on faith, if they can clarify just what it means. But that freedom is unavailing, because the context of the entire cosmological debate on divine creation is one of *argument* in natural rather than fideist theology. Thus, it would clearly be question begging, if not simply frivolous, to claim, in effect, that, within the steady-state cosmology, the perfect cosmological principle is tacitly predicated on the proviso that God *refrain* from *suspending* density conservation and/or from arresting the cosmic expansion. Neither Bondi nor Gold nor Hoyle—all reputedly atheists—would dream of such a proviso. And it is not they who are begging the question. *Besides,* the proposed deus ex machina of indirect divine creation is plainly ad hoc, since no evidence is offered for it at all.

As is now very clear, I trust, the steady-state theory radically belies the inveterate thesis that, *no matter what the physics of our world,* any matter–energy coming into being ex nihilo requires an external divine creative cause. And that alone, I claim, clearly discredits the received theistic view as articulated by Quinn. Indeed, it is, I claim, one of the gravest and *most insidious* of errors in the *entire history* of philosophy to legislate the need for external causes independently of what the actual physics of our world may be. I shall now articulate this major moral historically before turning to the Big Bang cosmology. In order to do so, let me now first refine my earlier published statement of the generalized fundamental lesson I draw from the history of science for the issues before us.

■ 3. THE IMPORT OF THE HISTORY OF SCIENCE FOR THE POSTULATION OF EXTERNAL CAUSES

Important episodes in the history of science have shown that new evidence or new theoretical insights have warranted fundamental changes in dealing with the following major question: Is it justified, in a given context, to postulate causes *external* to physical or biological systems as *intervening* in them, in order to explain some observed behavior of these systems? The historical evolution of the answers to this question bears directly on the legitimacy of inferring an *external* cause to account for the behavior of the universe as a whole, *or even for its very existence.* Let us see just how.

According to Aristotle, a force is needed as the external cause of a sublunar body's nonvertical motion, even if it moves horizontally with constant velocity. In his physics, the demand for such a disturbing external dynamical cause to explain any such motion arises from the following assumption: When a sublunar body is not acted on by an external force, its *natural,* spontaneous, dynamically unperturbed behavior is to be at rest at its "proper place," or—if it is not already there—to move vertically toward it.

Yet, as we know, Galileo's analysis of the motions of spheres on inclined planes, among other things, led him to conclude that the empirical evidence speaks against just this Aristotelian assumption. As Newton's First Law of Motion tells us, uniform motion *never* requires any external force as its cause; only accelerated motion does. Thus, Galileo and Newton *eliminated* a *supposed external* dynamical cause on *empirical* grounds, explaining that uniform motion can occur spontaneously without such a cause.

But, if so, then the Aristotelian demand for a causal explanation of *any* motion whatever by reference to an external perturbing force is predicated on a *false underlying assumption*.

Clearly, the Aristotelians begged the question by tenaciously continuing to ask: What net external force, pray tell, keeps a uniformly moving body going? Thus, scientific and philosophical questions can be anything but innocent by loading the dice with a petitio principii!

A brief example from the history of biology, starting with Louis Pasteur but including Alexander I. Oparin and Harold C. Urey, likewise illustrates a change as to the hypothesized need for external causes in the debate on the feasibility of the spontaneous generation of life from nonliving substances.[27]

I have adduced these examples in addition to the steady-state world to show that a scientific or philosophical theory may be fundamentally mistaken in calling for some sort of external cause to explain certain states of affairs. No physicist or philosopher can be justly criticized for failing to answer a causal question inspired by that mistaken demand for an external cause.[28] Incidentally, I do not deny that, in *other* cases, physical evidence may show the need for an external cause where none was theretofore suspected, as noted by the historian Lorraine Darden.

Now let me argue that the stated moral from the particular examples I adduced from the history of science spells a salutary caveat for the purported problem of creation.

■ 4. THE QUESTION OF THE RATIO ESSENDI AS A PSEUDO-PROBLEM

I claim that the question, "Why is there anything at all rather than just nothing?" is a misguided query, at least to the extent that it calls for a cause external to the universe. Thus, it is wrong-headed, I shall now contend, to ask for the external cause or reason of the bare existence and persistence of the world, its so-called ratio essendi. But it is vital to distinguish such a supposed *creative* cause or reason, as Aquinas did, from a merely *transformative* cause, which just produces *changes* in things that *already exist* in *some* form, or generates new entities from previously existing objects.

There is a crucial underlying assumption that animates the theological creationist and conservationist ratio essendi given by an array of famous theists. They take it to be axiomatic that *if* there is a physical world *at all,* then its spontaneous, undisturbed or natural state is one of *utter nothingness,* whatever that is. Those many theists who make this dubious assumption have thereby generated grounds for claiming that the very existence of matter, energy, or whatever constitutes a *deviation* from the alleged spontaneity of nothingness. And that supposed deviation must then have a suitably potent *external* cause.

Just this assumption of spontaneous nothingness is at least insinuated by the biblical story of Genesis. But Aquinas and Leibniz, among others, make it explicit. Aquinas used the loaded, question-begging word "creature" to refer to any contingent entity and declared: "the being of *every* creature depends on God, so that not for a moment could it subsist, *but would fall into nothingness* were it not kept in being by the operation of Divine power."[29] Thus, here we have the fateful crucial presupposition: There would be no world at all or just nothingness, whatever that is, were it not for divine creative and conservative *intervention.*

But what, I must ask, is the evidence for this philosophically fateful assumption of the spontaneity of nothingness? Why, in the absence of an external supernatural (creative) cause, *should* there be just nothing, even if we are clear what that would mean? Leibniz and Richard Swinburne have offered a defense of the spontaneity of nothing by arguing from *simplicity* that the nonexistence of the world (nothingness) is *its* most probable state. But I argue in my forthcoming "A New Critique of Theological Interpretations of Physical Cosmology"* that this defense is completely unavailing.

The baseless tacit presupposition of spontaneous nothingness also contributed to Leibniz's demand for a necessary being to provide a *sufficient reason* for the existence and persistence of contingent things. Yet *I deny that the mere logical or empirical contingency of the existence of any given particulars can support the spontaneity of utter nothingness* and the need for a logically necessary being as the creator. It will emerge that the theological presupposition of the spontaneity of nothingness lacks even the most rudimentary plausibility. Moreover, some philosophers, such as Henri Bergson, have asserted the unintelligibility of the notion of absolute Nothingness.

As I have just argued, the seminal question as to the ratio essendi of the world of contingent beings, far from being innocent and imperative, has forfeited the rationale that animated it at the hands of such major figures as Aquinas and Descartes. Their problem turns out to have been a *pseudo*problem. And their proposed theological resolution of it is a pseudo-*explanation.* One cannot overestimate, I believe, the extent to which the dubious rationale for a ratio essendi unconsciously insinuates itself to confer spurious plausibility on that pseudo-explanation. This point must be borne in mind as *prophylaxis* against the insidious temptation to ask for a *creative cause* of the very existence of the *entire world.*

---

* *Editor's Note:* This essay is included in the present volume as chapter 9.

■ 5. THE BIG BANG UNIVERSE OF THE GENERAL THEORY OF RELATIVITY

Two subtopics will concern us:

What is the Big Bang in the event ontology of the general theory of relativity? For brevity, I shall hereafter speak of the general theory of relativity as the "GTR" (5.1)

I shall contend that *physical* energy conservation *rules out* divine creatio continuans (5.2).

5.1 What Is the Big Bang in the Event Ontology of the General Theory of Relativity?

In my earlier writings,[30] I had discussed two Big Bang models, which I called Case (i) and Case (ii), respectively, and which I am about to characterize. Yet, as I noted then and will see shortly, Case (i) is *not* a bona fide model of the GTR for reasons given in the *event ontology* of that theory. In the putative Case (i) model, the Big Bang is *supposedly* the temporally first physical event of the space–time and is said to occur at the instant $t = 0$. But the Big Bang does not meet the requirements for being a bona fide physical event in the GTR. Instead, there is a hole in the space–time manifold at the putative $t = 0$, such that *at unboundedly ever earlier moments* of time before, say, 14 billion years ago, the space–time metric of the GTR becomes degenerate, and the so-called scalar curvature as well as the density approach infinity.[31] The locution "Big Bang" is a shorthand *façon de parler* for this mathematical behavior of the four-metric and scalar curvature at *regressively* earlier times.

The physicist John Stachel[32] has justified the view that this singular status robs the Big Bang of its *event status* in the GTR. As he showed, points of the *theoretical* manifold first *acquire* the physical significance of being *events,* when they stand in the chrono-geometric relations specified by the space–time metric, which familiarly does double duty as the gravitational field in the GTR.

Thus, in the GTR, it turns out that "the notion of an event makes physical sense only when [both] manifold and metric structure are [well] defined around it."[33] And in that theory, space–time is taken to be "the collection of all [physical] events."[34] But the Big Bang does *not* qualify as a physical point event of the space–time to which one could assign three spatial coordinates and one time coordinate. Therefore, contrary to the Case (i) model, which features a *first* physical event, the past cosmic time interval is *open* or unbounded rather than closed or bounded by a first moment, although its metrical duration in years is only *finite.*

Despite the ontological illegitimacy of the Case (i) model, I have discussed it because Pope Pius XII, Sir Bernard Lovell, and William Craig[35] each claimed support from it for divine creation ex nihilo. Besides, the Case (i) model had figured in the astrophysicist Narlikar's *secular* creationism with which I took issue elsewhere.[36]

But, as we just saw, the Big Bang is actually excluded as not being a physical event occurring at an actual moment of time. Thus understood the relativistically bona fide Big Bang models *differ* from those in Case (i) by being temporally *unbounded* (open) in the past. And hence the past physical career of the Big Bang universe *did not include a first physical event or state at which it could be said to have begun.* I designated the bona fide temporally unbounded models as Case (ii) models.

However, in either Case (i) or Case (ii), the current *age* of the Big Bang universe is *metrically of finite* duration, whose numerical value is under dispute, depending numerically on the time rate of its expansion.[37] Moreover, there are good reasons in the GTR for claiming that *no instants of physical time whatever* existed *before* that finite time interval in either Case (i) or Case (ii).[38] Thus, even if the singular Big Bang *were* included as an event having occurred at a bona fide moment of time t = 0, this hypothetical instant *had no temporal predecessor.* A fortiori, it could *not* have been *preceded* by a state of nothingness, even if the notion of such a state were well-defined.

As we now see, physical processes of some sort *already* existed at every *actual* instant of past time. After all, despite the finite duration of the past, *there was no time* at all at which the physical world did not exist *yet.* Thus, we can say that the Big Bang universe *always* existed, although its age is only, say, somewhere between 8 or 15 times 10^9 years. Here, the word "always" means "for all actual times," but it does not guarantee that time, past or future, is of infinite duration in years.

As we saw, in the Case (i) world, there did not exist any instants of cosmic time before $t = 0$. Therefore, no supposed earlier cause, either creative or transformative, could possibly have been operative before $t = 0$. For that reason alone, the Big Bang could not have had any temporally prior creative or transformative cause. Nor could "it" have had a simultaneous cause, creative or otherwise, because there simply was no "it" or instantaneous event that could have been the momentary effect of such a cause. And in the face of the groundlessness of the spontaneity of nothingness, there is no basis for a creative cause of the Big Bang as construed in the Pickwickian sense of a *façon de parler* I mentioned already.

Let me take for granted the altogether reasonable view that only *events* can qualify as the momentary *effects* of other events or of the action of an agency. As I just argued, the Big Bang is a nonevent, and $t = 0$ is not at all a bona fide time of "its" occurrence. Thus, the Big Bang *cannot be the effect of any cause in the case of either event causation or agent causation alike.* By the same token, a nonexistent event at the putative $t = 0$ cannot have a cause, *either earlier or simultaneous!* Besides, it cannot have an *earlier* cause, either creative or transformative, if only because there was no earlier time at all. And recall (from Section 4) that I have already undercut the entire rationale for any creative ratio essendi anyway by discrediting its assumed spontaneity of nothingness.

5.2 Physical Energy Conservation Versus Divine *Creatio Continuans*

We are ready now to examine Quinn's contention that purportedly *required* divine creation and conservation are consistent with the Big Bang models in both Case (i) and Case (ii). Indeed, Quinn asserts such consistency in all those cases in which *the GTR or any other physical theory* features a *physical* energy-conservation law.

It is very important to bear in mind that the theistic tradition which Quinn tries to defend has insisted aprioristically on the necessity of divine preservation of matter or energy against annihilation, regardless of the particular forms of matter or energy that populate the physical ontologies of successive scientific theories. Thus, he is concerned to argue strenuously that the necessity of perpetual divine conservation is logically *compatible* with the old matter-conservation law dating from Lavoisier, and also with such energy conservation as is valid in GTR universes. Indeed, Quinn and his fellow theists insist quite generally on the logical compatibility of the necessity of divine conservation with whatever *physical* matter–energy conservation law is presumed to be true at any given stage of science. Each such stage features a specific technical physical ontology of matter or energy. But I shall argue that, instead, there is *incompatibility* between the physical and divine conservation scenarios.

Quinn[39] offers the following definitions of divine creation and conservation, which I find very obscure: (i) God *creates* x at $t =_{def}$ God willing that x-exists-at-t brings about x-existing-at-t, and there is no t' prior to t such that x exists at t'; and (ii) God *conserves* x at $t =_{def}$ God willing that x-exists-at-t brings about x-existing-at-t, and there is some t' prior to t such that x exists at t'.

Quinn points out that his formulations deliberately leave open whether God's volitions or willings "are timelessly eternal by not building into this locution [of divine volitions] a variable ranging over times of occurrence of divine willings."[40] But I submit that the notion of *timelessly eternal acts of willing* is obscure and elusive to the point of making such divine willings altogether nonexplanatory as causes of the existence of our world. Quinn's use of the concept of "willing" clearly draws on the acts of volition familiar from the conative life of humans. But such volitional states are *inherently temporal* rather than "timelessly eternal." Thus, Quinn's divine volitional creation scenario is conceptually elusive. Furthermore, insofar as it is analogous at all to human volitions, there is no evidence whatsoever for the occurrence of such Pickwickian volitions.

Nor do I understand what we are to make of the posited scenario that the instantaneously "ensuing" temporal *bringing about* of the existence-of-x-at-time-t is the *effect* of such an *atemporal* volition. Besides, all of the cases of instantaneous action-at-a-distance familiar from prerelativistic physics (e.g., gravitational attraction in Newton's law of universal gravitation) feature *causally symmetric laws of* coexistence (*inter*actions), whereas Quinn's instantaneous divine creative causation is claimed to be *causally asymmetric*. Furthermore, let me just recall anew the Jesuit Buckley's agnostic disclaimer that Catholic theology does *not* know *how* God brings about the existence of the world.

It must be borne in mind that the theists whom Quinn claims to vindicate assert the necessity of divine *creatio continuans* unqualifiedly for the lifetime of a tree, for the conservation of the energy in an isolated finite subsystem of the universe, and—when such conservation is defined for the universe as a whole—for the entire cosmos.

We can now turn to Quinn's treatment of the bona fide relativistic Big Bang models of Case (ii), featuring a temporally *unbounded* past. He describes his theological scenario for all authentic moments of time as follows:

> God *conserves* the sum total of matter-energy whenever it exists, but there is no time at which he creates it or brings it into existence after a prior period of its nonexistence.[41]

But, now, my thesis will be the following: Insofar as the GTR does license a matter–energy conservation law for a specified subclass of the Case (ii) Big Bang models or for isolated subsystems of the universe, the physics itself *rules out* Quinn's theological doctrine that physical energy conservation is only an *epiphenomenon* in the sense of Malebranche's occasionalism, *requiring* repeated divine creation ex nihilo at every instant. One form of the energy-conservation law tells us that the total energy content of an isolated or closed system remains constant naturally and spontaneously. Another form, which is even taught in freshman physics or chemistry, asserts tout court that *energy can be neither created nor destroyed.*

To be more specific concerning both cosmological and subcosmological energy conservation that is licensed by the GTR, consider the spatially closed (or "three-sphere") "Friedmann"–Big Bang universe, which exists altogether for only a finite span of time. It is clearly a physically closed system since there is nothing else. When the matter of that universe takes the form of "dust" (i.e., when the pressure in it vanishes), the *total rest mass of that universe is conserved for the entire time period of its existence.*[42]

Apart from the stated cosmological rest mass conservation law, Robert M. Wald[43] points out that "in general relativity ... a conserved total energy of an isolated system [i.e., subsystem of the universe such as a condensed star, immersed in an asymptotically flat spacetime[44]] can be defined." (That total energy is the so-called ADM energy.[45]) Note that for any particular physical theory T such as the GTR, a physical system passes muster as "closed" in the absence of any outflow or influx of the kinds of physical entities that qualify as mass or energy in the ontology of T.

In the present Big Bang context, my argument from *physical* energy conservation against the necessity of divine *creatio continuans* is as follows: *Given the pertinent mass- or energy-conservation law of the Friedmann–Big Bang dust world, it follows decisively that the physical closure of this universe is causally sufficient for the conservation of its particular mass–energy content.* But just that *physical causal sufficiency for energy conservation,* in turn, *rules out* the major claim of theistic creationism that such physical conservation *requires* perpetual divine creative intervention ab extra as a *necessary condition!*

Here, as in the steady-state world, Leibniz can get his sufficient reason for physical existence from the physics itself and would not need God. And, as I have already emphasized twice, Leibniz's *quest* for an external sufficient reason was ill grounded on the alleged spontaneity of nothingness.

It is of cardinal importance to note vis-à-vis Quinn that the causal sufficiency of the physics for energy conservation which I have claimed is licensed by the *conjunction* of the physical energy-conservation law with the physical closure of the universe, *not* by the physical closure alone. Mutatis mutandis for the stated *sub*cosmological systems for which the GTR licenses a conservation law. In short, my thesis of causal sufficiency relies on a solution to the initial value problem.

But Quinn's view of divine conservation as the *total* and *sole* cause of energy conservation turns this paramount physical process into a mere *epiphenomenon* in the spirit of Malebranche's *occasionalism*. Thereby, Quinn robs the physics of any causal role in energy conservation, just as he had made the physics causally irrelevant to the genesis of new matter in the steady-state cosmology. Yet, as I have just argued, the *physics* is, in fact, *causally sufficient* in each of the major rival physical cosmologies. And since Quinn claims to accept the physics, his demotion of it to causally ineffectual, and hence also to causally *non*explanatory factors is untenable. Moreover, if he is to be believed, a philosophically enlightened physics teacher ought to explain energy conservation to students by attributing it *solely* to divine intervention, since the physics does no causally explanatory work in Quinn's scenario.

The bizarre character of that scenario is thrown into still bolder relief, when we consider an *alternative* formulation of the energy-conservation law that is found in standard reference works, such as the *International Encyclopedia of Science*,[46] which *articulates* the statement: "*The mass-energy content of an isolated system remains constant.*" The articulation follows immediately upon it and reads: "*The energy can be converted from one form to another, but can neither be created nor destroyed.*" Hence, even if the system is open, a *change* in its energy content can occur only by the exportation or importation of energy, *not* by its creation ex nihilo or *annihilation*.

Thus, the *alternative* formulation of the energy-conservation law *applies alike* to physically open and closed systems. And, importantly, this formulation does not restrict at all the kinds of agencies or devices that are declared *unable* to create or destroy energy. Instead, it asserts the impossibility of its creation or annihilation tout court as a law of nature. Therefore, *if* the law is true *and* there is also a God, he is *not almighty*.

Furthermore, since the law declares the impossibility of the annihilation of the energy tout court, the energy *could* not *lapse into nothingness* in the absence of God. Therefore, contrary to the long theistic tradition of perpetual creation espoused by Quinn, God is clearly *not* needed to *prevent* such supposed spontaneous annihilation by creative intervention. *This is a conclusion of cardinal importance.*

Lastly, let me object to Augustine's version of creation ex nihilo. In Book XI of his *Confessions,* he considers a challenger's question: "What did God do before He made Heaven and Earth?" But Augustine *rejects* the answer of someone who replied that God was busy preparing hell for those who would ask this question! Instead, he tells us that there simply was no time before creation, because God first had to create *both* time *and* matter. As I remarked at the start, the British physicist C. J. Isham regards Augustine's reply that "time itself was made by God" as "profound."[47]

Yet I consider it very unsatisfactory. What are we to understand by Augustine's assertion that God "brings about" the existence of time itself or creates it? I submit that his claim is either unintelligible or, at best, uselessly circular and unillumi-nating. In any case, if Augustine means only that time and matter are *existentially coextensive* in the sense of a "relationalist" ontology of time, then, as I have been at pains to argue, they do not need any *external cause* or creator as the *ratio essendi* of their very existence, let alone a divine one. Furthermore, the locution "*was* made," as used in regard to the creation of *time itself,* must not be allowed to suggest that, like stars or atoms, time itself came into existence *in the course of time.* Such a notion would make *illicit* appeal to some fictitious *supertime.* Therefore, *pace* Isham, the locution *"time itself was made"* by God is senseless here.

Similar objections apply, in my view, to *Aquinas's* doctrine, reported by Quinn, "that one of two things God made in the beginning was a unique first *now* from which time began."[48]

■ 6. QUANTUM COSMOLOGIES

The so-called quantum cosmologies are quite speculative. And no self-consistent theory of quantum gravity—uniting quantum theory and general relativity—is currently available.[49] Thus, it may be premature to entrust one's philosophic for-tunes to the extant versions of quantum cosmology, let alone to invoke them as support for divine creationism.

Although it is probably the better part of wisdom to wait philosophically until the dust settles in physics, let me just suggest here why, in my view, the creation-ist cannot get support from quantum cosmology that was *unavailable,* as I have argued, from the pre-quantum Big Bang and steady-state theories. It will turn out that some of the arguments I gave against theistic creationist interpretations of the classical Case (i) and Case (ii) models carry over to the three quantum cosmolo-gies. And just like the Case (ii) models, the third quantum version does not even provide a point of application for an attempt to argue for *initial* divine creation. Nor does it lend itself to divine creatio continuans any more than the other two.

The relevant highlights of the three quantum cosmologies can be briefly described as follows:

1. The semiclassical *inflationary* Big Bang models pioneered by Alan Guth and subsequently modified by Linde, Albrecht, and Steinhardt. In Guth's version, the

model is a *modification of the Case (ii) Big Bang world of the GTR* such that (a) between 10^{-35} and 10^{-33} seconds, the expansion rate was inflationary or enormously higher than thereafter; and (b) the Big Bang universe itself originated in quantum fluctuations in *non*gravitational fields. There is a so-called true vacuum featuring quantum energy fluctuations during the first 10^{-35} seconds, which is succeeded by the so-called false vacuum of the inflationary period. In these models, Einstein's GTR field equations are used to derive the false vacuum.

During the inflationary period, *energy density* is conserved, which means that in analogy to the popping into existence of new *matter* in the old steady-state theory, additional energy pops into existence *during that period.* But it turns out that after this inflationary period, the energy value returns permanently to the status quo ante. Thus, except for the tiny inflationary period, the model exhibits such physical energy conservation as is present in the classical Case (ii) model.

Clearly, I can carry over to this semiclassical quantum model my objections to the *theological* interpretations of the Case (ii) models and of the steady-state theory.

2. A second version of quantum cosmology is furnished by the so-called wave-function models.[50] Whereas the semiclassical inflationary models quantize only *non*gravitational fields, the wave-function models quantize all fields. But, like the former, they also feature an inflationary episode. The *temporal* structure of the wave-function models is that of the Case (i) Big Bang model, but with the important difference that there is *no singularity* at the initial state $t = 0$. Thus, here there is a bona fide first state of the universe. But it cannot have an earlier cause, since there is no prior time. Nor is there any basis for thinking that its initial state has a *simultaneous* asymmetric cause supplied by divine volition. We have no empirical evidence at all for the existence of creative causes ex nihilo. The demand for such a cause of the very existence of the entire universe is inspired—as I showed in Section 4—by the *groundless* assumption of the spontaneity of nothingness. Moreover, there is no extant viable account of a criterion of asymmetric *instantaneous* causation such that divine volition would qualify under it as the creative cause of the universe. In any case, attributions of volitions to God are completely ex post facto and can be invoked unwarrantedly no matter what the facts of the world. Yet the physics of the wave-function model yields a *probability* for the existence of our world as one member of a set of alternative worlds.

Overall, my objections to a theological reading of the wave-function model can be stated by carrying over those I offered à propos of the classical Case (i) model and against divine conservation à propos of the Case (ii) Big Bang model.

3. The third set of quantum cosmologies, the *vacuum fluctuation* models, are quite distinct from the first two, although there are quantum fluctuations in the course of the careers of the other models as well. Quentin Smith[51] has lucidly outlined a series of these models, beginning with Tryon's in 1973, and including those of Brout, Englert, Gott, and others.

Their cardinal feature is that there is a preexisting background space in which our universe is embedded and that our world is a quantum fluctuation of the vacuum of this larger space. Yet our world is only one of many vacuum fluctuation worlds that emerge randomly from the embedding vacuum space. As Quentin Smith explains,[52] these models lend themselves to incorporation in Brandon Carter's theory of a World Ensemble explanation of our world, and especially of its "anthropic coincidences."

These models are of interest for various philosophic purposes. But the prior existence of their background space provides no point of application for an attempt to argue for initial divine creation. Nor do they lend themselves to divine creatio continuans, any more than any of the others we have considered.

■ 7. CONCLUSION

I conclude that, in the major cosmologies of the twentieth century, there is no scope at all for a creative role of the deity qua ratio essendi.

■ NOTES

The author thanks Allen I. Janis, Richard Gale, and Quentin Smith very much for the benefit of very helpful comments and reactions.

1. P. Quinn, "Creation, Conservation, and the Big Bang," in J. Earman et al. (eds), *Philosophical Problems of the Internal and External Worlds*, 1993, ch. 23, pp. 589–612.

2. Starting with A. Grünbaum, "The Pseudo-Problem of Creation in Physical Cosmology," *Philosophy of Science* 56 (1989), 373–94.

3. I have also argued for "The Poverty of Theistic Morality" in my contribution to K. Gavroglu et al. (eds.), *Science, Mind and Art*, 1995, pp. 203–242. A version of this essay is included in the present volume as chapter 6.

4. Grünbaum, "The Pseudo-Problem of Creation in Physical Cosmology," in J. Leslie (ed.) *Physical Cosmology & Philosophy*, 1990, pp. 92–112. This is a reprint of Grünbaum, "Pseudo-Problem," op cit. The paper was also reprinted in *Epistemologia* 12 (1989), 3–32; and *Free Inquiry* 9 (1990), 48–57. A German translation, "Die Schöpfung als Scheinproblem der physikalischen Kosmologie," appeared in H. Albert (ed.), *Wege der Vernunft*, 1991, pp. 164–191.

5. "Creation as a Pseudo-Explanation in Current Physical Cosmology," *Erkenntnis* 35 (1991), 233–254.

6. C. Isham, "Creation of the Universe as a Quantum Process," in R. J. Russell et al. (eds.), *Physics, Philosophy and Theology: A Common Quest for Understanding*, 1988, p. 387.

7. Quinn, "Creation, Conservation, and the Big Bang."

8. Cf. Grünbaum, "Pseudo-Problem of Creation," pp. 96–97.

9. M. Buckley, "Religion and Science: Paul Davies and John Paul II," *Theological Studies* 51 (1990), 310–324, here 314.

10. Pius XII, "Modern Science and the Existence of God," *Catholic Mind* 49 (1952), 188, 190.

11. Associated Press (April 24, 1992): "U.S. Scientists Find a 'Holy Grail': Ripples at Edge of the Universe," *International Herald Tribune*: p. 1.

12. F. Hoyle, "Light Element Synthesis in Planck Fireballs," *Astrophysics and Space Science* 198 (1992), 177–193.

13. B. Lovell, *The Individual and the Universe*, 1961; see also Lovell, "Reason and Faith in Cosmology" (*Ragione e Fede in Cosmologia*), *Nuovo Civilta Delle Macchine* 4(3–4) (1986), 101–108.

14. Lovell, *Individual and the Universe*, p. 117.

15. Ibid., p. 124 (my italics).

16. Lovell, "Reason and Faith in Cosmology."

17. Grünbaum, "Pseudo-Problem of Creation," 375.

18. Quinn, "Creation, Conservation and the Big Bang," p. 608 (my italics).

19. Ibid., p. 593.

20. Ibid., p. 590.

21. Ibid., pp. 591–592 (my italics).

22. Ibid., p. 597.

23. Ibid.

24. Ibid. (my italics). For an array of difficulties besetting the notion of divine volitional bringing about, see Grünbaum, "Origin Versus Creation in Physical Cosmology," in: L. Krüger and B. Falkenburg (eds.), *Physik, Philosophie und die Einheit der Wissenschaften*, 1995, pp. 221–254.

25. Quinn, "Creation, Conservation and the Big Bang," p. 602.

26. Ibid.

27. See Grünbaum, *Philosophical Problems of Space and Time*, 2nd ed., 1973, pp. 571–574.

28. Ibid., pp. 406–407.

29. See Quinn, "Creation, Conservation and the Big Bang," p. 593 (my italics), cited from Aquinas's *Summa Theologiae*, p. 511; trans. the English Dominican Fathers, 1981.

30. Besides the ones already mentioned, see especially "Pseudo-Creation of the Big Bang," *Nature* 344 (1990), pp. 821–822 (included in the present volume as ch. 10).

31. See R. Wald, *General Relativity*, 1984, pp. 99–100.

32. John Stachel, "The Meaning of General Covariance," in: J. Earman et al. (eds.), *Philosophical Problems of the Internal and External Worlds*, 1993, pp. 138–144.

33. Wald, *General Relativity*, p. 213.

34. S. Hawking and G. Ellis, *The Large-Scale Structure of Space-Time*, 1973, p. 56.

35. W. Craig, "Creation and Big Bang Cosmology," in *Philosophia Naturalis* 32 (1994), 217–224. But see my critical reply in the same issue of *Philosophia Naturalis*, pp. 225–236; and Craig's rejoinder there, pp. 237–249.

36. Grünbaum, "Narlikar's 'Creation' of the Big Bang Universe Was a Mere Origination," *Philosophy of Science* 60 (1993), pp. 638–646.

37. Wald, *General Relativity*, p. 99.

38. Ibid.

39. Quinn, "Creation, Conservation and the Big Bang," p. 598.

40. Ibid., p. 597.

41. Ibid., p. 601 (my italics).

42. Wald, *General Relativity*, p. 100, equation {5.2.19}.

43. Ibid., p. 70 n. 6.

44. Ibid., p. 269.

45. Ibid., p. 293.

46. "Conservation of Mass Energy," *International Encyclopedia of Science*, vol. 1, ed. J. R. Newman (Edinburgh: Thomas Nelson, 1965), p. 276.

47. Isham, "Creation of the Universe as a Quantum Process," p. 387.

48. Quinn, "Creation, Conservation and the Big Bang," p. 595.

49. P. Renteln, "Quantum Gravity," *American Scientist* 79 (1991), 508–527.

50. J. B. Hartle and S. W. Hawking, "Wave Function of the Universe," *Physical Review*, D28 (July–Dec.1983): 2960–2975; S. W. Hawking, "Quantum Cosmology," in: S. W. Hawking and W. Israel (eds.), *Three Hundred Years of Gravitation*, 1987, pp. 631–651; A. Vilenkin, "Creation of Universes from Nothing," *Physics Letters B* 117(1–2) (Nov. 4, 1982), 25–28; A. Vilenkin, "Birth of Inflationary Universes," *Physical Review*, D27 (June 1983), 2848–2855.

51. Q. Smith, "World Ensemble Explanations," *Pacific Philosophical Quarterly* 67 (1986), 81–84.

52. Ibid., pp. 73–86.

9 A New Critique of Theological Interpretations of Physical Cosmology

ABSTRACT

This paper is a sequel to my "Theological Misinterpretations of Current Physical Cosmology" (this volume, ch. 8). There I argued that the Big Bang models of (classical) general relativity theory, as well as the original 1948 versions of the steady-state cosmology, are each logically incompatible with the time-honored theological doctrine that perpetual divine creation (creatio continuans) is *required* in each of these two theorized worlds. Furthermore, I challenged the perennial theological doctrine that there must be a divine creative cause (as distinct from a transformative one) for the very existence of the world, a ratio essendi. This doctrine is the theistic reply to the question: "Why is there something, rather than just nothing?"

I begin my present paper by arguing against the response by the contemporary Oxford theist Richard Swinburne and by Leibniz to what is, in effect, my *counter*-question: "But why should there be just nothing, rather than something?" Their response takes the form of claiming that the a priori probability of there being just nothing, vis-à-vis the existence of alternative states, is maximal, because the nonexistence of the world is conceptually *the simplest*. On the basis of an analysis of the role of simplicity in scientific explanations, I show that this response is multiply flawed and thus provides no basis for their three contentions that (i) if there is a world at all, then its "normal," natural, spontaneous state is one of utter nothingness or total nonexistence, so that (ii) the very existence of matter, energy, and living beings constitutes a *deviation* from the allegedly "normal," spontaneous state of "nothingness," and (iii) that deviation must thus have a suitably potent (external) divine cause. Related defects turn out to vitiate the medieval Kalam Argument for the existence of God, as espoused by William Craig,

Next I argue against the contention by such theists as Richard Swinburne and Philip L. Quinn that (i) the specific content of the scientifically most fundamental laws of nature, including the constants they contain, requires supra-scientific explanation, and (ii) a satisfactory explanation is provided by the hypothesis that the God of theism *willed them* to be exactly what they are.

Furthermore, I contend that the theistic teleological gloss on the anthropic principle is incoherent and explanatorily unavailing.

Finally, I offer an array of considerations against Swinburne's attempt to show, *via Bayes's theorem,* that the existence of God is more probable than not.

222 ■ PHILOSOPHY OF 20TH-CENTURY PHYSICAL COSMOLOGIES

■ 1. INTRODUCTION

In Richard Gale's book *On the Nature and Existence of God* (1991), he devotes a very penetrating chapter (chapter 7) to a critique of cosmological arguments for the existence of God, after giving a generic characterization of all such arguments. As is well-known, there are different species of such arguments. But Gale reaches the following negative verdict on the genus:[1]

> My two arguments ... constitute ontological disproofs of the existence of the very sort of being whose existence is asserted in the conclusion of every version of the cosmo-logical argument, thereby showing that these arguments are radically defective. These ontological disproofs, however, do not pinpoint the defective spot in these arguments.

My *initial* aim in this paper is precisely to pinpoint the defects of the time-honored arguments for perpetual divine creation given by a succession of theists including Aquinas, Descartes, Leibniz, and Locke as well as by the present-day theists Richard Swinburne and Philip L. Quinn. One of these defects will also turn out to vitiate a pillar of the medieval Arabic Kalam argument for a creator.[2]

■ 2. THE NONEXISTENCE OF THE ACTUAL WORLD AS ITS PURPORTED "NATURAL" STATE

2.1 Swinburne and Leibniz on the Normalcy of Nothingness

In Richard Swinburne' s extensive writings in defense of (Christian) theism, he presents two versions of his argument for his fundamental thesis that "the most natural state of affairs of the existing world and even of God is *not* to exist *at all!*" As he put it: "It is extraordinary that there should exist anything at all. Surely the most natural state of affairs is simply nothing: no universe, no God, nothing. But there is something."[3] It will be expeditious to deal first with the more recent version of his case[4] and then with his earlier substantial articulation of Leibniz's argument from a priori simplicity.[5]

Surprisingly, Swinburne deems the existence of something or other to be "extraordinary," that is, literally out of the ordinary. To the contrary, surely, the most pervasively ordinary feature of our experience is that we are immersed in an ambiance of existence. Swinburne's initial assertion here is, at least prima facie, a case of special pleading in the service of a prior philosophical agenda. Having made that outlandish claim, Swinburne builds on it, averring that "surely the most natural state of affairs is simply nothing." Hence, he regards the cosmic existential question, "Why is there anything at all, rather than just nothing?" as paramount.

As we know, the Book of Genesis in the Old Testament starts with the assertion that, in the beginning, God created Heaven and Earth from scratch. And, as John Leslie pointed out:[6] "When modern Western philosophers have a tendency to ask it [i.e., the aforementioned existential question], possibly this is only because they are heirs to centuries of Judaeo-Christian thought." This conjecture derives added poignancy from Leslie's observation that "to the general run of Greek thinkers the mere existence of a thing [or of the world] was nothing remarkable. Only their changing patterns provoked [causal] inquisitiveness." And Leslie mentions Aristotle's views as countenancing the acceptance of "reasonless existence."

Yet there is a long history of sometimes emotion-laden, deep puzzlement, even on the part of atheists such as Heidegger, about the mere existence of our world.[7] Thus, Wittgenstein acknowledged the powerful *psychological* reality of wondering at the very existence of the world. Yet, logically, he rejected the question altogether as "nonsense," because he "cannot imagine its [the world's] not existing,"[8] by which he may perhaps have meant not only our world but, more generally, as Rescher points out, some world or other.[9] Wittgenstein could be convicted of a highly impoverished imagination, if he could not imagine the nonexistence of just our particular world.

Before turning to the logical aspects of the cosmic existential question, let me mention a psychological conjecture as to why not only theists but also some atheists find that question so pressing. For example, Heidegger deemed, "Why is there anything at all, rather than just nothing?" the most fundamental question of metaphysics.[10] Yet he offered no indication of an answer to it, and he saw its source in our facing nothingness in our existential anxiety.

I gloss this psychological hypothesis as surmising that our deeply instilled fear of death has prompted us to wonder why we exist so precariously. And we may then have extrapolated this precariousness, more or less unconsciously, to the existence of the universe as a whole.

Psychological motivations aside, let me recast Swinburne's aforecited statement, "The most natural state of affairs is simply nothing," to read instead, "The most natural state of the existing world is to *not* exist at all." This reformulation avoids the hornet's nest inherent in the question as to the sheer intelligibility of utter nothingness qua purportedly *normal state* of our world.[11]

Yet my reformulation is still conceptually troublesome: How can *non*existence at all be *coherently* a state, natural or otherwise, of the actual, existing world? Swinburne speaks vaguely of "the most natural state of affairs," leaving it unclear whether his "state of affairs" pertains only to our actual world or also to any other logically possible world that might have existed instead. But it is clear that he has in mind at *least* our actual world, in which case my reformulation of his claim is incoherent and not helpful. The stronger claim pertaining to any alternative world as well was perhaps intended by Derek Parfit, who wrote: "Why is there a Universe at all? It might have been true that nothing ever existed: no living beings, no stars, no atoms, not even space or time .... No question is more sublime than why there is a Universe: why there is anything rather than nothing."[12]

No matter whether one is considering Swinburne's original formulation or Parfit's, "It might have been true that nothing existed," it is surely epistemically appropriate to ask for the grounds on which Swinburne and Parfit, respectively, rest their assertions. Parfit does not tell us, whereas Swinburne does. Therefore, I shall scrutinize Swinburne's argument for it, and also Leibniz's.

I shall offer my own reasons for endorsing Henri Bergson's injunction as follows: We should never assume that the "natural thing" would be the existence of nothing. He rested this proscription on grounds radically different from mine, when he declared: "The presupposition that *de jure* there should be nothing, so that we must explain why *de facto* there is something, is pure illusion."[13] But Bergson's reasons for charging illusoriness are conceptual and a priori, whereas mine will turn out to be empirical.

As we know, a long theistic tradition has it that this de jure presupposition is correct *and* that there must therefore be an explanatory cause *external* to the world for its very existence; furthermore, it is argued that this external cause is an omnipotent, omnibenevolent and omniscient personal God.

But, in outline, my challenge to this reasoning will be as follows: (i) In this context, the question, "What is the *external* cause of the very existence of the universe?" is avowedly predicated on the doctrine that, in Swinburne's words, "Surely the most natural state of affairs is simply nothing"; yet (ii) as I shall argue in detail, just this doctrine is ill founded, contrary to the arguments for it offered by Leibniz and Swinburne; and (iii) therefore, the question calling for an external cause of the very existence of the world is a *nonstarter*, that is, it poses a *pseudo-problem*. By the same token, the answer that an omnipotent God is that cause will turn out to be ill founded.

What are the appropriate grounds for gleaning what is indeed the natural, spontaneous, normal state of the world in the absence of an intervening external cause? In opposition to an a priori conceptual dictum of naturalness, I have previously argued from the history of science that changing evidence makes the verdict inevitably *empirical* rather than a priori.[14] Here, a summary will have to suffice.

I welcome Swinburne's use of the phrase "natural state of affairs," which dovetails with the parlance I used, when I elaborated on the notion of "natural state" by speaking of it as the "spontaneous, externally undisturbed, or normal" state. In essence, Swinburne's claim that "the most natural state of affairs is simply nothing" had been enunciated essentially by Aquinas, Descartes, Leibniz, and a host of other theists. Hereafter, I shall designate this thesis as asserting "the spontaneity of nothingness," or "SoN" for brevity.

In my parlance, the terms "natural," "spontaneous," "normal," and "externally unperturbed" serve to characterize the historically dictated *theory-relative* behavior of physical and biological systems, when they are not subject to any *external* influences, agencies or forces. In earlier writings, I called attention to the *theory relativity* of such naturalness or spontaneity by means of several examples from physics and biology.

Thus, I pointed out[15] that the altogether "natural" behavior of suitable subsystems in the now defunct original Bondi and Gold steady-state world is as follows: Without any interference by a physical influence external to the subsystem, let alone by an external matter-*creating* agency or God, matter pops into existence spontaneously in violation of Lavoisier's matter conservation. This spontaneous popping into existence follows deductively from the conjunction of the theory's postulated matter-*density* conservation with the Hubble law of the expansion of the universe. For just that reason, I have insisted on the use of the *agency-free* term "matter accretion" to describe this process and have warned against the use of the *agency-loaded* term "matter *creation.*"

In the same vein, I emphasized that, according to Galileo and to Newton's First Law of Motion, it is technically "natural" that a *force-free* particle moves uniformly and rectilinearly, whereas Aristotle's physics asserted that a force is required as the external cause of any sublunar body's nonvertical uniform rectilinear motion. In short, Aristotle clashed with Galileo and Newton as to the "natural," spontaneous, dynamically unperturbed behavior of a body, which Aristotle deemed to be one of rest at its proper place. Thus, Galileo and Newton *eliminated* a supposed external dynamical cause on *empirical* grounds, explaining that uniform motion can occur spontaneously without such a cause.

But, if so, then the Aristotelian demand for a causal explanation of any nonvertical motion whatever by reference to an external perturbing force is predicated on a false underlying assumption. Clearly, the Aristotelians then begged the question by tenaciously continuing to ask: "What net external force, pray tell, keeps a uniformly moving body going?" Thus, scientific and philosophical questions can be anything but innocent by loading the dice with a petitio principii!

An example from biology yields the same lesson. It has been said that Louis Pasteur "disproved" the "spontaneous" generation of life from nonliving substances. Actually, he worked with sterilized materials over a cosmically minuscule time interval and showed that bacteria in an oxidizing atmosphere would not grow in these sterilized materials. From this he inferred that the natural, unperturbed behavior of *non*living substances *precludes* the spontaneous generation of living things. That was in 1862. But in 1938 Alexander I. Oparin in the then Soviet Union, and in 1952 Harold C. Urey in the United States rehabilitated the hypothesis of the spontaneous generation of life to the following effect: Life on Earth originated by spontaneous generation under favorable conditions prevailing some time between 4.5 billion years ago and the time of the earliest fossil evidence 2.7 billion years ago. I have summarized this rehabilitation as follows:

> When the earth was first formed, it had a reducing atmosphere of methane, ammonia, water, and hydrogen. Only at a later stage did photochemical splitting of water issue in an oxidizing atmosphere of carbon dioxide, nitrogen and oxygen. The action of electric discharges or of ultra-violet light on a mixture of methane, ammonia, water, and hydrogen yields simple organic compounds such as amino acids and urea, as shown by work done since 1953 [footnote omitted]. The first living organism originated by a series of non-biological steps from simple organic compounds which reacted to form structures of ever greater complexity until producing a structure that qualifies as living.[16]

Indeed, in a new book, Paul Davies has argued persuasively that progress in biology and astronomy is transforming the one-time mystery of the origin of life into a soluble problem.[17] The clash between the inferences drawn by Pasteur, on one hand, and by Oparin and Urey, on the other, provides a biological illustration of the theory dependence of the "natural," spontaneous behavior of a system, just as the theory shifts from Aristotle to Galileo and from matter-energy conservation to matter accretion provide vivid illustrations from physics. And in each case, *empirical evidence* was required to justify the avowed naturalness.

As illustrated by the ill-conceived question put to Galileo by his Aristotelian critics, it is altogether misguided to ask for an external cause of the *deviations* of a system from the pattern that an empirically discredited theory tenaciously affirms to be the "natural" one.[18]

The proponents of SoN have not offered any empirical evidence for it. Yet the lesson of the history of science appears to be that just such evidence is required. However, some of the advocates of SoN have offered an a priori conceptual argument in its defense. I now turn to their defense.

2.2 Leibniz's and Swinburne's Simplicity Argument for the Normalcy of Nothingness

The imposition of a priori notions of naturalness is sometimes of-a-piece with the imposition of tenaciously held criteria of the mode of scientific explanation required for understanding the world. The demise of Laplacean determinism in physics, and its replacement by irreducibly stochastic, statistical models of microphysical systems, is a poignant case of the empirical discreditation of a tenacious demand for the satisfaction of a previously held ideal of explanation: It emerges a posteriori that the universe just does not accommodate rigid prescriptions for explanatory causal understanding that are rendered otiose by a larger body of evidence.

Relatedly, the stochastic theory of radioactive decay in nuclear physics, for example, runs counter to Leibniz's demand for a "sufficient reason" for all logically contingent states of affairs. Thus, since the existence of our actual world is logically contingent, he insisted that there must be a sufficient reason for its existence.

2.2.1 Leibniz's Simplicity Argument

I now cite this demand from Leibniz's essay "The Principles of Nature and of Grace Based on Reason":[19]

> Sec. 7. Thus far we have spoken as simple *physicists*: now we must advance to *metaphysics*, making use of the great principle, little employed in general, which teaches that *nothing happens without a sufficient reason;* that is to say that nothing happens without it being possible for him who should *sufficiently understand* things, to give a

reason sufficient to determine why it is so and not otherwise. This principle laid down, the first question which should rightly be asked will be, *why is there something rather than nothing?*

To justify this question, he then resorts to an a priori argument from simplicity:

For *nothing is simpler and easier than something*. Further, suppose that things must exist, we must be able to give a reason *why they must exist so* and not otherwise.

And hence he concludes in section 8:

Sec. 8. Now this sufficient reason for the existence of the universe cannot be found *in the series of contingent things* ... [italics in original except for the one sentence "nothing is simpler and easier than something"].

Contextually, Leibniz implicitly enunciated SoN, when he declared that "nothing is simpler and easier than something." Having thus assumed SoN by recourse to a priori simplicity, he is in a position to reason as follows about the different states of the world:

Every subsequent state is somehow copied from the preceding one (although according to certain laws of change). No matter how far we may have gone back to earlier states, therefore, we will never discover in them a full reason why there should be a world at all, and why it would be such as it is. Even if we should imagine the world to be eternal, therefore, the reason for it would clearly have to be sought elsewhere ....[20]

As for Leibniz's claim that "nothing is *simpler* and *easier* than something," I ask: But why is this *conceptual* claim, if granted, mandatory for what is the ontologically spontaneous, externally undisturbed state of the *actual* world? Alas, Leibniz does not tell us here. Yet, as I argued in Section 2.1, according to our best scientific knowledge, spontaneity is relative to changing *empirically based* scientific theories.

Furthermore, Philip Quinn has, in effect, issued an important demurrer (private communication): Let us suppose that the purported state of nothingness is conceptually nonelusive and the simplest. Then it would still not follow from this maximum conceptual simplicity that SoN is the simplest *hypothesis* within the set of all logically possible hypotheses, a set that we do not encompass intellectually. In short, conceptual simplicity does not necessarily bespeak theoretical simplicity, as Quinn has illustrated for this context by the following example:

Suppose the purpose of a hypothesis is to explain how the observed universe is produced from a postulated initial state. Perhaps the hypothesis that postulates the simplest initial state will be forced to postulate a complex productive mechanism in order to achieve its explanatory purpose, while the same purpose can be achieved by a rival that postulates a slightly less simple initial state together with a vastly simpler productive mechanism. In such a case the hypothesis that postulates the simplest initial state will not be the hypothesis with the greatest overall simplicity (private communication again).

The moral of this sketchy history is twofold: (i) The character of just what behavior of the actual world and of its subsystems is "natural" is an empirical a posteriori matter rather than an issue that can be settled a priori; yet (ii) SoN has no *empirical* credentials at all, as acknowledged, in effect, by the purely conceptual arguments for it which have been offered by its recent defenders.

Given this empiricist moral, I must dissent from Leslie's view that beliefs about the natural state of the universe are matters of "intuition." Says he: "Intuitions about what should be viewed as a universe's 'natural state'—where this means something not calling for explanation by a divine person or any other external factor—can be defended or attacked only very controversially."[21] As I have argued, however, the naturalness or spontaneity of the states of physical and biological systems or of the cosmos is epistemologically a matter of empirical evidence and not of the conflict of personal intuitions regarding naturalness.

But the question could be and has been asked why this form of "scientism" should be mandatory. Friedrich von Hayek[22] and his acolytes have characterized scientism as a doctrine of explanatory scientific imperialism with utopian pretensions. Much more precisely, Richard Gale and Alexander Pruss[23] defined scientism as implying that everything that *is* explained is explained by either science or some kind of explanation having strong affinities to actual scientific explanation. Thus, in their construal, scientism is not taken to assert that everything is explained by science tout court but only that everything that is actually explained is explained by science.

It is easy enough, as theists like Leibniz, Swinburne, and Philip L. Quinn have done, to disavow scientism as just defined, although Swinburne insists that his version of theism is methodologically of-a-piece with various modes of scientific inference, such as the use of Bayes's theorem to credibilify scientific hypotheses. And he then marshals that theorem to aver that God probably exists.

But such a theistic disavowal of scientism calls for a potent justification of the theistic explanatory alternative. The most prominent alternative that theists have proffered is modeled on volitional agency explanations of human actions, as distinct from ordinary event causation.

Yet in Section 3 I shall argue that a divine volitional explanation of the actual topmost or most fundamental laws of nature, of their constants, and of the pertinent boundary or initial conditions founders multiply.

2.2.2 Swinburne's Simplicity Argument

With no help from Leibniz toward a cogent defense of SoN by reference to simplicity, I turn to Swinburne's quite general argument from simplicity for which he claims multiple sanction from science.[24] And I note first that he claims probity for his appeal to simplicity by maintaining that it has "the same structure" as the use of simplicity in scientific theorizing. In his words:

> The structure of a cumulative case for theism was thus, I claimed [in *The Existence of God*], the same as the structure of a cumulative case for any unobservable entity, such as a quark or a neutrino. Our grounds for believing in its existence are that it is an entity of a simple kind with simple modes of behavior which leads us to expect the more complex phenomena which we find.[25]

Furthermore, having argued that an infinite capacity is simpler than any one finite capacity, Swinburne contends that in a rank ordering of graduated properties, God's omnipotence, omniscience, and (presumed) omnibenevolence are the simplest.[26] Hereafter, I shall refer to this triad as "God's triplet of *omnis*." Swinburne puts his case for the simplicity of this triplet as follows:

> The postulation of God ... is the postulation of *one* entity of a simple kind, the simplest kind of person there could be, having no limits to his knowledge, power, and freedom.[27]

Yet Swinburne had also told us that a natural state of nothingness *without God* is simpler than a world containing God.[28] And furthermore, he deems the cardinal number zero of entities simpler than the number 1 for which he just claimed simplicity *vis-à-vis* a larger cardinal.

Occam's injunction, as symbolized by his razor, is to abstain from postulating entities beyond necessity. Mindful of this prescription, Swinburne characterizes the simplicity and complexity of hypotheses in terms of the *number* of entities, the sorts of entities, and the kinds of relations among entities that they postulate.[29] But clearly, in scientific theorizing, the regulative ideal of Occam's razor is subject to the crucial proviso of heeding the total available evidence, including its complexity.

Thus, it now turns out that there were important episodes in the history of actual science, in which increasingly greater theoretical faithfulness to the facts required the *violation* of Swinburne's a priori criterion of simplicity with respect to the number of postulated entities. Thus, such numerical simplicity as can be achieved while explaining the phenomena is an empirical matter and not subject to Swinburne' s legislative a priori *conceptual* simplicity.

Ironically, this lesson is spelled by one of his own examples. In his view, the putative infinite speed of a particle is simpler than some finite value. Yet the velocity of light is known to be finite, and the special theory of relativity tells us that this finite velocity is an upper bound for the transmission of any causal influence. Other examples from actual science that violate Swinburne's mandate of conceptual simplicity abound. Let me enumerate some of them.

(i)   By Swinburne's normative criterion of numerical simplicity, the pre-Socratic Thales's monistic universal hydro-chemistry of the world's substances is about a hundred-fold simpler than the empirically

discovered periodic table of the elements. And there are even two iso-
topes of Thales's water, one heavier than the other. Furthermore, in
organic chemistry, isomerism is a complication. Moreover, the single
frequency of monochromatic light is simple for Swinburne, but ubiquitous white light is composed of a whole range of spectral frequencies.

(ii)   Yet again, suppose that fundamental physics were reduced to the
well-known quadruplet of forces, then these four forces are numerically
less simple than a single such force, not to speak of Swinburne's much
simpler number zero of them. *Evidently, our world does not accommodate* a priori *conceptual decrees of simplicity!* Thus, it is unavailing for
Swinburne to tell us: "If there is to exist something, it seems impossible
to *conceive* of anything simpler (and therefore a priori more probable)
than the existence of God" [italics added].[30] But Swinburne has not
shown cogently that greater *conceptual* simplicity automatically makes
for greater a priori probability.

Indeed, this claim boomerangs, as Keith Parsons has pointed out as
follows:

A demon, for instance [especially Satan], is a single entity, it is a spiritual being and
hence not composed of parts; it presumably exercises its power over persons and physical objects in some direct and simple way, and it is in all its deeds actuated by a single
motivating drive—malevolence. Hence, explanation of a case of psychosis in terms of
demon[ic] possession seems much simpler than any of the current psychological or neurological explanations. The simplicity and untestability (How could it ever be shown
that demons *do not* cause psychoses?) of such hypotheses gives them great obscurantist
potential.[31]

(iii)   Among laws of nature, van der Waals's laws for gases are more complicated than the Boyle-Charles law for ideal gases. Again, in the Newtonian
two-body system of the earth and the sun, Kepler's relatively simple laws
of planetary motion are replaced by more complicated ones that take
account of the sun's own acceleration. Third, Einstein's field equations are
awesomely complicated nonlinear partial differential equations, and as
such are enormously more complicated than the ordinary second order
differential equation in Newton's law of universal gravitation. Remarkably,
Swinburne himself mentions this greater complexity of Einstein's gravitational field equations, but his comment on it does not cohere with his
demand for a priori simplicity. He says:

Newton's laws ... are (probably) explained by Einstein's field equations of General
Relativity [as special approximations under specified restrictive conditions]. In passing
from Newton's laws to Einstein's there is I believe a considerable loss of [a priori] simplicity .... But there is some considerable gain in explanatory power.[32]

Note, however, that the sacrifice of a priori simplicity for the sake of greater explanatory power is dictated by *empirical* constraints. Thus, Swinburne seems to admit, in effect, that empirical facts override his a priori simplicity qua the governing heuristic criterion of theory formation. In sum, epistemologically, all of the more complicated laws I have mentioned were of course prompted by empirical findings.

(iv)   And finally, simplicity enters into curve fitting to a finite number of data points. But just how? Clark Glymour, in effect, answers this question tellingly as follows:

> It is common practice in fitting curves to experimental data, in the absence of an established theory relating the quantities measured, to choose the "simplest" curve that will fit the data. Thus linear relations are preferred to polynomial relations of higher degree, and exponential functions of measured quantities are preferred to exponential functions of algebraic combinations of measured quantities, and so on.[33]
>
> The trouble is that it is just very implausible that scientists typically have their prior degrees of belief distributed according to any plausible simplicity ordering, and still less plausible that they would be rational to do so. I can think of very few simple relations between experimentally determined quantities that have withstood continued investigation, and often simple relations are replaced by relations that are infinitely complex: consider the fate of Kepler's laws.[34]

I presume that Glymour's remark about Kepler's three laws does not pertain just to the complication arising from the two-body problem, which I already mentioned under (iii), but a fortiori to the *ten-body* problem of the Newtonian gravitational interaction of the sun with all of the nine planets. The solutions of these equations of motion are infinitely complex in the sense that they take the form of infinite series rather than featuring a much simpler closed, finite form. Besides, Richard Feynman has pointed out that this full planetary system is "chaotic" in the technical sense of modern chaos theory: Very slight differences in the initial velocities or accelerations issue after a while in very large orbital differences.

It emerges that empirical facts as to how much or little "simplicity" there is in the world undermine Leibniz's and Swinburne's notion that the conceptual deliverances of epistemically a priori simplicity—even if they were coherent—can at all be mandatory for what is ontologically the case.

In a perceptive critical review of Swinburne's *Is There a God?* (1996), Quentin Smith[35] examines his argument that theism is the simplest hypothesis, since God is infinite, while infinity and zero are the simplest notions employed by scientists. And Swinburne's reason for claiming that a state of nothing, *excluding* God, is the most natural state of the world is likewise that such a presumed state is conceptually the simplest.

Smith points out, however, that Swinburne equivocates on three different senses of "infinity" which need to be distinguished. Briefly, Smith explains, these three senses are the following: (i) "Infinite" refers to Georg Cantor's lowest transfinite cardinal number Aleph-zero; (ii) "infinite" refers to a speed, as in an instantaneous transmission of an effect, which is familiar from Newtonian gravitational inter-action but is clearly different from the transfinite cardinal Aleph-zero; and (iii) a third sense, different from the first two, pertains to the maximum degree of a grad-uated qualitative property. In this sense, God is infinite, because he is presumed to have the maximum degree of power, knowledge and goodness.

Parsons[36] and Michael Martin[37] had offered other objections to Swinburne's notion of simplicity.

Recall Swinburne's contention that "if there is to exist something, it seems impossible to *conceive* of anything simpler (and therefore a priori more proba-ble) than the existence of God." Recall also that this claim does not heed Quinn's aforestated caveat not to slide unsupportedly from being the simplest concept to being the simplest hypothesis. Then we can see that, for Swinburne, conceptual simplicity has *ontological* significance by being legislative for what does exist. Thus, for him, conceptual simplicity is not, at least in the first instance, a methodological, pragmatic, or inductive criterion.

Accordingly, Swinburne's writings do not, I believe, bear out the following sug-gestion as a counter to me: What he really had in mind was not a criterion of absolute simplicity based on concept simplicity alone but rather an injunction to "always accept that theory which is the simplest one consistent with the data." But this reading would turn Swinburne's thesis into a rather commonplace version of Occamite methodology. Thus construed, he would then be defending the hypoth-esis that God exists as the simplest explanation of the world's existence and content consistent with all known data. Admittedly—so the suggestion runs—this retort would not save Swinburne's espousal of SoN, but it might allow him to parry a number of my animadversions.

To this I say: I doubt that his philosophical framework could compatibly incor-porate this suggestion. Besides, one basic part of that framework is the supposed divine volitional explanation of the existence of the world, and of its contents. But, as I shall argue in Section 3, that explanation fails on several counts.[38]

Evidently, Swinburne presents us with a misdepiction of the use of simplicity criteria in actual science, although he claims continuity with actual scientific the-ory construction for his conceptual standard of simplicity. Just as the lesson spelled out by scientific theoretical progress undermined his a priori conceptual avowal of SoN as a basis for external divine creation, so also his pseudo-Occamite argument for normative a priori simplicity fails. Thus, Swinburne's attempt to underwrite SoN by recourse to simplicity is abortive. But, in the absence of SoN, the logical contingency of the existence of the world does not jeopardize its existence one iota! Accordingly, the claim that a divine external cause is needed to prevent the world from lapsing into nothingness is baseless.

2.3 The Role of the Normalcy of Nothingness in the Medieval Arabic Kalam Argument

To conclude my contention that the appeal to SoN wrought philosophical mischief in several of the major theistic cosmological arguments, let me consider the medieval Arabic Kalam argument for a creator, as articulated and championed by William Craig.[39]

We shall see that, contrary to Craig's assertion, the so-called Kalam version of the cosmological argument, which he defends, is likewise predicated, though only quite tacitly and insidiously, on the baseless SoN. The Kalam argument was put forward by such medieval Arab philosophers as al-Kindi and others.

In Craig's 1994 "Response to Grünbaum on Creation and Big Bang Cosmology," he wrote:

> Grünbaum conflates three versions of the cosmological argument. The Kalam version, which I have defended, says nothing about a *causa/ratio essendi*. The Thomist version, as it comes to expression in Aquinas's *Tertia Via*, argues for a *causa essendi* on the basis of the real distinction between essence and existence in contingent things, *a distinction which disposes them to nothingness*. The Leibnizian version in no way presupposes a disposition toward nothingness in contingent things, but seeks a *ratio* for the existence of anything, even an eternal thing which has no disposition to nothingness, in a being which is metaphysically necessary .... *Thus Grünbaum's demand for evidence of the spontaneity of nothingness is not in every case a relevant demand* [italics added].[40]

But I claim that I am not guilty of any conflation of the three versions of the cosmological argument for the existence of God. In the case of Aquinas, Craig acknowledges my demand for evidence supporting SoN as "a relevant demand." But since he gives no hint as to how this demand could be met, I presume that he has no response to my argument against SoN. As for Leibniz, I have documented already that, contrary to Craig, Leibniz's cosmological version does presuppose SoN, so that Craig's denial that it does so is just incorrect. Yet I allow that Leibniz's sundry publications may not be coherent on this issue.

As for Craig's endorsement of the Kalam version, I can now show that, malgré lui, it derives its spurious plausibility from a tacit, though subtle, appeal to SoN. In his attack on my views, which was replete with red herrings, Craig claimed that the old Kalam cosmological argument justifies a creationist theological interpretation of the Big Bang world.[41] Specifically, in my paraphrase, he offers the following Kalam proposition to be metaphysically necessary: "Anything that begins to exist but does not have a transformative cause *must have* a creative cause *ex nihilo*, rather than no cause at all."[42]

Yet in the Big Bang universe, we have an unbounded interval of past time that is only metrically finite in years. This means that there is no first moment of time. Ordinally and topologically, the past-open time interval is isomorphic with a time that is metrically infinite in years. But Craig untutoredly declares an infinite past time to be logically impossible! And he speaks of the Big Bang universe as "beginning to exist" by misdepicting the Big Bang singularity as a genuine first moment of time.[43]

But why, I ask, does a Big Bang universe that "begins to exist," in the special sense that there were no instants of time preceding all of the moments in the metrically finite past, require an external creative cause at all in the absence of a transformative cause? Is it not because Craig tacitly embraces SoN and uncritically assumes that the externally uncaused, natural state of the world is one of nothingness? What else makes it *psychologically compelling* to Craig and some others that an externally uncaused physical universe is "metaphysically" impossible tout court? Would Craig's intuition of metaphysical necessity not dissipate once its tacit reliance on the baseless SoN and its misextrapolation from cases of warranted external causation are made explicit?

SoN as a source of Craig's avowed "metaphysical intuition that something cannot come out of absolutely nothing"—which is akin to the scholastic dictum ex nihilo nihil fit—seems also to be subtly present in his quasi-Leibnizian argument from the supposed potentiality of the universe to exist. That potentiality, Craig tells us, is causally but not temporally prior to the Big Bang. And he relies on it to buttress his stated metaphysical intuition as follows:

> A pure potentiality cannot actualize itself .... On the theistic hypothesis, the potentiality of the universe's existence lay in the power of God to create it. On the atheistic hypothesis, there did not even exist the potentiality for the existence of the universe. But then it seems inconceivable that the universe should become actual if there did not exist any potentiality for its existence. It seems to me therefore that a little reflection leads us to the conclusion that the origin of the universe had a cause.[44]

Here Craig is telling us that an external cause is required to effect the realization of "the [mere] potentiality of the universe's existence" and that, if the latter potentiality did not exist, "then it seems inconceivable that the universe should *become* actual." But what reason is there in the temporally unbounded big bang model for claiming that the big bang universe ever "*became* actual"? The most immediate reason seems to be the ill-founded SoN, and the question-begging supposition that "the potentiality of the universe's existence lay in the power of God to create it," a potentiality, which then required divine creation to be actualized.

Yet Craig insists that since the singularity of the Big Bang model avowedly had no earlier cause, it must have had a simultaneous one, because it is metaphysically impossible that it be uncaused or "come out of absolutely nothing." And he charges me with having overlooked this "obvious alternative" of a simultaneous cause, claiming that the Big Bang singularity and its purported divine cause "both occur coincidentally (in the literal sense of the word), that is, they both occur at t_o."[45] But surely the *temporal* coincidence of events is not tantamount to *literal* coincidence. And, as is well known to physical cosmologists, if t_o is used as a label for the singularity, it does *not* designate a bona fide instant of physical time.[46] Instead, the term "the Big Bang" is short, in this instance, for the behavior of the universe during its unbounded early temporal past.

Now I must ask anew: What, other than the insidious SoN, could make *psychologically compelling* Craig's avowed "metaphysical intuition that something cannot [spontaneously] come out of absolutely nothing"?[47] I answer: Once we abandon his misleading language of "coming *out of* nothing," we can describe the situation as follows: The Big Bang models feature a world whose past time is unbounded (open) but metrically finite in years. Absent the tacit presupposition of the baseless SoN, there is just no cogent reason for requiring an external creative cause for the existence of that world! We must be ever mindful to extirpate the baseless SoN from our cognitive (unconscious) awareness.

John Earman, when presumably speaking of the Kalam argument, writes:

> A seemingly more sophisticated but not essentially different response is that something cannot begin to exist without a cause so if there is no physical cause of the beginning to exist, there must be a metaphysical one. Here I am in complete agreement with Professor Grünbaum in that the standard big bang models … imply that for every time *t* there is a prior *t'* and that the state at *t'* is a cause (in the sense of causal determinism) of the state at *t* (fn. 7 omitted here).[48]

Samuel Clarke, Leibniz, and, in their wake, other philosophers have asked for an explanation of the existence of this *set* of states *as whole*, and indeed of the conjunction of all facts,[49] over and above the explanation of each individual state *t* by some prior state or other *t'*. But why is it thought that the entire series of states requires an *external* cause, instead of being a fundamental, logically contingent brute fact? If the Clarkians envision divine volition as providing the explanation they demand,[50] then I argue, as I am about to do, that such a theological explanation fails multiply. Besides, Quentin Smith, in a perceptive paper "Internal and External Causal Explanations of the Universe" (1995), contended that contemporary discussions of the Clarke and Leibniz challenge "are vitiated by an inadequate understanding of the relation between a cause external to a whole and the whole itself (in the broad sense of 'whole' that includes sets, mereological sums, aggregates and organic unities)."[51] Smith argues that, regardless of what kind of whole the universe may be, it cannot be *externally* caused by the God of classical theism, who supposedly created it ex nihilo.

I conclude from the foregoing that Craig has failed to show cogently that the universe ever "became actual" at the phantom time t_o, let alone that the atheist, anticreationist position is damaged by not countenancing a corresponding potentiality.

■ 3. CRITIQUE OF THE "EXPLANATION" OF THE MOST FUNDAMENTAL LAWS OF NATURE BY DIVINE CREATIVE VOLITION

Philip Quinn wrote:

The conservation law for matter-energy is logically contingent. So if it is true, the question of why it holds rather than not doing so arises. If it is a fundamental law and only scientific explanation is allowed, the fact that matter-energy is conserved is an inexplicable brute fact. For all we know, the conservation law for matter-energy may turn out to be a derived law and so deducible from some deeper principle of symmetry or invariance. But if this is the case, the same question can be asked about this deeper principle because it too will be logically contingent. If it is fundamental and only scientific explanation is allowed, then the fact that it holds is scientifically inexplicable. Either the regress of explanation terminates in a most fundamental law or it does not. If there is a deepest law, it will be logically contingent, and so the fact that it holds rather than not doing so will be a brute fact. If the regress does not terminate, then for every law in the infinite hierarchy there is a deeper law from which it can be deduced. In this case, however, the whole hierarchy will be logically contingent, and so the question of why it holds rather than some other hierarchy will arise. So if only scientific explanation is allowed, the fact that this particular infinite hierarchy of contingent laws holds will be a brute inexplicable fact. Therefore, on the assumption that scientific laws are logically contingent and are explained by being deduced from other laws, there are bound to be inexplicable brute facts if only scientific explanation is allowed.

There are, then, genuine explanatory problems too big, so to speak, for science to solve. If the theistic doctrine of creation and conservation is true, these problems have solutions in terms of agent-causation. *The reason why there is a certain amount of matter-energy and not some other amount*[52] *or none at all is that God so wills it, and the explanation of why matter-energy is conserved is that God conserves it.* Obviously nothing I have said proves that the theistic solutions to these problems are correct. I have *not* shown that it is *not* an *inexplicable brute fact* that a certain amount of matter-energy exists and is conserved. For all I have said, the explanatory problems I have been discussing are insoluble. But an insoluble problem is not a pseudoproblem; it is a genuine problem that has no solution. So Grünbaum's claim that creation is a pseudoproblem for big bang cosmogonic models misses the mark.[53]

In the same vein, Quinn[54] cites Swinburne's book *The Existence of God.*[55] Speaking of the laws of nature L, Swinburne declared: "L would explain why whatever energy there is remains the same; but what L does not explain is why there is just this amount of energy."

My response is twofold: (i) I contend that Quinn offers a non sequitur in his conclusion: "So Grünbaum's claim that [the problem of] creation is a pseudoproblem for big bang cosmogonic models misses the mark"; and (ii) the theistic volitional explanations for the existence and nomic structure of the world championed by Quinn and Swinburne are inherently defective.

(i)   In Quinn's argument for his complaint that I had leveled an unsound charge of pseudo-problem, he conflates two different problems, only one of which I had indicted as a pseudo-issue. In a passage that he himself[56] had adduced from Leibniz, that philosopher had lucidly stated the pertinent two distinct questions when he demanded "a full reason why there should be a world at all, and why it should be such as it is." Quinn reasoned fallaciously[57] that if the latter question is a "genuine explanatory problem" even when addressed to the most fundamental laws and facts of nature—as he claims—then so also the former question, "Why is there a world at all?" must be genuine. But in my complaint of pseudo-problem, I had targeted only the question, "What is the *external cause* of the very existence of the universe?" It is *this* problem that is at issue when Quinn speaks of my dismissal of "the problem of creation."

I had rejected it as misbegotten, because it is avowedly or tacitly predicated on the SoN doctrine, a tenet that I have been at pains to discredit as ill based. And I favor the use of the pejorative term "pseudo-problem" to derogate a question that rests on an ill-founded or demonstrably false presupposition, yet, in so doing, I definitely do not intend to hark back to early positivist indictments of "meaninglessness."

Assuming that the most fundamental laws and facts of the world are logically contingent, one can clearly allow the question why they are what they are, as contrasted with logically possible alternatives to them, even as one rejects the different existential question, "Why is there anything at all, rather than just nothing?" Thus, when I indicted the latter as a pseudo-problem, I did not thereby disallow the former. Yet Quinn reasons illicitly that since the former question is genuine, my rejection of the latter "misses the mark."

In the opening paragraph of his lengthy critique of my rejection of theological interpretations of physical cosmology, Quinn[58] had declared that the aim of his critique is to refute my charge of pseudo-problem. I claim that he failed, when he conflated Leibniz's two questions.

(ii)   This brings me to the theological answer given by Swinburne and Quinn to their question why the nomological structure and content of the world are what they are.

To set the stage for my array of animadversions against their divine volitional answer, let me mention Richard Gale's view that "… ultimate disagreements between philosophers are due to their rival sentiments of *rationality* as to what constitutes a rationally satisfying explanation of reality."[59] Two such rival views of rationality, Gale points out, are the scientific worldview, on one hand, and the man-centered one, which employs anthropomorphisms, on the other. Theistic advocates of "natural religion" champion the anthropomorphic perspective of personhood in their proffered explanations of everything via divine creative volition, a standpoint rejected by Santayana, Bertrand Russell, and a host of others.

Let me defer, for now, adjudicating the merits of these two competing worldviews and first set forth some fundamental epistemological and methodological differences between them. These deep differences exist, despite Swinburne's claim of solid methodological continuity between the two worldviews and their respective criteria of rationality. Says he: "The very same criteria which scientists use to reach their own theories lead us to move beyond those theories to a creator God who sustains everything in existence."[60] Moreover, he asserts theistic *pan*-explainability, declaring, "... using those same [scientific] criteria, we find that the view that there is a God explains *everything* we observe, not just some narrow range of data. It explains the fact that there is a universe at all [via SoN], that scientific laws operate within it."[61]

Note, however, that Swinburne and others who offer divine volitional explanations would offer precisely such an explanation, if the facts of our world were radically different, or even if, in a putative world ensemble of universes, each of them had its own laws, vastly different from the respective laws in the others. Their schema of theistic volitional explanations relies on roughly a model of intentional action affine to Aristotle's practical syllogism (hereafter "PS") for intentional action.

As we just saw, Swinburne maintains that the hypothesis of divine creation "moves beyond" scientific explanations via the very same epistemological criteria. As against that contention, let me now set forth the substantial explanatory discrepancies between them.

Neither Swinburne nor Quinn spelled out the provision of a deductive theistic volitional explanation, which they claim for the hypothesis of divine creation. I now offer a reconstruction of essentially the deductive explanatory reasoning that, I believe, they had in mind. And I am glad to report that Quinn (private communication) authenticated my reconstruction, at least in regard to himself. It reads:

Premise 1:
God freely willed that the state of affairs described in the *explanandum* ought to materialize.

Premise 2:
Being omnipotent, he was able to cause the existence of the facts in the *explanandum* without the mediation of other causal processes.

Conclusion:
Our world exists, and its contents exhibit its most fundamental laws.

It is to be understood that this reconstruction is only schematic, since the actual specifics of the most basic laws are not stated in the Conclusion.

But two basic considerations jeopardize the epistemic viability of this proffered volitional theological explanation:

(i) Epistemically, it will succeed *only if* the theist can produce cogent evidence, *independent of the* explanandum, for the very content of the volition that the proffered explanation imputes to the Deity; failing that, the deductive argument here is not viable epistemically. But where has the theist produced such independent evidence? Moreover, Premise 2 unwarrantedly assumes the availability of a successful cosmological argument for the existence of the God of theism.

(ii) Relatedly, the explanation is conspicuously ex post facto, because the content of the volition imputed to God is determined retrospectively, depending entirely on what the specifics of the most fundamental laws have turned out to be.

William James has beautifully encapsulated the ex post facto character of the relevant sort of theological explanation, in which God is Hegel's Absolute. Speaking of the facts of the world, James declared:

> Be they what they may, the Absolute will father them. Like the sick lion in Esop's fable, all footprints lead into his den, but *nulla vestigia retrosum* [i.e., no trace leads back out of the den]. You cannot redescend into the world of particulars by the Absolute's aid, or deduce any necessary consequences of detail important for your life from your idea of his nature.[62]

But no such ex post facto deficiency is found in typical explanations in physics or biology such as (i) the Newtonian gravitational explanation of the orbit of the moon; (ii) the deductive-nomological explanations of optical phenomena furnished by Maxwell's equations, which govern the electromagnetic field, or in a statistical context; and (iii) the genetic explanations of hereditary phenotypic human family resemblances.

But according to familiar scientific evidential criteria which Swinburne tirelessly professes to employ—as in his appeal to Bayes's theorem—his and Quinn's deductive argument appears to be *epistemologically frivolous* by being altogether ex post facto. Nonetheless, Swinburne, speaking of the explanandum e, is explicitly satisfied with such an ex post facto mode of explanation: "... clearly whatever e is, God being omnipotent, has the power to bring about e. He will do so, if he chooses to do so."[63] Yet since we obviously have no independent evidential access to God's choices, Swinburne has to *infer* completely ex post facto whether God's choices included e from whether or not e is actually the case.

In ordinary action explanations on the model of Aristotle's PS, we often, if not typically, do have independent evidence—or at least access to independent evidence—as to the content of the agent's motives. That is to say, we have evidence for the imputed motives *other than* the action taken by the agent. And, absent such evidence, we reject the proffered action explanation as viciously circular. Thus, if

240 PHILOSOPHY OF 20TH-CENTURY PHYSICAL COSMOLOGIES

I explain an unreasonable reprimand of an academic colleague by the department chairman as vindictive, I may do so on the basis of independent evidence that the chairman harbors aggressive feelings toward the colleague, because of the colleague's repeated expressions of disrespect for the chairman. By the same token, we reject imputations of motives for given actions as facile, when there is no independent evidence for the agent's possession of the attributed motives.

In this sense, I claim that the attribution of the existence of the world to God's willing that it exist is unacceptably ex post facto. Let me emphasize that, as I lodge it, the complaint that ex post facto explanations afford no independent empirical check on their premises is *not* focused on their being nonpredictive. Nonpredictiveness is not tantamount to untestability, since a theory may be retrodictive without being predictive. For example, (neo)-Darwinian evolutionary theory is essentially unpredictive in regard to the long-term evolution of species, but it retrodicts numerous previously unknown past facts in the fossil record. Similarly, being an untreated syphilitic is not predictive of affliction with neurologically degenerative paresis. Yet, since only untreated syphilitics become paretics, being paretic testably *retrodicts* having been an untreated syphilitic. But an explanation that is neither retrodictive nor predictive and whose premises have no corroboration by evidence independent of the given explanandum is paradigmatically ex post facto.

There are further difficulties in the previous theistic explanation: God's omnipotence will now serve to show that Swinburne's and Quinn's deductive volitional explanation does not meet Leibniz's aforecited demand for a "full reason why," if there is a world at all, "it should be such as it is." Having harped on God's omnipotence earlier, Swinburne develops it further:

> God, being omnipotent, cannot rely on causal processes outside his control to bring about effects, so his range of easy control must coincide with his range of direct control and *include all states of affairs which it is logically possible for him to bring about.*[64]

Precisely because God is omnipotent, however, he could clearly have chosen any one of the logically possible sets of fundamental laws to achieve his presumed aims—goals that are outlined by Swinburne—rather than the actual laws. Yet, if so, then exactly that latitude shows that, if the stated epistemic defects are to be avoided, the theological explanatory scenario fails to satisfy Leibniz's demand. Swinburne himself concedes that the theistic explanation is wanting: "It is compatible with too much. There are too many different possible worlds which a God might bring about."[65] Thus, God's supposed choice to create the actual world is presumably a matter of brute fact.

How, then, does Swinburne justify that his previous theological explanation improves upon a scientific system in which explanation is envisioned as departing from the most fundamental laws of nature, which are themselves taken to hold as a matter of brute fact? In the face of the epistemic flaws I have set forth, Swinburne's and Quinn's theological superstructure appears to be an explanatorily misguided step.

Furthermore, the details of the proposed theological explanation, employing the aforementioned modified version of the PS, is beset by difficulties of its own. Swinburne opines that "... although certain physical conditions of the brain need to occur if human agents are to have intentions which are efficacious, the human model suggests a simpler model in which such limitations are removed."[66] Leaving aside his hapless a priori simplicity, let me recall that he and Quinn rely on direct unmediated divine volitional creation of the world ex nihilo.[67] Yet Quinn cautions us: "I leave open the question of whether God and his volitions are timelessly eternal by not building into this locution [of direct bringing about] a variable ranging over times of occurrence of divine willings."[68] On the other hand, Swinburne dissociates himself from the notion of divine timeless eternity, and says: "I understand by God's being eternal that he always has existed and always will exist."[69]

Like many others, I find it unintelligible to be told by Quinn that any mental state, especially a volitional one that creates ex nihilo, can be "timelessly eternal." A fortiori it defies comprehension how such a timeless state can "bring about" a state of existence at any one ordinary *worldly* time. The reply that this complex event happens in *metaphysical* time merely mystifies further an already conceptually elusive situation.

Thus, the best I can do to make the supposed creative process intelligible is to construe their direct divine causation as taking the following form: God is in the injunctive mental state "let there be the existing world," including the biblical "let there be light." And this mental state instantaneously causes the world to exist.

But in all of our ordinary and scientific reasoning, it would be regarded as magical thinking to suppose that any *mere thought* could bring about the actual existence of the thought object, let alone out of nothing. Hence I can only welcome the assertion of the Jesuit theologian Michael Buckley that, as for divine volitional creation, "We really do not know how God 'pulls it off.'"[70] But then Buckley continues in an apologetic mode: "Catholicism has found no great scandal in this admitted ignorance." While I accept this account of the attitude of the exponents of Catholic doctrine, I regard the admitted explanatory gaping lacuna as a "scandal" of unintelligibility.

As a further cardinal methodological difference between scientific and creationist reasoning, I now consider Quinn's attempt to reconcile the following two claims: (i) the theistic doctrine that divine perpetual creation (or recreation) is causally necessary to prevent the universe from lapsing into nothing at any one moment, on one hand, and (ii) the assertion of the scientific mass–energy conservation law, on the other. I had argued for their logical incompatibility.[71]

Yet in discussing this issue of logical compatibility, we must be mindful of the fact that the technical scientific concept of energy is highly *theory relative* by depending on the energy ontology of the particular pertinent scientific theory.

And, furthermore, we must heed the caveat emphasized by John Leslie[72] that the application of the energy concept to the entire universe is quite problematic in contemporary physical theory, notably in Big Bang and quantum cosmology, a caveat not honored by either Swinburne or Quinn.

How, then, does Quinn reason specifically that he can reconcile the supposed causal necessity of divine underwriting of the physical conservation law with its formulation in physics or chemistry textbooks and in scientific encyclopedias? One such encyclopedia reads: "The mass-energy content of an isolated system remains constant. The energy can be converted from one form to another, but can neither be created nor destroyed."[73] Note importantly that this statement asserts the impossibility of either the creation or annihilation of energy tout court as a law of nature. Furthermore, since the law declares the impossibility of the annihilation of energy, the energy *could not lapse into nothingness* anyway in the absence of God's supposed ontological support. Therefore, contrary to the long theistic tradition of perpetual creation espoused by Quinn, God is clearly *not* needed to *prevent* such supposed spontaneous annihilation by creative intervention. *This is a conclusion of cardinal importance.*[74]

Quinn had laid the groundwork for his response, when he maintained that God is the only and the total cause of energy conservation. Then he replied to my charge of incompatibility as follows:

> Grünbaum has argued that what I have said so far does not get to the heart of the matter. His key thesis, he says, is that the mere physical closure of a system is causally sufficient for the conservation of its matter-energy because the conservation of matter-energy is a matter of natural law. However, this thesis rests on an understanding of the conservation law that theists of the sort I have been discussing would reject. They would insist that the sum total of matter-energy in a physically closed system remains constant from moment to moment only if God acts to conserve it from moment to moment. Because they hold that divine conserving activity is causally necessary for the conservation of matter-energy even in physically closed systems, such theists would deny that the mere physical closure of a system is causally sufficient for the conservation of its matter-energy. So they would take the true conservation law to contain an implicit *ceteris paribus* clause about God's will. When spelled out in full detail, the law is to be understood as implying that if a system is physically closed, then the sum total of matter-energy in it remains constant if and only if God wills to conserve that sum total of matter-energy.
>
> It is important to realize that it does not lie within the competence of empirical science to determine whether Grünbaum's understanding of the conservation law is rationally preferable to the theistic understanding I have sketched. The empirical methods of science could not succeed in showing that divine activity does not conserve the matter-energy in physically closed systems.[75]

I have several critical comments on this passage. And I use the term "energy conservation" as short for mass–energy conservation.

(i)   Recall that the doctrine that perpetual divine creation is causally neces-
        sary for physical energy conservation is tacitly predicated on SoN, which
        I have shown to be baseless.

(ii)   Quinn, though not Swinburne, told us that physical energy conservation
        is only an epiphenomenon, presumably in the sense of Malebranche's
        occasionalism.[76] And he rests that thesis on the purported causal exclu-
        sivity and totality of God's conservationist role. Thus, he lays down the
        proviso that the physical conservation law holds "if and only if God wills
        to conserve that sum total of matter-energy." This contention, however,
        makes that physical law totally irrelevant, both causally and *explanato-
        rily*, to physical energy conservation.

(iii)   Oddly, Quinn peremptorily shifts the probative burden of evidence from
        his own shoulders to mine. In lieu of himself giving *positive* evidence for
        his theistic proviso, which he ought to have done—over and above tell-
        ing us that his theistic confrères avow it—he illegitimately calls on me to
        refute it. But surely it is not incumbent on me to disprove such a wholly
        untestable proviso any more than Quinn is required to confute a *rival*
        proviso asserting that only Satan's intervention preserves the total energy
        with the help of poltergeists who refrain from making noises!

(iv)   Having thus improperly shifted the burden of proof from himself to me,
        Quinn concludes that "it does not lie within the competence of empiri-
        cal science to determine whether Grünbaum's understanding of the con-
        servation law is rationally preferable to the theistic understanding." But
        since Quinn did not justify his proviso by evidence, on what grounds, I
        ask, does *he* feel rationally entitled to espouse it?

(v)   Last, but by no means least, I argue that Quinn's imposition of his pro-
        viso, *in epistemic effect*, turns the physical conservation law into an empty
        tautology. The proviso amounts to the claim of creatio continuans that
        the physical conservation law is true only if God does not make it false by
        "suspending" it. And this claim, in turn, is tantamount to the following
        conjunction: (a) The *tautology* that the law is true only if it is not false;
        and (b) The law will cease to hold if God "suspends" it at some time.

Yet plainly, we have no independent evidential access to whether and when God
decides to discontinue his supposed ontological support for the physical law, so
that God's "suspension" in (b) is epistemically otiose. Evidentially, God's "suspension" amounts to the empirical falsity of the law. Thus, *epistemologically*, Quinn's
avowal reduces to the *tautological* averral that the conservation law holds except
when it doesn't, which is the empirically empty logical truth that the law is either
true or false!

What physicist, I ask, would *nowadays* countenance such an epistemic trivialization of the law, be he or she religious or not? And, since the physical law asserts
tout court that energy can neither be created nor destroyed, why should everyone

else *not* regard Quinn's imposition of the theistic proviso on the physical law as a mere ad hoc rescuing maneuver needed to meet my charge of incompatibility?

(vi) Let us recall the challenge I issued to the intelligibility of the process of instantaneous divine creation; and also the cognate admission by the Jesuit Michael Buckley that we really do not know how God "pulls it off." Then one further moral of the epistemic demise of Quinn's proviso is that, contrary to his and Swinburne's contention, their version of divine volitional explanation provides no epistemically viable account of why the physical energy conservation law holds, let alone of why the magnitude of the total energy is what it is, presumably to within a choice of units and of a zero of energy.

As for the putative divine "suspension" of one or more of the laws of nature, Swinburne has offered criteria for a divine miracle qua bona fide "Violation of a Law of Nature,"[77] which we certainly don't need to discuss here any further. Instead, we can turn to Swinburne's argument from religious experience of a transcendent God endowed with the triad of "omnis."[78] And here, again, I confine myself to just deeming it impossible that the human cognitive apparatus could ever experience such a God. How, for example, can a human being possibly receive cognizance of an omnipotent transcendent being via perceptual or nonperceptual direct religious experience? Surely this is a matter of daredevil *inference* rather than of direct human experience. Hence I consider the reports of such alleged transcendent experiences to be fundamental epistemic misinterpretations of them, whatever else they might deliver.[79]

I have provided an only occasionally evaluative account of some of the fundamental differences between the scientific and the anthropomorphic worldviews sketched previously by Richard Gale. Note that theism relies on the anthropomorphism of attributing personhood to God, and it is *anthropocentric* as well, by holding that God chose a world in which human beings would play an important role. Let me now provide my promised appraisal of the rivalry between these competing worldviews.

Gale (personal communication) has expressed the following view concerning that rivalry:

> That a belief results from wish fulfillment and lacks evidential support is not alone a reason for charging it with being irrational. For our belief that our senses and memories are in general reliable results from wish fulfillment and has no non-circular evidential support, but is not irrational for that reason.

To this, I respond in several ways:

1. It is both unreasonable and utopian to demand of an epistemic system that it provide an *internal* justification of its fundamental epistemic points of departure, if such there be. As Gale surely knows, such foundationalism has been ably challenged. In any case, just as every non-a

priori explanatory system has to start from some unexplained explainers, so also any epistemic system must, in effect, contain some internally unjustified precepts as to what is to count as evidence. No epistemic system can completely pick itself up by its own bootstraps.

2. I see no grounds for Gale's claim that our epistemic dependence on sensory experience and memories is, in the first instance, wish driven. Instead, such epistemic dependence is precisely the mechanism of our very existence and biological survival. And within our ordinary and scientific epistemic systems, both our sensory perceptions and memories are relentlessly *self-correcting.* By the same token, they tell us that wishful thinking more often than not leads to painful frustration and disappointment or even disaster, especially in some psychoses. Thus, Freud observed: "Experience teaches us that the world is not a nursery."[80] For just such reasons, Donald Davidson characterized wishful thinking as "a model for the simplest kind of irrationality."[81]

3. The issue of rationality posed by Quinn apropos his theistic proviso, and surely also by Swinburne's invocation of scientific evidence in his reliance on Bayes's theorem, is set within a framework that completely bypasses Gale's demand for an epistemic system's self-justification: The appeal to self-correcting sensory experience, including memory, is *common ground* between these theists and myself in our debate on rational preferability. Therefore, I believe that this commonality and my other foregoing responses to Gale as well as to Quinn spell the following moral: I am fully entitled to conclude, contra Quinn, that my standard scientific construal of the conservation law, without his epistemic trivialization of it by his theistic proviso, is rationally preferable to Quinn's.

Finally, the theistic explanatory scenario for our world is abortive, because it is ethically incoherent: As Hume has emphasized, no omnibenevolent and omnipotent God would ever create a world with so overwhelmingly much gratuitous and uncompensated natural evil such as cancer, evil that is not due to human decisions and actions. In particular, evil comprises both moral and natural evil. Thus, even if God could be exonerated from moral evil via the so-called free will defense, the strong challenge to God's omnibenevolence from natural evil remains. This egregious difficulty is attested by wide agreement, even among theists, that no extant theodicy has succeeded in neutralizing it.

True enough, Swinburne offered his own theodicy. But Quentin Smith has discredited his earlier efforts,[82] while Gale undermined Swinburne's later ones.[83] For my part, I claim that Swinburne sees the world through rose-colored glasses, preparatory to enlisting this glowing view of the world in the service of his theistic agenda. Thus, he opined:

> The world contains much evil, but the evil is not endless and it is either evil brought
> about by men, or evil of a kind which is necessary if men are to have knowledge of

the evil consequences of possible actions (without that knowledge being given in ways which will curtail their freedom), and which provides the other benefits described.[84]

But this apologetic scenario does not meet the more inclusive challenge that David Hume issued through the interlocutor Philo in his *Dialogues Concerning Natural Religion*. As we shall see in Section 5, Swinburne uses Bayes's theorem from the probability calculus to claim that the existence of God is more probable than not. Yet Swinburne makes no mention of Wesley Salmon's telling 1978 paper "Religion and Science: A New Look at Hume's *Dialogues*."[85] In that article, Salmon endorsed Philo's stance as part of casting Hume's [Philo's] case into Bayesian form. A year later, Salmon further articulated his pro-Humean stance.[86]

■ 4. THE "ANTHROPIC PRINCIPLE"

4.1 The Scientific Status of the Anthropic Principle

The so-called weak anthropic principle (WAP) has been construed in a number of naturalistic, nontheological ways, whose nub is the following: *Given the currently postulated laws of nature, very sensitive physical conditions, going back to the earliest stages of a big bang universe, are causally necessary for the cosmic evolution and existence of carbon-based humanoid life*. And these very delicate initial or boundary conditions are a priori exceedingly improbable, where the a priori probabilities are presumably defined on the set of all logically possible values of the pertinent physical conditions, which include the physical constants in the laws of nature. Various authors speak of these sensitive initial conditions as being "fine-tuned" for life. But John Leslie, an exponent of the design interpretation of the "fine-tuning," issued the disclaimer "talk of 'fine tuning' does not presuppose that a divine Fine Tuner, or Neoplatonism's more abstract God, must be responsible."[87] And the Roman Catholic theist Ernan McMullin rightly cautions: "*Fine tuning* has something of the ambiguity of the term *creation*; if it be understood as an action, then the existence of a 'fine tuner' seems to follow. Perhaps a more neutral term would be better."[88] Indeed, for just that reason, I suggest the use of the term "biocritical values" in lieu of the locution "fine-tunings." The following scientific statements support my point.

In an admirably thorough and cogent article, John Earman gave an account and appraisal of WAP, but without scrutiny of a theistic design construal of it, and he concludes in part:

> The litany of the many ways the universe is fine tuned for life falls into two parts. First, for example, a tiny change in the strong nuclear force would mean the absence of complex chemical elements needed for life .... Second, for example, a change in the energy density at Planck time by an amount as small as 1 in 10^{-5} as compared with the critical density (corresponding to a flat universe) would mean either that the universe would have been closed and would have recollapsed millions of years ago or else that it would

have been open with a presently negligible energy density. The second category does not call for an attitude of agog wonder-at-it-all. Rather, it points to a potential defect, in the form of a lack of robustness of explanation, of the standard hot big bang scenario, a defect which the new inflationary scenario promises to overcome by showing how exponential expansion in the early universe can turn fairly arbitrary initial conditions into the presently observed state .... Nor is it evident that puzzlement is the appropriate reaction to the first category. A mild form of satire may be the appropriate antidote. Imagine, if you will, the wonderment of a species of mud worms who discover that if the constant of thermometric conductivity of mud were different by a small percentage they ould not be able to survive.

Even if puzzlement as to the fine tuning of constants is appropriate, it does not follow that we must look for enlightenment either to Design or to worlds-within-worlds ....

Insofar as the various anthropic principles are directed at the evidentiary evaluation of cosmological theories they are usually interpretable in terms of wholly sensible ideas, but the ideas embody nothing new, being corollaries of any adequate account of confirmation. And insofar as anthropic principles are directed at promoting man or consciousness to a starring role in the functioning of the universe, they fail, for either the promotion turns out to be an empty tease or else it rests on woolly and ill-founded speculations.[89]

To give pause to a teleological construal of WAP, it is well to remember that the universe is not particularly hospitable to humans. To emphasize this restricted hospitality, T. Schick quotes the renowned attorney Clarence Darrow as follows:

Admitting that the earth is a fit place for life, and certainly every place in the universe where life exists is fitted for life, then what sort of life was this planet designed to support? There are some millions of different species of animals on this earth, and one-half of these are insects. In numbers, and perhaps in other ways, man is in a great minority. If the land of the earth was made for life, it seems as if it was intended for insect life, which can exist anywhere. If no other available place can be found they can live by the million on man, and inside of him. They generally succeed in destroying his life, and if they have a chance, wind up by eating his body.[90]

And in a very fine 1997 paper on "The Lessons of the Anthropic Principle," John Worrall wrote:

It is notoriously easy to play the coincidence game. Coincidences—"massively improbable" events can be created at will—especially if you are happy not to think too sharply about what you treat as the random variables underlying the probabilities.[91]

Even in cases where probability talk is justified, it's just plain silly to say that some event has such a low probability of happening that it "cannot have happened by chance"—chance is what governs both high and low probability events: low probability events tend to occur less often.

But if the best of all possible worlds would be one in which this perceived intolerable coincidence had been explained scientifically—that is, shown to be the deductive consequence of a further deeper theory that had independently testable empirical consequences that turned out to be correct, then surely the worst of all possible worlds is the one in which, by insisting that some feature of the universe cannot just be accepted as "brute fact," we cover up our inability to achieve any deeper, testable description in some sort of pseudo-explanation—appealing without any independent warrant to alleged a priori considerations or to designers, creators and the rest.[92]

4.2 Critique of the Theistic Design Interpretation of the Weak Anthropic Principle

In his *Dialogues on Natural Religion*, David Hume famously warned against sliding from an ordered universe to one featuring theistic design. In this subsection, I shall examine critically the theistic design interpretation of WAP, with which Earman did not deal and which Worrall dismissed.

Recall (from Section 3) that divine omnipotence has the defect of failing to explain the explanandum by allowing too much. So also, I shall now argue, Swinburne and Leslie undermine their teleological construal of WAP by invoking the omnipotent God's design to explain the a priori very improbable biocritical values ("fine-tuning"). It is crucial to note that, as a point of departure of their theistic argument, both Swinburne and McMullin take the laws of nature as given and fixed, no less than philosophical naturalists do. Thus, Swinburne wrote: "Given the actual laws of nature or laws at all similar thereto, boundary conditions will have to lie within a narrow range of the present conditions if intelligent life is to evolve."[93] Elsewhere, Swinburne had made the same claim.[94] Moreover, he averred: "The peculiar values of the constants of laws and variables of initial conditions are substantial evidence for the existence of God, *which alone can give a plausible explanation of why they are as they are*."[95] And divine action supposedly does so by teleologically transforming the a priori very low probability of the biocritical conditions into probable ones.

Clearly, Swinburne operates with the assumption that the laws of nature are *given* as fixed. And he does so twice: (i) in his initial inductive inference of divine teleology from the biocritical values, whose critical role is predicated on a given set of laws; and (ii) in the theistic teleological explanation of why these values "are as they are." And my challenge will be that this explanation does not cohere with divine omnipotence, which negates the fixity of the laws. Relatedly, though far more cautiously than Swinburne, McMullin wrote:

> According to the Biblical Account of creation, God chose a world in which human beings would play an important role, and would thus have been committed to whatever else was necessary in order for this sort of universe to come about. If fine tuning was needed this would present no problem to the Creator.[96]

Although McMullin is careful to speak of fine-tuning conditionally, bear in mind that the pertinent fine tuning would be needed *only* in the context of the *givenness* of the actual laws of nature.

The logical structure of Swinburne's explanation of the existence of the bio-critical values can be schematized essentially as follows: the premises are (i) the laws of nature are given, (ii) God wants to create human life, and (iii) under the constraints of the given laws, "fine-tuning" is necessary for the existence of human life. And the conclusion is that God selected the biocritical values in his creation, and therefore they materialized.

But this explanation founders on the shoals of divine omnipotence, just as did the theistic volitional explanation of the ultimate laws of nature (cf. Section 3). As noted earlier, Swinburne's account of omnipotence included the following creative nomological latitude: "God, being omnipotent ... his range of easy control must ... include all states of affairs [including laws] which it is *logically possible* for him to bring about."[97] Yet this conclusion undermines the theistic teleological explanation of the biocritical values, a philosophical account that was predicated on the *contrary* assumption that God confronts fixed, rather than disposable laws of nature. Instead, God can achieve any desired outcome by any laws of his choosing. This result demonstrates the logical incoherence of the theistic anthropic account. Moreover, divine omnipotence makes the causal necessity of the biocritical values *irrelevant* to the divine teleological scenario: In the context of suitably different natural laws, relating their corresponding variables to the existence of humanoid life, the a priori very low probability of the standard biocritical values is no longer an issue at all, since the critical role of these values is relative to a specified fixed set of laws and is not played by them per se.

Again, as in the case of the theistic explanation of the laws of nature by divine volition (Section 3), the doctrine of divine omnipotence has boomeranged!

Moreover, in the teleological scenario, God is held to be omnibenevolent. But, given his omnipotence, his choice to let human and animal life evolve *via* Darwinian evolution clashes head-on with his omnibenevolence. How could the God of these two omnis possibly choose the mindless, unbelievably cruel and wasteful Darwinian process—Tennyson's "nature red in tooth and claw"—as his way of bringing intelligent life onto the earth? After all, an untold number of far more benevolent mechanisms were available to him, even including the biblical Garden of Eden arrangement of Adam and Eve. Besides, as Leslie has reported: "There have been certainly five and maybe well over a dozen mass extinctions in Earth's biological history."[98] The Darwinian mechanism befits a cruel monster.

In an argument quite different from mine, Quentin Smith has contended[99] that (i) there is a huge range of values of the constants of nature and initial conditions which do *not* issue in intelligent life but which are also a priori overwhelmingly improbable, and (ii) Swinburne's claim[100] that the biocritical values specially cry out for explanation is unjustified.

It turns out that the so-called anthropic coincidences, which are purely physical, are *coextensive* with those that are critical for the formation and primordial existence of stars and galaxies. As Earman remarked, "the selection function [of initial–boundary conditions] is served just as well by the existence of stars and planetary systems supporting a carbon-based chemistry but no life forms."[101] Thus, since the "fine-tunings" are not distinctively anthropic after all, we must beware that the so-called weak anthropic principle not be allowed to give the misleading impression that the "fine-tunings" are *anthropically* unique, qua being necessary for *our* existence. In this sense, the label "anthropic principle" is pseudonymous. I refer the reader to a *superb* critique of teleological interpretations of WAP, just published by the Nobelist physicist Steven Weinberg.[102]

■ 5. CRITIQUE OF SWINBURNE'S BAYESIAN ARGUMENT FOR THE EXISTENCE OF GOD

5.1 The Incoherence of Swinburne's Apologia

Quinn has given a concise summary of Swinburne's Bayesian argument for the existence of God:

> He [Swinburne] argues that the hypothesis of theism is more probable given the exis-
> tence over time of a complex physical universe than it is on tautological evidence alone,
> and he further contends that this argument is part of a cumulative case for theism
> whose ultimate conclusion is that "on our total evidence theism is more probable than
> not."[103]

But Swinburne hedged his Bayesian plaidoyer, declaring: "Certainly one would not expect too evident and public a manifestation [of the existence of God] .... If God's existence, justice and intentions became items of evident common knowl- edge, then man's freedom to choose [belief or disbelief] would in effect be vastly curtailed."[104] In short, in Swinburne's view, the requirements of human freedom of choice allegedly require God to play a kind of hide-and-seek game with us.

Richard Gale clarified this feature of Swinburne's view as follows:

> While Swinburne's overall aim is to establish that the [Bayesian] probability that God
> exists is greater than one-half, he does not want the probability to be too high, for he
> fears that this would necessitate belief in God on the part of whoever accepts the argu-
> ment, thereby negating the accepter's freedom to choose not to believe.[105]

Yet, assuming this clarification of Swinburne's argument here, the argument is flatly incoherent: He appeals to the need for free choice of belief to justify God's *not* giving us evidence for a high probability of his existence, on the grounds that a high probability would *cause* us to believe in God willy-nilly! But why, oh why, would a high probability *necessitate* our belief at all, *if* we have free choice to believe or not in the first place when we are confronted with evidence, however strong? In the process of saving our freedom to believe, Swinburne inconsistently assumes the causal determination of our beliefs.[106] His argument fares no better, if the pertinent freedom to choose is instead between good and evil. To suppose that this freedom would be vastly curtailed if God's existence, justice and intentions became items of evident common knowledge, postulates the *causal dependence* of our moral conduct on our putative theological knowledge. But how can that cohere with Swinburne's libertarian view of human action?

Gale cites Swinburne as having asserted that "S believes that *p* if and only if he believes that *p* is more probable than any alternative," where the alternative is usually not-*p*. But Gale rightly disputes that claim: He gives examples from Tertullian to Kierkegaard, and from his own life, to the following effect: "It certainly is possible for someone to believe a proposition while believing that it is improbable, even highly improbable."[107]

In sum, Swinburne's apologia for God's evidential coyness is deeply incoherent. A cognate appeal to God's elusiveness has been made by the old apologetic doctrine of deus absconditus, such as Martin Buber's thesis of "the eclipse of God."[108] The characterization of God as self-concealing is found repeatedly in the Old Testament and was also espoused by Martin Luther, for instance.

5.2 Swinburne's Bayesian Argument for the Existence of God

Swinburne takes it for granted that Bayes's theorem, which is derived from the formal (Kolmogorovian) probability axioms, is applicable to the probability of *hypotheses* and can thus serve as a paradigm for probabilifying scientific and theological hypotheses. This kind of use of Bayes's theorem has been challenged and even rejected, if only because of the well-known problems besetting the determination of *non*subjective values of the so-called prior probability of the hypothesis at issue, as will be clear subsequently. Yet Swinburne asserts exaggeratedly that "Bayes's Theorem … is a crucial principle at work for assessing hypotheses in science, history and all other areas of inquiry."[109]

But if we do use Bayes's theorem to probabilify hypotheses, we must be mindful of a crucial distinction made by Carl Hempel yet unfortunately not heeded by Swinburne. Wesley Salmon has lucidly articulated it as follows:

> Bayes's theorem belongs to the context of confirmation, not to the context of explanation …. This is a crucial point. Many years ago, Hempel made a clear distinction between two kinds of why-questions, namely, *explanation-seeking* why questions and *confirmation-seeking* why-questions. Explanations-seeking why-questions solicit answers to questions about why something occurred, or why something is the case. Confirmation-seeking why-questions solicit answers to questions about why *we believe* that something occurred or something is the case. The characterization of non-demonstrative inference as inference to the best explanation serves to muddy the waters—not to clarify them—by fostering confusion between these two types of why-questions. Precisely this confusion is involved in the use of the "cosmological anthropic principle" as an explanatory principle.[110]

Swinburne likewise muddies the waters. He tries to use Bayes's theorem both to probabilify (i.e., to increase the confirmation of) the existence of God, on one hand, and, on the other, to show that theism offers the best explanation of the known facts, assuming that God exists. And his account of the notation he uses in his statement of the theorem reveals his failure to heed the Hempel–Salmon distinction.

Thus, Swinburne tells us the following: h represents the hypothesis to be probabilified incrementally; "k is our general background knowledge of what there is in the world and how it works"; "e is our *phenomena to be explained* and other relevant observational evidence"; $p(h/e \cdot k)$ is "the [posterior] probability of a hypothesis h on empirical evidence e and background knowledge k"[111]; $p(h/k)$ is the prior probability of h on k; $p(e/h \cdot k)$ is the "likelihood" of e on the conjunction of h and k; and $p(e/k)$ is the "expectedness" of e on k. But note importantly that Swinburne speaks of e both as the *explanandum* and as the "relevant observational evidence," thereby rolling them into one, and running afoul of Salmon's caveat that "Bayes's theorem belongs to the context of confirmation, not to the context of explanation."

In this notation, the short form of Bayes's theorem asserts:

$$p(h/e \cdot k) = \frac{p(h/k) \times p(e/h \cdot k)}{p(e/k)}$$

Here the terminology is as follows:

$p(h/e \cdot k) =_{\text{Def.}}$ The "posterior" probability of h on $e \cdot k$.
$p(h/k) =_{\text{Def.}}$ The "prior" probability of h on k.
$p(e/h \cdot k) =_{\text{Def.}}$ The "likelihood" of e on $h \cdot k$.
$p(e/k) =_{\text{Def.}}$ The "expectedness" of e on k.

It is vital to distinguish the *absolute* confirmation of h from its *incremental* confirmation.[112] The former obtains iff the probability of h is fairly close to 1; the latter, also called "relevant confirmation," is defined by the condition that the posterior probability of h be greater than its prior probability.

Upon dividing both sides of Bayes's theorem by the prior probability $p(h/k)$ it is evident that h is incrementally confirmed by e, iff $p(e/h \cdot k) > p(e/k)$, i.e. iff the likelihood exceeds the expectedness.

It is vital to be clear on what Swinburne takes to be the hypothesis h in his Bayesian plaidoyer for the existence of the God of theism. He tells us explicitly: "Now let h be our hypothesis—'God exists' .... Our concern is with the effect of various pieces of evidence on the proposition in which we are interested—'God exists.'"[113] And then he argued for the following conclusion: "On our total evidence theism is more probable than not."[114] Yet to reach this conclusion, he proceeds as follows: "... to start without any factual background knowledge (and to feed all factual knowledge gradually into the evidence of observation), and so to judge the prior probability of theism solely by a priori considerations, namely, in effect, simplicity."[115] However, in Section 2.1, I have already contended that the conceptual deliverances of a priori simplicity, even if they were coherent, cannot be at all mandatory for what is probably actually the case.

Above, I also advanced considerations that undermine Swinburne's theodicy. And I now add that the weakness of such a theodicy is shown further by the content of the Roman Catholic Exorcist Rite. It brings in Satan as a counterweight to God in order to reconcile divine omnibenevolence with such natural evils as death, declaring that Satan "hast brought death into the world."[116] As Salmon interprets Hume's argument in the *Dialogues*, in the face of all the evil in the world, and of the hypothesis of an omnipotent omnibenevolent God, the likelihood of the evidence in the numerator of Bayes's theorem is less than the expectedness in the denominator.[117] It then follows immediately from the theorem that the ratio of the posterior to the prior is less than one; that is, the total evidence *disconfirms* the existence of God.

Yet, even disregarding this devastating consequence, Swinburne's proposed additive agglomeration of posterior probabilities, starting with tautological evidence k, has a burden of proof that he did not address: The avoidance of some highly counterintuitive results of Rudolf Carnap's, which bedevil incremental confirmation. Salmon has clearly summarized an array of them.[118] One of them threatens Swinburne's assumption of additivity of probabilities: Two separate items of evidence e_1 and e_2 can *each* separately provide positive incremental confirmation for a given h, while their conjunction $e_1 \cdot e_2$ incrementally *disconfirms* h! Swinburne assumes unwarrantedly that the posterior probability is a monotomically increasing function under the conjunctive enlargement of e_n (n = 1, 2, ...).

Again, even leaving both of these great hurdles aside, there is a yet further major difficulty that Swinburne likewise does not address, let alone overcome. And the latter alone would suffice to undermine his agglomeration argument.

5.3 The Problem of "Old Evidence"

Old evidence is constituted by facts already known. But if a hypothesis (e.g., Darwinian evolutionary theory) *retrodicts* past events that were not previously known, then evidence of the occurrence of these events does count as new evidence no less than successfully predicted events. More generally, as Salmon has put it so well (personal communication), Bayes's theorem is a device for updating the appraisal of a hypothesis on the basis of new or previously unavailable, or unconsidered evidence.

Now, Swinburne tells us regarding his agglomerative program of adding posterior probabilities that "any division of evidence between e and k will be a somewhat arbitrary one. Normally, it is convenient to call the latest piece of observational evidence e and the rest k; but sometimes it is convenient to let e be all observational evidence and let k be mere tautological evidence."[119] Yet in the case of old evidence as defined, that is, facts already known, how can Swinburne *avoid* conceding that the expectedness in the denominator is equal to 1 and argue effectively that it is less than 1? The circumvention of an expectedness equal to 1 is crucial if there is to be incremental confirmation of h: As noted in Section 5.2, the condition for such confirmation is that the likelihood in the numerator exceed the expectedness in the denominator. But since no probability value or product of such values can exceed the value 1, this condition for incremental confirmation cannot be met if the expectedness equals 1.

What, then, does Swinburne need to accomplish in order to avoid this untoward result? He absolutely must be able to *split off* a given piece of old evidence e from what remains from the old background knowledge k, such that k does not still entail e.

Deborah Mayo reports that several authors have insisted that probability assignments should have been relativized to current knowledge *minus e*. Or, as urged by Paul Horwich, that although the expectedness of old evidence is actually 1, a "Bayesian should assess how much e would alter our degree of belief assignment to h relative to 'our epistemic state prior to the discovery' of e, when its probability was not yet 1."[120] If such splitting off could succeed, then the expectedness would no longer be prima facie 1. Philip Quinn has emphasized a caveat for me (private correspondence): Unless we allow such splitting off of e, Bayes's theorem cannot be useful in the confirmation of a hypothesis by known facts. Thus the question now becomes: Can the required splitting off of e succeed? Let me give my own reasons for a negative answer by means of two concrete illustrations.

As we saw, Swinburne offers the existence of laws of nature as evidence e supporting the existence of God. And he also adduces that hypothesized existence, in turn, as the sole explanation of natural lawfulness, an explanatory claim in which he is joined by Quinn.[121] But, assuming that the laws e have been split off, consider the complement subset of the already known background knowledge, which supposedly now *excludes* the laws of nature. Surely, that complement class will still include a vast array of practical knowledge by means of which we are able to control our environment and survive at all. Yet just that practical knowledge *inextricably involves the laws of nature*. Thus, when we go ice sledding on a (partially) frozen lake, we count on the lawful anomalous expansion of water as we do when we expect ice cubes to float upon being dropped into water at room temperature. Other examples in point are legion. For instance, our technological knowledge of the use of color filters depends essentially on our cognizance of the lawful spectral decomposition of white light. It emerges that the attempt to split off the laws of nature, as required by Swinburne's program fails. Philip Quinn likewise did not spell out how he would achieve the splitting off he advocated.

Marek Druzdzel (private communication) has retorted to these objections that, if they were sound, a jury of Bayesians could never convict a murderer who is caught with a smoking gun on the basis of that old evidence. To this, Salmon (again in private communication) has offered the cogent rejoinder that, in this juridical context, the jury is duty-bound to start with the presumption of innocence, and that it seems unfeasible to factor this normative legal point into a de facto prior probability. Thus, this high-profile crime example does not gainsay my claim that Swinburne's handling of old evidence fails.

Quite generally, in the chapter of Earman's *Bayes or Bust?* (1992) entitled "The Problem of Old Evidence," he drew the following conclusion regarding the status of old evidence: "... the Bayesian account of confirmation retains a black eye."[122]

I conclude that Swinburne and other Bayesians have failed to solve the problem of old evidence. Even more importantly, as we saw, Swinburne's avowed program to probabilify the existence of God cumulatively as exceeding one-half has likewise been unsuccessful: He has failed to establish that the posterior probability of the existence of God exceeds one-half. A fortiori he has failed to show, in turn, that the hypothesis of the existence of God can serve as a warranted premise to provide explanations.

■ 6. CONCLUSION

None of the cross section of diverse theists, past and present, whose arguments I have considered, has presented cogent evidence for the existence of their God.

■ NOTES

I am greatly indebted for very helpful discussions to my colleagues and friends (in alphabetical order) Richard Gale, Allen Janis, John Leslie, Philip Quinn, Wesley Salmon, and Quentin Smith. I likewise benefited from the comments of an anonymous referee. And I owe my colleague and friend Gerald Massey the reference to Schellenberg's book *Divine Hiddenness and Human Reason* (1993).

1. Gale, *On the Nature and Existence of God*, 1991, p. 284.

2. See Craig, W., *The Kalam Cosmological Argument*, 1979.

3. Swinburne, R., *Is There a God*, 1996, p. 48.

4. Ibid.

5. Swinburne, *The Existence of God*, 1979; see also 3rd ed., 1991.

6. Leslie, J., "Efforts to Explain All Existence," *Mind* 87 (1978), 185.

7. Edwards, P., "Why," in: Edwards, *The Encyclopedia of Philosophy*, Vol. 8, 1967, pp. 296–302.

8. Wittgenstein, L., "Lecture on Ethics," in: Klagge and Nordmann (eds.), *Philosophical Occasions*, 1993, p. 41–42.

9. Rescher, N., *The Riddle of Existence*, 1984, p. 5.

10. Heidegger, M., *Einführung in die Metaphysik*, 1953, p. 1.

11. In 1931, Rudolf Carnap explained in a major paper ("*Überwindung der Metaphysik durch Logische Analyse der Sprache*") that the noun "nothingness" is a product of logical victimization by the grammar of our language.

12. Parfit, D., "Why Anything? Why This?" *London Review of Books* 20(2) (1998), 24.

13. Quoted in Leslie, "Efforts to Explain All Existence," p. 181, from Bergson's *The Two Sources of Morality and Religion*, Part 2.

14. Grünbaum, A., "Theological Misinterpretations of Current Physical Cosmology," *Philo* 1(1) (1998), 15–34; see also the previous version of that paper in *Foundations of Physics* 26 (1996), 523–543.

15. Grünbaum, "Theological Misinterpretations," sections 2–4.

16. Grünbaum, *Philosophical Problems of Space and Time*, 2d ed., 1973, pp. 573–574.

17. Davies, P., *The Fifth Miracle: The Search for the Origin and Meaning of Life*, 1999.

18. Grünbaum, *Philosophical Problems*, pp. 406–407.

19. Leibniz, G. W., "The Principles of Nature and of Grace Based on Reason," (1714), in: P. P. Wiener (ed.), *Leibniz Selections*, 1951, p. 525.

20. Cited in Quinn, "Creation, Conservation, and the Big Bang," in: J. Earman et al. (eds.), *Philosophical Problems of the Internal and External Worlds*, 1993, p. 605, from Leibniz's *Philosophical Papers and Letters*.

21. Leslie, "Cosmology and Theology," *Stanford Encyclopedia of Philosophy*, http://plato.stanford.edu/entries/cosmology-theology (version 1998).

22. Hayek, F. von, *The Counter-Revolution of Science: Studies on the Abuse of Reason*, 1952.

23. Gale, R. and Pruss, A., "A New Cosmological Argument," *Religious Studies* 35(4) (Dec. 1999), 461–476.

24. Swinburne, *Is There a God?*

25. Quoted in Parsons, K. M., *God and the Burden of Proof*, 1989, p. 81.

26. Swinburne, *Existence of God*, 3d ed., ch. 5; Swinburne, *Is There a God*, ch. 3

27. Swinburne, *Existence of God*, 3d ed., p. 322.

28. Swinburne, *Is There a God?* p. 48.

29. Swinburne, *Existence of God*, 3d ed., p. 84.

30. Ibid., pp. 283–284.

31. Parsons, *God and the Burden of Proof*, p. 84.

32. Swinburne, *Existence of God*, 3d ed., p. 79.

33. Glymour, C., *Theory and Evidence*, 1980, p. 78.

34. Ibid., 79.

35. Smith, Q., "Swinburne's Explanation of the Universe," *Religious Studies* 34 (1998), 91–102.

36. Parsons, *God and the Burden of Proof*, ch. 2.

37. Martin, M., *Atheism*, 1990, pp. 110–118.

38. As I was putting the finishing touches to this paper, Swinburne made me aware of his Aquinas Lecture *Simplicity as Evidence of Truth* (1997). I regret that it was too late to deal with it here.

39. Craig, W., *Kalam Cosmological Argument*. 1979.

40. Craig, "A Response to Grünbaum on Creation and Big Bang Cosmology," *Philosophia Naturalis* 31 (1994), 247.

41. Craig, "The Origin and Creation of the Universe: A Reply to Adolf Grünbaum," *British Journal for the Philosophy of Science* 43 (1992), 233–240.

42. See Craig, *Kalam Cosmological Argument*, 1979, pp. 141–148; see also, besides the other aforecited work of Craig's, Craig and Quentin Smith, *Theism, Atheism, and Big Bang Cosmology*, Oxford: Clarendon 1993, pp. 147, 156.

43. See Grünbaum, "Some Comments on William Craig's 'Creation and Big Bang Cosmology,'" *Philosophia Naturalis* 31 (1994), 225–236; see also my "Theological Misinterpretations of Current Physical Cosmology," *Philo* 1 (1998), sec. 5A, 25–26 (this volume, ch. 8, sec. 5.1).

44. Craig, "Pseudo-Dilemma?" Letter to the Editor, *Nature* 354 (December 5, 1991), p. 347.

45. Craig, "Creation and Big Bang Cosmology," *Philosophia Naturalis* 31(1994), 218, 222, fn. 1.

46. Grünbaum, "Theological Misinterpretations," 25–26.

47. Craig, "Pseudo-Dilemma?" p. 347.

48. Earman, J., *Bangs, Crunches, Whimpers, and Shrieks: Singularities and Acausalities in Relativistic Spacetimes*, 1995, p. 208.

49. See Gale, *On the Nature and Existence of God*, 1991.

50. Quinn, "Creation, Conservation, and the Big Bang," 1993.

51. Smith, "Internal and External Causal Explanations of the Universe," *Philosophical Studies* 79 (1995), 283–310.

52. Quinn and Swinburne quantify the "amount of matter energy," which clearly depends, however, on the choice of units, as well as on the zero of energy in the pertinent physical theory. Does God's supposed creative decree contain such mundane specifications?

53. Quinn, "Creation, Conservation, and the Big Bang," 607–608, italics added.

54. Ibid., p. 606.

55. Swinburne, *Existence of God*, 3d ed., pp. 123–125.

56. Quinn, "Creation, Conservation, and the Big Bang," p. 605

57. Ibid., p. 607.

58. Ibid., p. 589.

59. Gale, "Santayana's Bifurcationist Theory of Time," *Bulletin of the Santayana Society* 16 (Fall 1999).

60. Swinburne, *Is There a God?* p. 2.

61. Ibid.; cf. Swinburne, *Existence of God*, 3rd ed., 1991, ch. 4, "Complete Explanation."

62. James, W., *Pragmatism*, 1975, p. 40.

63. Swinburne, *Existence of God*, 3d ed., 1991, p. 109.

64. Ibid., p. 295, italics added.

65. Ibid., p. 289.

66. Ibid., p. 296.

67. Ibid., 294; Quinn, "Creation, Conservation, and the Big Bang," p. 602.

68. Ibid., p. 597.

69. Swinburne, *Existence of God*, 3d ed., p. 8.

70. Buckley, M., "Religion and Science: Paul Davies and John Paul II," *Theological Studies* 51 (1990), 314.

71. Grünbaum, "Theological Misinterpretations," pp. 15–34; see also the other version of this paper in *Foundations of Physics* 26 (1996), 523–543.

72. Leslie, *The End of the World*, 1996, p. 131.

73. Newman, J. R. (ed.), "Conservation of Mass-Energy," *International Encyclopedia of Science*, vol. 1, 1965, 276.

74. See Grünbaum, "Theological Misinterpretations," p. 29; see also the version *Foundations of Physics*, p. 539.

75. Quinn, "Creation, Conservation, and the Big Bang," p. 603.

76. Swinburne (*Existence of God*, 3d ed., p. 103) rejects occasionalism as an "untenable view" of the relation between scientific and theological explanation.

77. Swinburne, "Violation of a Law of Nature," in: R. Swinburne (ed.), *Miracles*, 1989, ch. 8, p. 79.

78. Swinburne, *Existence of God*, 3d ed., ch.13.

79. For criticisms directed at the particulars of Swinburne's recourse to religious experience, see Martin, *Atheism*, 1990, ch. 6; and Gale, "Swinburne's Argument from Religious Experience," in: A. Padgett (ed.), *Reason and the Christian Religion*, 1994, ch. 3.

80. Freud, S., "The Question of a Weltanschauung," *Standard Edition* 22 (1933), p. 168.

81. Davidson, D., "Paradoxes of Irrationality," in R. Wollheim and J. Hopkins (eds.), *Philosophical Essays on Freud*, 1982, p. 298.

82. Smith, "An Atheological Argument from Evil Natural Laws," *International Journal for the Philosophy of Religion* 29 (1991), pp. 165–168; see also his "Anthropic Coincidences, Evil and the Disconfirmation of Theism," *Religious Studies* 28 (1992), pp. 347–350, and his *Ethical and Religious Thought in Analytic Philosophy of Language*, 1997, pp. 137–157.

83. See Swinburne, *Providence and the Problem of Evil*, 1998; and Gale's "Review of *Providence and the Problem of Evil* by Richard Swinburne," *Religious Studies* 36 (2000), 209–219.

84. Swinburne, *Existence of God*, 3d ed., p. 284.

85. Salmon, W., "Religion and Science: A New Look at Hume's Dialogues," *Philosophical Studies* 33 (1978), 143–176.

86. Salmon, "Experimental Atheism," *Philosophical Studies* 35 (1979), 101–104.

87. Leslie, *Universes*, p. 3.

88. McMullin, E., "Cosmology and Religion," in: N. Hetherington (ed.), *Cosmology*, 1993, p. 602.

89. Earman, "The SAP Also Rises: A Critical Examination of the Anthropic Principle," *American Philosophical Quarterly* 24 (1987), 314–315.

90. Schick, T., Jr., "The 'Big Bang' Argument for the Existence of God," *Philo* 1 (1998), 98.

91. Worrall, J., "Is the Idea of Scientific Explanation Unduly Anthropocentric? The Lessons of the Anthropic Principle," in *Discussion Paper Series from the London School of Economics*, 1997, DP25/96, p. 11.

92. Ibid., p. 13.

93. Swinburne, *Existence of God*, 3d ed., p. 306.

94. Swinburne, "Argument from the Fine Tuning of the Universe," in: J. Leslie (ed.), *Physical Cosmology and Philosophy*, 1990, p. 160.

95. Swinburne, *Existence of God*, 3d ed., p. 312, italics added.

96. McMullin, "Cosmology and Religion," p. 603.

97. Swinburne, *Existence of God*, 3d ed., p. 295, italics added.

98. Leslie, *The End of the World*, 1996, p. 81.

99. Smith, "Anthropic Explanations in Cosmology," *Australasian Journal of Philosophy* 72 (1994), 372.

100. Swinburne, "Argument from the Fine Tuning of the Universe," p. 154.

101. Earman, "SAP Also Rises," p. 309. For numerous relevant details, see Leslie, *Universes*, 1990, ch. 1, section 1.4.

102. Weinberg, S., "A Designer Universe?" *New York Review of Books* 46(16), October 21, 1999, 46–48.

103. Quinn, "Creation, Conservation, and the Big Bang," p. 622. See Swinburne, *Existence of God*, 1979, p. 291, for the internal citation.

104. Swinburne, *Existence of God*, 3d ed., ch. 13, p. 244.

105. Gale, "Swinburne's Argument from Religious Experience," in: A. Padgett (ed.), *Reason and the Christian Religion*, 1994, p. 39.

106. Cf. Grünbaum, "Free Will and Laws of Human Behavior," in: H. Feigl, W. Sellars, and K. Lehrer (eds.), *New Readings in Philosophical Analysis*, 1972, section 2.2 and 2.3 on the role of causality in belief formation. This paper is included as ch. 4 of the present volume.

107. Gale, "Swinburne's Argument from Religious Experience," p. 40.

108. Grünbaum, "The Poverty of Theistic Morality," in: K. Gavroglu, J. Stachel, and M. W. Wartofsky (eds.), *Science, Mind and Art*, 1995, pp. 211–212. This paper is included in the present volume as ch. 6: "In Defense of Secular Humanism." Cf. Schellenberg, *Divine Hiddenness and Human Reason*, 1993.

109. Swinburne, "Argument from the Fine Tuning of the Universe," p. 155.

110. Salmon, "Explanation and Confirmation; A Bayesian Critique of Inference to the Best Explanation," in: G. Hon and S. Rakover (eds.), *Explanation: Theoretical Approaches and Applications*, 2001, pp. 85–86.

111. Swinburne, *Existence of God*, 3d ed., pp. 64–65, 281, 289 (italics added).

112. Salmon, "Confirmation and Relevance," in: G. Maxwell and R. M. Anderson (eds.), *Induction, Probability, and Confirmation*, 1975, p. 6.

113. Swinburne, *Existence of God*, 3d ed., p. 16, 18.

114. Ibid., p. 291.

115. Ibid., p. 294, p. 63n.

116. Entry: "Exorcism," *Encyclopedia Britannica*, 1929.

117. Salmon, "Religion and Science: A New Look at Hume's Dialogues," *Philosophical Studies* 33 (1978), pp. 143–176; and personal communication.

118. Salmon, "Confirmation and Relevance," pp. 14–16.

119. Swinburne, *Existence of God*, 3d ed., p. 65.

120. Mayo, D., "The Old Evidence Problem," in: Mayo, *Error and the Growth of Experimental Knowledge*, 1996, p. 334 and fn. 10 there.

121. Quinn, "Creation, Conservation, and the Big Bang," pp. 607–608.

122. Earman, "The Problem of Old Evidence," in: Earman, *Bayes or Bust?* 1992, ch. 5, p. 135.

10 Pseudo-Creation of the Big Bang

Sir John Maddox[1] has rejected the Big Bang model of the universe as "philosophically unacceptable," and Jean-Marc Lévy-Leblond[2] has replied: "It need not be as 'philosophically unacceptable' as he contends." But Maddox expects, and Lévy-Leblond allows, that scientific evidence will turn out to justify the abandonment of the model.

I claim that, insofar as both its classical and quantum versions become unacceptable, they will do so only on scientific rather than on philosophical grounds. For example, the very recently discovered "great wall" and "great attractor," the so-called dark matter, the newly observed most distant and oldest quasars[3] and the role played by plasma in cosmic evolution[4] pose a theoretical challenge that the Big Bang framework may, in due course, be unable to meet.

But there are basic logical flaws in the assumptions that prompted Maddox's philosophical rejection of the cataclysmic cosmic scenario as well as in Lévy-Leblond's claim that his linear time has enhanced the conceptual palatability of the Big Bang. They either beg the question or illicitly impose a priori demands on physical cosmology.

The "philosophical difficulty" that Maddox sees in the Big Bang model is "that an important issue, that of the ultimate origin of our world, cannot be discussed." His argument is predicated on the occurrence of the cosmic cataclysm at a "well-defined instant" t_0 which had no temporal predecessor. But as Roberto Torretti has explained,[5] general relativity theory does not countenance t_0 as a bona fide instant of the Big Bang space–time.

Yet the alleged philosophical difficulty would be spurious, I contend, even if the model did feature a time interval that is closed, rather than open, for $t = t_0$. By hypothesis, there simply did not exist any instants before it. But precisely this total absence of times earlier than t_0 also rules out the very existence of an earlier cause of any event that does occur at the hypothesized instant t_0. Hence, if the Big Bang is taken to have occurred at the putative t_0, that initial event is causally sui generis. It just cannot have any cause at all in the universe of the given model, nor, of course, can that paramount occurrence be the effect of any prior cause.

Therefore, it just begs the question to insist, as Maddox does peremptorily, on characterizing the supposedly initial event as "an effect." By thus illicitly requiring the existence of an earlier cause within the assumed model, this characterization also requires the existence of at least one instant before t_0 after all, which saddles the cosmological model with a temporal inconsistency. Maddox objects that, qua *purported* effect, the Big Bang "is an effect whose cause cannot be identified or even discussed," but the elusiveness of the phantom earlier cause is due to its sheer nonexistence in the face of the demand that it must exist nonetheless.

Evidently, it is imperative to distinguish the sound question, "Does the universe have a temporal beginning?" from the quite different alleged problem, "If the universe did have a bounded past of finite duration, what was the cause of its initial event at t_0?" Maddox takes the latter query to be the "important issue ... of the ultimate origin of our world." But, absent the phantom cause or "ultimate origin," the craving for it—and for the causal explanation of the classical Big Bang at the putative instant t_0—is a seductive pseudo-problem rather than a riddle eluding scientific solution. A question cannot be regarded as a well-posed challenge to a theory merely because the questioner finds it psychologically insistent and finds the answer to it elusive after introducing a temporal inconsistency into the theory a priori.

Though the Big Bang model is beset by some scientific liabilities, it is not vitiated by the purported difficulty. A fortiori, there is no warrant at all for Maddox's contention that "creationists ... have ample justification in the doctrine of the Big Bang." Indeed, the putative cosmological model poses no challenge whatever to atheism.[6] It is the a priori philosophical aversion to a bounded, metrically finite past that plays into the hands of the creationists. This aversion goes back to Aristotle, who indirectly, as it were, ruled out the supposed Big Bang model a priori by deeming an instant without temporal predecessor to be inconceivable (*Physics*, Book VIII, 251b). But such a moment is quite conceivable, and the verdict as to its actual existence must be reserved to the mathematics of our best physical theories.

As noted by Lévy-Leblond, general relativity excludes t_0 from the set of bona fide physical instants. Assuming also that there is no final future crunch, he points out that the cosmic time interval of the Robertson–Walker space–time metric is open in both directions ($t_0 < t < \infty$) but patently does not extend before t_0. Relying on an a priori prejudice, however, Lévy-Leblond finds it philosophically discomfiting that, even though the past on this open interval is ordinally unbounded, the age of the Big Bang universe in the given metric is finite, presumably somewhere between twelve and twenty billion years.

Citing Misner,[7] he aims "to send back the birth of the Universe [metrically] to (minus) infinity where, or rather when, it seems to belong." Hence, he was pleased to introduce his new linear time metric, which confers an infinite duration on the ordinally unbounded past of the Big Bang universe.

True, this alternative time metrization is quite legitimate. Interestingly, it is physically realized, as Lévy-Leblond explains, by a clock geared to the expansion of the universe. But just why must the birth of the universe "belong" to a past of metrically infinite duration? Since the philosophical malaise experienced by those who shrink *a priori* from a metrically finite age of the universe is baseless, the ability to allay this discomfiture does not add merit to the remetrization.

Pace Maddox and Lévy-Leblond, the Big Bang model featuring a finite age of the universe on the standard cosmic timescale is not philosophically disquieting at all.

In one of the versions of current quantum cosmology,[8] the Big Bang is no longer held to comprise all of the earliest instants; instead, the inflationary expansion is preceded by two so-called vacuum states in a metrically finite past. Within that theoretical framework, it is no longer misguided to ask: "What caused the Big Bang?" Given the model's random quantum fluctuations in its "false" vacuum, the general theory of relativity tells us why there was an inflationary expansion. Here as well, the model lends no support to the doctrine of divine creation ex nihilo,[9] but that teaching lingers on in the misleading overtones of the terms "creation" and "nothing." The former connotes the operation of a creator or external causal agency, and the latter a completely unstructured state.

Alas, the recent literature on some versions of quantum cosmology contains inappropriate uses of these locutions which may suggest that this theory abets creationism. For example, such physicists as James B. Hartle and Stephen Hawking[10] as well as Alexander Vilenkin speak misleadingly of certain primordial physical states as "nothing," even though, as Vilenkin says, these states are avowedly only "a realm of unrestrained quantum gravity," which is "a state with no classical space-time."[11]

Recently, plasma cosmology[12] has posed a major challenge to the gravity-dominated Big Bang models by assigning a critical cosmic role to hot electrically charged gases. The plasma model evolves without any beginning, being metrically infinite in both time directions on the standard cosmic scale. But this feature does not make the plasma universe philosophically preferable to any of its Big Bang rivals. The merits of their competing claims turn instead on their scientific credentials, which include the adequacy with which they fit observational findings.

■ NOTES

1. Maddox, J., "Down With the Big Bang," *Nature* 340 (1989), p. 425.

2. Lévy-Leblond, J.-M., "The Unbegun Big Bang," *Nature* 342 (1989), p. 23; "Did the Big Bang Begin?" *American Journal of Physics* 58 (1990), pp. 156–159.

3. Wilford, N.W., "The Big Bang Survives an Onslaught of New Cosmology," *New York Times*, Ideas and Trends section E.21, January 21, 1990, p. 5.

4. Peratt, A. L., "Not With a Bang: The Universe May Have Evolved from a Vast Sea of Plasma," *Sciences*, January–February 1990, pp. 24–32.

5. Torretti, R., *Relativity and Geometry*, 1983, pp. 210–219.

6. Grünbaum, A., "The Pseudo-Problem of Creation in Physical Cosmology," *Philosophy of Science* 56 (1989), p. 373.

7. Misner, C. W., "Absolute Zero of Time," *Physical Review* 186 (1969), pp. 1328–1333.

8. Weisskopf, V., "The Origin of the Universe," *Bulletin of the American Academy of Arts and Sciences* 42 (January 1989), p. 22.

9. Grünbaum, A., *Proceedings of the International Congress "Truth in Science,"* *Accademia Nazionale Del Lincei*, Rome, Italy, October 13–14, 1989; see also Grünbaum, "Pseudo-Problem of Creation."

10. Hartle, J. B. and S. W. Hawking, "Wave Function of the Universe," *Physical Review* D28 (1983), p. 2960.

11. Vilenkin, A., "Birth of Inflationary Universes," *Physical Review* D27 (1983), p. 2848.

12. Peratt, "Not With a Bang."

■ BIBLIOGRAPHY

Alexander, Harry L. [1955]: *Reactions With Drug Therapy*, Philadelphia: W.B. Saunders.
Aquinas, Thomas [1981]: *Summa Theologiae*; trans. the English Dominican Fathers, New York: Benziger Bros.
Bartley III, William [1982]: "The Philosophy of Karl Popper: Part III: Rationality, Criticism, and Logic," *Philosophia* (Israel) 11, nos. 1–2, pp. 121–221.
Bergin, Allen. E. [1970]: "The Evaluation of Therapeutic Outcomes," in A. E. Bergin & S. L. Garfield (eds.), *Handbook of Psychotherapy and Behavior Change*, New York: John Wiley and Sons, pp. 217–270.
Bergson, Henri [1974, originally published 1935]: *The Two Sources of Morality and Religion*, trans. R. A. Audra & C. Brereton, with the assistance of W. Carter. Westport, CT: Greenwood Press.
Berofsky, Bernard (ed.) [1966]: *Free Will and Determinism*, New York: Harper Collins College Div.
Bertocci, Peter A. [1968/1973]: "Creation in Religion," in P. Wiener (ed.), *Dictionary of the History of Ideas: Studies of Selected Pivotal Ideas, Vol.1: Abstraction in the Formation of Concepts to Design Argument*, New York: Scribner, p. 571.
Blodgett, Stephen W. [1967]: "Conservation of Mass-Energy," in J. R. Newman (ed.), *The Harper International Encyclopedia of Science*, New York: Harper & Row, pp. 276–277.
Bondi, Hermann [1960]: *Cosmology*, 2nd ed., Cambridge Monographs on Physics Series, Cambridge: Cambridge University Press.
Buber, Martin [1952]: *Eclipse of God*, New York: Harper.
—— [1967]: "The Dialogue between Heaven and Earth," in N. Glatzer (ed.), *On Judaism*, New York: Schocken Books, pp. 214–226.
Buckley, Michael [1990]: "Religion and Science: Paul Davies and John Paul II," *Theological Studies* 51, pp. 310–324.
Burtt, E. A. (ed.) [1939]: *The English Philosophers From Bacon to Mill*, New York: Random House (Modern Library Series).
Cain, Seymour [1993/1994]: "In Response to Grünbaum's Defense," *Free Inquiry* 14, no. 1, pp. 55–57.
Campbell, Charles Arthur [1967]: *In Defence of Free Will*, London: Allen & Unwin.
Carlson, Erik, and Olsson, Erik J. [2001]: "The Presumption of Nothingness," *Ratio* 14, no. 3, pp. 203–221.
Carnap, R. [1931]: "Überwindung der Metaphysik durch Logische Analyse der Sprache," in H. Schleicher (ed.), *Logischer Empirismus—der Wiener Kreis. Ausgewählte Texte mit einer Einleitung*, Munich: W. Fink, 1975, pp. 149–171.
—— [1936/37]: "Testability and Meaning," *Philosophy of Science* 3 (1936), pp. 419–471; *Philosophy of Science* 4 (1937), pp. 2–40.
—— [1963]: "Intellectual Autobiography," in Paul A. Schilpp (ed.), *The Philosophy of Rudolf Carnap*, La Salle, IL: Open Court, pp. 3–84.
Carter, Stephen L. [1993]: *The Culture of Disbelief*, New York: Basic Books.

Cassirer, Ernst [1956]: *Determinism and Indeterminism in Modern Physics*, New Haven, CT: Yale University Press.

Cioffi, Frank [1985]: "Psychoanalysis, Pseudo-Science and Testability," in G. Currie & A. Musgrave (eds.), *Popper and the Human Sciences*, Dordrecht: Martinus Nijhoff, pp. 13–44.

—— [1988]: "Exegetical Myth-Making in Grünbaum's Indictment of Popper and Exoneration of Freud," in P. Clark & C. Wright (eds.), *Mind, Psychoanalysis and Science*, Oxford: Blackwell, pp. 61–87.

Craig, William [1979]: *The Kalam Cosmological Argument*, New York: Harper & Row.

—— [1991]: "Pseudo-Dilemma?" Letter to the Editor, *Nature* 354 (5 December), p. 347.

—— [1992]: "The Origin and Creation of the Universe: A Reply to Adolf Grünbaum," *British Journal for the Philosophy of Science* 43, pp. 233–240.

—— [1993]: "The Theistic Cosmological Argument," in W. Craig & Q. Smith, *Theism, Atheism, and Big Bang Cosmology*, Oxford: Clarendon Press, pp. 3–76.

—— [1994]: "Creation and Big Bang Cosmology," *Philosophia Naturalis* 31, pp. 217–224.

—— [1994]: "A Response to Grünbaum on Creation and Big Bang Cosmology," *Philosophia Naturalis* 31, no. 2, p. 237–249.

Craig, William, and Smith, Quentin [1993]: *Theism, Atheism, and Big Bang Cosmology*, Oxford: Clarendon Press.

Davidson, Donald [1982]: "Paradoxes of Irrationality," in R. Wollheim & J. Hopkins (eds.), *Philosophical Essays on Freud*, New York: Cambridge University Press, pp. 289–305.

Descartes, René [1967]: "Meditation III. Of God: That He Exists," in *The Philosophical Works of Descartes*, vol. 1, trans. E. S. Haldane & G. R. T. Ross, Cambridge: Cambridge University Press, pp. 157–171.

Duhem, Pierre [1954]: *The Aim and Structure of Physical Theory*, Princeton, NJ: Princeton University Press.

Earman, John [1974]: "An Attempt to Add a Little Direction to the 'Problem of the Direction of Time,'" *Philosophy of Science* 41, pp. 15–47.

—— [1987]: "The SAP Also Rises: A Critical Examination of the Anthropic Principle," *American Philosophical Quarterly* 24, no. 4, pp. 307–317.

—— [1992]: "The Problem of Old Evidence," in J. Earman, *Bayes or Bust: A Critical Examination of Bayesian Confirmation Theory*, Cambridge, MA: MIT Press, sec. 5, pp. 119–136.

—— [1993]: "The Cosmic Censorship Hypothesis," in J. Earman et al. (eds.), *Philosophical Problems of the Internal and External Worlds: Essays on the Philosophy of Adolf Grünbaum*, Pittsburgh: University of Pittsburgh Press, ch. 3, pp. 45–82.

—— [1995]: *Bangs, Crunches, Whimpers, and Shrieks: Singularities and Acausalities in Relativistic Spacetimes*, New York: Oxford University Press.

Eccles, John C. [1974]: "The World of Objective Knowledge," in P. A. Schilpp (ed.), *The Philosophy of Karl Popper*, LaSalle, IL: Open Court, pp. 349–370.

Edwards, Paul [1967]: Entry: "Why," in P. Edwards (ed.), *The Encyclopedia of Philosophy*, Vol. 8, New York: Macmillan Company, pp. 296–302.

—— [1967]: Entry: "Atheism," in P. Edwards (ed.), *The Encyclopedia of Philosophy*, Vol. 1, New York: Collier Macmillan Ltd., pp. 174–189.

—— [1973]: "Buber, Fackenheim and the Appeal to Biblical Faith," in P. Edwards & A. Pap, *A Modern Introduction to Philosophy*, 3rd ed., New York: Free Press, pp. 394–398.

Einstein, Albert [1941]: "Science and Religion," in D. J. Bronstein & H. M. Schulweis (eds.), *Approaches to the Philosophy of Religion*, New York: Prentice Hall, 1954, pp. 68–72.

—— [1949]: "Reply to Criticisms," in P. A. Schilpp (ed.), *Albert Einstein: Philosopher-Scientist*, LaSalle, IL: Open Court, pp. 663–688.

Eisenhart, Luther P. [1949]: *Riemannian Geometry*, Princeton, NJ: Princeton University Press.

Eliade, Mircea [1992]: *Essential Sacred Writings from Around the World*, San Francisco, CA: Harper Collins.

Eysenck, Hans Jürgen [1965]: "The Effects of Psychotherapy," *International Journal of Psychiatry* 1, pp. 97–178.

—— [1966]: *The Effects of Psychotherapy*, New York: International Science Press.

Eysenck, G. D. Wilson [1973]: *The Experimental Study of Freudian Theories*, London: Methuen.

Fackenheim, Emil [1973]: "On the Eclipse of God," in P. Edwards & A. Pap, *A Modern Introduction to Philosophy*, 3rd ed., New York: Free Press, pp. 523–533.

Fisher, Seymour, and Greenberg, Roger P. [1977]: *The Scientific Credibility of Freud's Theory and Therapy*, New York: Basic Books.

Földesi, Tamas [1966]: *The Problem of Free Will*, Budapest: Akadémiai Kiadó.

Fox, Robin L. [1992]: *The Unauthorized Version: Truth and Fiction in the Bible*, New York: Albert A. Knopf.

Frank, Philipp [1947]: *Einstein, His Life and Times*, New York: Albert A. Knopf.

Freud, Sigmund [1953–1974]: *Standard Edition of the Complete Psychological Works of Sigmund Freud*, trans. J. Strachey et al., London: Hogarth Press (cited as S.E.).

Fromm, Erich [1950]: *Psychoanalysis and Religion*, New Haven, CT: Yale University Press.

—— [1970]: *The Crisis of Psychoanalysis*, New York: Holt, Rinehart and Winston.

Gale, Richard M. [1976]: *Negation and Non-Being*, American Philosophical Quarterly Monograph Series, no. 10–0084–6422, Oxford: Blackwell.

—— [1991]: *On the Nature and Existence of God*, New York: Cambridge University Press.

—— [1994]: "Swinburne's Argument from Religious Experience," in A. Padgett (ed.), *Reason and the Christian Religion: Essays in Honor of Richard Swinburne*, Oxford: Oxford University Press, ch. 3.

—— [1999]: "Santayana's Bifurcationist Theory of Time," *Bulletin of the Santayana Society* 16, pp. 1–13.

—— [2000]: "Review of *Providence and the Problem of Evil* by Richard Swinburne," *Religious Studies* 36, pp. 209–219.

Gale, Pruss A. [1999]: "A New Cosmological Argument," *Religious Studies* 35, pp. 461–476.

Giere, Ronald [1970]: "An Orthodox Statistical Resolution of the Paradox of Confirmation," *Philosophy of Science* 37, pp. 354–362.

Glymour, Clark [1974]: "Freud, Kepler and the Clinical Evidence," in R. Wollheim (ed.), *Freud*, Anchor Books: New York, pp. 285–304. An *Afterword* appears in R. Wollheim & J. Hopkins (eds.), *Philosophical Essays on Freud*, Cambridge University Press, New York, 2nd ed., 1982.

Good, Irving. J. [1967]: "The White Shoe Is a Red Herring," *British Journal for the Philosophy of Science* 17, p. 322.

Grünbaum, Adolf [1952]: "Causality and the Science of Human Behavior," *American Scientist* 40, pp. 665–676.

—— [1954]: "Science and Ideology," *Scientific Monthly* 79 (July), pp. 13–19.

—— [1955]: "Time and Entropy," *American Scientist* 43, pp. 550–572.

—— [1957]: "Determinism in the Light of Recent Physics," *Journal of Philosophy* 54, no. 23, pp. 713–741.

—— [1960]: "The Duhemian Argument," *Philosophy of Science* 27, pp. 75–87.

—— [1962]: "Geometry, Chronometry and Empiricism," *Minnesota Studies in the Philosophy of Science*, vol. 3, Minneapolis: University of Minnesota Press.

—— [1962]: "Science and Man," *Perspectives in Biology and Medicine* 5, pp. 483–502.

—— [1968]: *Modern Science and Zeno's Paradoxes*, George Allen and Unwin: London.

—— [1972]: "Free Will and Laws of Human Behavior," in H. Feigl, W. Sellars, & K. Lehrer (eds.), *New Readings in Philosophical Analysis*, New York: Appleton-Century-Crofts, pp. 605–627.

—— [1973]: *Philosophical Problems of Space and Time*, 2nd ed., Dordrecht, The Netherlands: D. Reidel Publishing Co.

—— [1974]: "Popper's Views on the Arrow of Time," in Paul A. Schilpp (ed.), *The Philosophy of Karl Popper*, LaSalle, IL: Open Court, pp. 775–797.

—— [1976]: "Is Falsifiability the Touchstone of Scientific Rationality? Karl Popper Versus Inductivism," in R. S. Cohen, P. K. Feyerabend, & M. W. Wartofsky (eds.), *Essays in Memory of Imre Lakatos*, Boston Studies in the Philosophy of Science 38, Dordrecht, The Netherlands: D. Reidel, pp. 213–252.

—— [1976]: "Can a Theory Answer More Questions Than One of Its Rivals?" *British Journal for the Philosophy of Science* 27, pp. 1–24.

—— [1979]: "Is Freudian Psychoanalytic Theory Pseudo-Scientific by Karl Popper's Criterion of Demarcation?" *American Philosophical Quarterly* 16, pp. 131–141.

—— [1984]: *The Foundations of Psychoanalysis: A Philosophical Critique*, Berkeley: University of California Press.

—— [1986]: "Author's Response," *Behavioral and Brain Sciences* 9, pp. 266–281.

—— [1988]: "The Role of the Case Study Method in the Foundations of Psychoanalysis," in L. Nagl & H. Vetter (eds.), *Die Philosophen und Freud*, Vienna: R. Oldenbourg Verlag, pp. 134–174. Reprinted in *Canadian Journal of Philosophy* 18, 623–658.

—— [1989]: "Why Thematic Kinships Between Events Do *Not* Attest Their Causal Linkage," in J. R. Brown & J. Mittelstrass (eds.), *An Intimate Relation: Studies in the History and Philosophy of Science*, Boston Studies in the Philosophy of Science, Dordrecht, The Netherlands: D. Reidel, pp. 477–493.

—— [1989]: "The Pseudo-Problem of Creation in Physical Cosmology," *Philosophy of Science* 56, 373–394.

—— [1989]: "Truth in Science," *Proc. Int. Congr., Accademia Nazionale Del Lincei*, Rome, Italy, October 13–14, 1989.

—— [1990]: "The Pseudo-Problem of Creation in Physical Cosmology," in J. Leslie (ed.), *Physical Cosmology and Philosophy*, Philosophical Issues Series, New York: Macmillan Publishing, pp. 92–112.

——[1990]: "Pseudo-Creation of the Big Bang," *Nature* 344, pp. 821–822.

—— [1991] "Die Schöpfung als Scheinproblem der physikalischen Kosmologie," in H. Albert, A. Bohnen, & A. Musgrave (eds.), *Wege der Vernunft, Festschrift zum Siebzigsten Geburtstag von Hans Albert*, Tübingen: J.C.B. Mohr, pp. 164–191.

—— [1991]: "Creation as a Pseudo-Explanation in Current Physical Cosmology," *Erkenntnis* 35, pp. 233–254.

—— [1992]: *Validation in the Clinical Theory of Psychoanalysis*, Madison, CT: International Universities Press.

—— [1993]: "Narlikar's 'Creation' of the Big Bang Universe was a Mere Origination," *Philosophy of Science* 60, pp. 638–646.

—— [1994]: "Some Comments on William Craig's Creation and Big Bang Cosmology," *Philosophia Naturalis* 31, pp. 225–236.

—— [1995]: "The Poverty of Theistic Morality," in K. Gavroglu, J. Stachel, & M. W. Wartofsky (eds.), *Science, Mind and Art: Essays on Science and the Humanistic Understanding in Art, Epistemology, Religion and Ethics, in Honor of Robert S. Cohen*, Boston Studies in the Philosophy of Science, vol. 165, Dordrecht, The Netherlands: Kluwer Academic Publishers, pp. 203–242.

—— [1995]: "Origin Versus Creation in Physical Cosmology," in L. Krüger & B. Falkenburg (eds.), *Physik, Philosophie und die Einheit der Wissenschaften*, Heidelberg, Germany: Spektrum Akademischer Verlag, pp. 221–254.

—— [1998]: "Theological Misinterpretations of Current Physical Cosmology," *Philo* 1, no. 1 (Spring–Summer), pp. 15–34.

—— [2000]: "A New Critique of Theological Interpretations of Physical Cosmology," *British Journal for the Philosophy of Science* 51, pp. 1–43.

—— [2008]: "Is Simplicity Evidence of Truth?" *American Philosophical Quarterly* 45, no. 2, pp. 179–189.

Grunberg, Emile, and Modigliani, Franco [1954]: "The Predictability of Social Events," *Journal of Political Economy (USA)* 62, pp. 465–478.

Habermas, Jürgen [1971]: *Knowledge and Human Interests*, trans. J. J. Shapiro, Boston: Beacon Press.

Harper, William L. [1975]: "Rational Belief Change, Popper Functions and Counterfactuals," *Synthese* 30, pp. 221–262.

Hartle, James B., and Hawking, Stephen [1983]: "Wave Function of the Universe," *Physical Review* D28, pp. 2960–2975.

Hasker, Robert W. [1998]: "Religious Doctrine of Creation and Conservation," in E. Craig (ed.), *Routledge Encyclopedia of Philosophy*, vol. 2, London: Routledge, pp. 695–700.

Hawking, Stephen [1987]: "Quantum Cosmology," in S. W. Hawking & W. Israel (eds.), *Three Hundred Years of Gravitation*, Cambridge: Cambridge University Press, pp. 631–651.

Hawking, Stephen, and Ellis, George F. R. [1973]: *The Large-Scale Structure of Space-Time*, New York: Cambridge University Press.

Hayek, Friedrich von [1952]: *The Counter-Revolution of Science: Studies on the Abuse of Reason*, Glencoe, IL: Free Press.

Heidegger, Martin [1953]: *Einführung in die Metaphysik*, Tübingen: Niemeyer.

Hilpinnen, Risto [1970]: "On the Information Provided by Observations," in J. Hintikka & P. Suppes (eds.), *Information and Inference*, Dordrecht: Reidel.

Hintikka, Jaakko [1975]: "Carnap and Essler Versus Inductive Generalization," *Erkenntnis* 9, pp. 235–244.

Holmes, Oliver Wendell [1915]: "Ideas and Doubts," *Illinois Law Review* 10, p. 2.

Hook, Sidney (ed.) [1961]: *Determinism and Freedom in the Age of Modern Science*, New York: Collier Books.

—— [1978]: "Solzhenitsyn and Secular Humanism: A Response," *Humanist*, Nov.–Dec., p. 6.

—— [1987]: "A Common Moral Universe?" *Free Inquiry* 7, no. 4, pp. 29–31.

Howson, Colin [1973]: "Must the Logical Probability of Laws Be Zero?" *British Journal for the Philosophy of Science* 24, pp. 153–160.

Hoyle, Fred [1992]: "Light Element Synthesis in Planck Fireballs," *Astrophysics and Space Science* 198, pp. 177–193.

Isham, Christopher [1988]: "Creation of the Universe as a Quantum Process," in R. J. Russell et al. (eds.), *Physics, Philosophy and Theology: A Common Quest for Understanding*, Rome: Vatican Observatory.

James, William [1975]: *Pragmatism*, Cambridge, MA: Harvard University Press.

Jaspers, Karl [1974]: *Allgemeine Psychopathologie*, New York: Springer Verlag.

Joseph, Horace. W. B. [1916]: *An Introduction to Logic*, 2nd rev. ed., Oxford: Oxford University Press.

Kline, George L. [1955]: "A Philosophical Critique of Soviet Marxism," *Review of Metaphysics* 9, pp. 90–105.

Kochen, Simon B., and Specker, Ernst P. [1967]: "The Problem of Hidden Variables in Quantum Mechanics," *Journal of Mathematics and Mechanics* 17, pp. 59–87.

Kolakowski, Leszek [1968]: *Toward a Marxist Humanism*, New York: Grove Press.

Kristol, Irving [1992]: "Quotable," *Chronicle of Higher Education*, April 22.

—— [1992]: "The Future of American Jewry," *Commentary* 92, no. 2, pp. 21–26.

Lakatos, Imre [1974]: "The Role of Crucial Experiments in Science," *Studies in History and Philosophy of Science* 4, pp. 309–325.

—— [1974]: "Popper on Demarcation and Induction," in P. A. Schilpp (ed.), *The Philosophy of Karl Popper*, The Library of Living Philosophers, Book I, LaSalle: Open Court, pp. 241–273.

Lamont, Corliss [1967]: *Freedom of Choice Affirmed*, New York: Horizon Press.

Lastick, Ian S. [1988]: *For the Land and the Lord: Jewish Fundamentalism in Israel*. New York: Council on Foreign Relations.

Lehrer, Keith (ed.) [1966]: *Freedom and Determinism*, New York: Random House.

Leibniz, G. W. [1697]: "On the Ultimate Origination of Things," in G. H. R. Parkinson (ed.), *Leibniz: Philosophical Writings*, trans. G. H. R. Parkinson & M. Morris; London: J.M. Dent & Sons, 1973.

—— [1714]: "Principles of Nature and of Grace Founded on Reason," in G. H. R. Parkinson, (ed.), *Leibniz: Philosophical Writings*, trans. G. H. R. Parkinson & M. Morris; London: J.M. Dent & Sons, 1973.

—— [1714]: "The Principles of Nature and of Grace Based on Reason," in P. P. Wiener (ed. and trans.), *Leibniz Selections*, New York: Charles Scribner's Sons, 1951.

—— [1956]: *Vernunftprinzipien der Natur und der Gnade; Monadologie = Principes de la Nature et de la Grace fondés en Raison; Monadologie*, French/German, Hamburg: Felix Meiner.

Leslie, John. [1978]: "Efforts to Explain All Existence," *Mind* 87, pp. 181–194.

—— [1990]: *Universes*, New York: Routledge.

—— [1996]: *The End of the World*, London: New York: Routledge.

—— [1998]: "Cosmology and Theology," *Stanford Encyclopedia of Philosophy*, http://plato.stanford.edu/entries/cosmology-theology.

Lévy-Leblond, Jean-Marc [1989]: "The Unbegun Big Bang," *Nature* 342, p. 23.

—— [1990]: "Did the Big Bang Begin?" *American Journal of Physics* 58, pp. 156–159.

Loveley, E. [1967]: "Creation: 1. In the Bible," in Editorial Staff at Catholic University of America, (eds.), *New Catholic Encyclopedia*, vol. 4, New York: McGraw-Hill, pp. 417–419.

Lovell, Bernard [1961]: *The Individual and the Universe*, New York: New American Library.

—— [1986]: "Reason and Faith in Cosmology," (*Ragione e Fede in Cosmologia*), *Nuovo Civilta Delle Macchine* 4, pp. 101–108.

Lütkehaus, Ludger [1999]: *Nichts: Abschied vom Sein, Ende der Angst*, Zürich: Haffmans Verlag.

MacKay, Donald M. [1967]: *Freedom of Action in a Mechanistic Universe*, Cambridge: Cambridge University Press.

Maddox, John [1989]: "Down With the Big Bang," *Nature* 340, p. 425.

Margenau, Henry [1968]: *Scientific Indeterminism and Human Freedom*, Latrobe, PA: Archabbey Press.

Margalit, Avishai [1994]: "The Uses of the Holocaust," *New York Review of Books* 41, no. 4, February 17.

Martin, Michael [1990]: *Atheism*, Philadelphia: Temple University Press.

Maxwell, Grover [1974]: "Corroboration Without Demarcation," in P. A. Schilpp (ed.), *The Philosophy of Karl Popper*, Book II, La Salle, IL: Open Court, pp. 292–321.

May, Gerhard [1978]: *Schöpfung aus dem Nichts: Die Entstehung der Lehre von der Creatio Ex Nihilo*, Berlin: de Gruyter.

Mayo, Deborah [1996]: "The Old Evidence Problem," in D. Mayo, *Error and the Growth of Experimental Knowledge*, Chicago: University of Chicago Press.

McMullin, Ernan [1993]: "Cosmology and Religion," in N. Hetherington (ed.), *Cosmology*, New York: Garland Publishing, ch. 31.

Meltzoff, Julian, and Kornreich, Melvin [1970]: *Research in Psychotherapy*, New York: Atherton Press.

Merton, Robert K. [1949]: *Social Theory and Social Structure: Toward the Codification of Theory and Research*, Glencoe, IL: Free Press.

Mill, John Stuart [1887]: *A System of Logic*, 8th ed., New York: Harper and Brothers.

Misner, Charles W. [1969]: "Absolute Zero of Time," *Phys. Rev.* 186, pp. 1328–1333.

Murray, Thomas E. [1952]: "Some Limitation of Science," *Chemical Engineering Progress* 48, pp. 20–22.

Musgrave, Allen [1975]: "Popper and 'Diminishing Returns from Repeated Tests,'" *Australasian Journal of Philosophy* 53, pp. 248–253.

Noonan, John T., Jr. [1993]: "Development in Moral Theology," *Theological Studies* 54, no. 4, December, pp. 662–677.

Northrop, Filmer S. C. [1947]: *The Logic of the Sciences and the Humanities*, New York: Meridian.

Norton, John [1992]: "Philosophy of Space and Time," in: Merrilee H. Salmon et al. (eds.), *Introduction to the Philosophy of Science*, Upper Saddle River, NJ: Prentice-Hall, pp. 179–231.

Nozick, Robert [1981]: *Philosophical Explanations*, Cambridge, MA: Harvard University Press.

Parfit, Derek [1998]. "The Puzzle of Reality: Why Does the Universe Exist?" in P. van Inwagen & D. Zimmerman (eds.), *Metaphysics: The Big Questions*, Malden, MA: Blackwell, pp. 418–27. (Originally published in *Times Literary Supplement*, July 3, 1992.)

—— [1998]: "Why Anything? Why This?" *London Review of Books* vol. 20, no. 2, January 22, pp. 24–27.

Park, James L., and Margenau, Henry [1968]: "Simultaneous Measurability in Quantum Theory," *International Journal of Theoretical Physics* 1, pp. 211–283.

Parsons, Keith M. [1989]: *God and the Burden of Proof*, Buffalo, NY: Prometheus Books.

Peratt, Anthony L. [1990]: "Not With a Bang: The Universe May Have Evolved From a Vast Sea of Plasma," *Sciences*, Jan.–Feb., pp. 24–32.

Pius XII [1952]: "Modern Science and the Existence of God," *Catholic Mind* 49, pp. 188–190.

Poincaré, Jules Henri [1946]: *The Foundations of Science*, Lancaster, PA: Science Press.

Popper, Karl R. [1950]: *The Open Society and Its Enemies*, Princeton, NJ: Princeton University Press.

—— [1950]: "Indeterminism in Quantum Physics and in Classical Physics," *British Journal for the Philosophy of Science* 1, no. 1, pp. 117–133; no. 2, pp. 173–195.

—— [1959]: *The Logic of Scientific Discovery*, London: Hutchinson.

—— [1962]: *Conjectures and Refutations*, New York: Basic Books.

—— [1973]: *Objective Knowledge*, Oxford: Oxford University Press.

—— [1974]: "Autobiography of Karl Popper," and "Replies to My Critics," in P. A. Schilpp (ed.), *The Philosophy of Karl Popper*, La Salle, IL: Open Court Book I, pp. 2–184, and Book II, pp. 961–1197.

—— [1983]: *Realism and the Aim of Science*, ed. W. W. Bartley III, Totowa, NJ: Rowman and Littlefield.

Putnam, Hilary [1957]: "Three-Valued Logic," *Philosophical Studies* 8, pp. 73–80.

Quine, Willard V. O. [1961]: *From a Logical Point of View*, 2nd ed., Cambridge: Cambridge University Press.

Quinn, Philip L. [1993]: "Creation, Conservation, and the Big Bang," in J. Earman et al. (eds.), *Philosophical Problems of the Internal and External Worlds: Essays on the Philosophy of Adolf Grünbaum*, Pittsburgh: University of Pittsburgh Press, ch. 23, pp. 589–612.

—— [2005]: "Cosmological Contingency and Theistic Explanation," *Faith and Philosophy* 22, no. 5, pp. 581–600.

Rachman, Stanley [1971]: *The Effects of Psychotherapy*, New York: Pergamon Press.

Renteln, Paul [1991]: "Quantum Gravity," *American Scientist* 79, pp. 508–527.

Rescher, Nicholas [1984]: *The Riddle of Existence: An Essay in Idealistic Metaphysics*, Lanham, MD: University Press of America.

——— [2003]: "Contingentia Mundi: Leibniz on the World's Contingency," in N. Rescher, *On Leibniz*, Pittsburgh: University of Pittsburgh Press, pp. 45–67.

Ricoeur, Paul [1970]: *Freud and Philosophy*, New Haven, CT: Yale University Press.

—— [1981]: *Hermeneutics and the Human Sciences*, trans. J. B. Thompson, New York: Cambridge University Press.

Roshwald, Mordecai [1955]: "Value-Judgments in the Social Sciences," *British Journal for the Philosophy of Science* 6, pp. 186–208.

Russell, Bertrand [1948]: *Human Knowledge*, New York: Simon and Schuster.

Ryle, Gilbert [1954]: *Dilemmas*, Cambridge: Cambridge University Press.

Samuelson, Norbert [2000]: "Judaic Theories of Cosmology," in J. Neusner, A. Avery-Peck, & W. Green (eds.), *The Encyclopedia of Judaism*, vol. 1, Leiden: Brill, pp. 126–136.

Salmon, Wesley C. [1959]: "Psychoanalytic Theory and Evidence," in S. Hook (ed.), *Psychoanalysis, Scientific Method and Philosophy*, New York: New York University Press, pp. 252–267.

—— [1966]: *The Foundations of Scientific Inference*, Pittsburgh: University of Pittsburgh Press.

—— [1971]: "Statistical Explanation," in W. C. Salmon, *Statistical Explanation and Statistical Relevance*, Pittsburgh: University of Pittsburgh Press, pp. 29–87.

—— [1975]: "Confirmation and Relevance," in G. Maxwell & R. M. Anderson (eds.), *Induction, Probability, and Confirmation*, Minnesota Studies in the Philosophy of Science, vol. 6, Minneapolis: University of Minnesota Press, pp. 3–36.

—— [1978]: "Religion and Science: A New Look at Hume's Dialogues," *Philosophical Studies* 33, pp. 143–176.

—— [1979]: "Experimental Atheism," *Philosophical Studies* 35, pp. 101–104.

—— [1998]: *Causality and Explanation*, New York: Oxford University Press.

—— [2001]: "Explanation and Confirmation; A Bayesian Critique of Inference to the Best Explanation," in G. Hon & S. Rakover (eds.), *Explanation: Theoretical Approaches and Applications*, Dordrecht, The Netherlands: Kluwer Academic Publishers, pp. 61–91.

Schellenberg, John L. [1993]: *Divine Hiddenness and Human Reason*, Ithaca, NY: Cornell University Press.

Schick, Theodore, Jr. [1998]: "The 'Big Bang' Argument for the Existence of God," *Philo* 1, pp. 95–104.

Schopenhauer, Arthur [1958, 1969]: *The World as Will and Representation*, Vols. 1 & 2, trans. E. F. J. Payne, New York: Dover.

Scriven, Michael [1965]: "An Essential Unpredictability in Human Behavior," in B. B. Wolman & E. Nagel (eds.), *Scientific Psychology*, New York: Basic Books, pp. 411–425.

Sellars, Wilfrid [1966]: "Fatalism and Determinism," in K. Lehrer (ed.), *Freedom and Determinism*, New York: Random House, pp. 141–174.

—— [1969]: "Metaphysics and the Concept of a Person," in J. F. Lambert (ed.), *The Logical Way of Doing Things*, New Haven, CT: Yale University Press, pp. 219–252.

Smith, Quentin [1986]: "World Ensemble Explanations," *Pacific Philosophical Quarterly* 67, no. 1, 73–86.

—— [1991]: "An Atheological Argument from Evil Natural Laws," *International Journal for the Philosophy of Religion* 29, pp. 159–174.

—— [1992]: "Anthropic Coincidences, Evil and the Disconfirmation of Theism," *Religious Studies* 28, pp. 347–350.

—— [1994]: "Anthropic Explanations in Cosmology," *Australasian Journal of Philosophy* 72, no. 3, pp. 371–382.

—— [1995]: "Internal and External Causal Explanations of the Universe," *Philosophical Studies* 79, pp. 283–310.

—— [1997]: *Ethical and Religious Thought in Analytic Philosophy of Language*, New Haven, CT: Yale University Press.

—— [1998]: "Swinburne's Explanation of the Universe," *Religious Studies* 34, pp. 91–102.

Sokolnikoff, Ivan S. [1946]: *Mathematical Theory of Elasticity*, New York: McGraw-Hill.

Stachel, John [1993]: "The Meaning of General Covariance," in: J. Earman et al. (eds.), *Philosophical Problems of the Internal and External Worlds: Essays on the Philosophy of Adolf Grünbaum*, Pittsburgh: University of Pittsburgh Press, pp. 129–160.

Suppes, Patrick [1964]: "On an Example of Unpredictability in Human Behavior," *Philosophy of Science* 31, no. 2, pp. 143–148.

Swinburne, Richard [1979]: *The Existence of God*, New York: Oxford University Press.

—— [1989]: "Violation of a Law of Nature," in R. Swinburne (ed.), *Miracles*, New York: Macmillan Publishing, ch. 8.

—— [1990]: "Argument from the Fine Tuning of the Universe," in J. Leslie (ed.), *Physical Cosmology and Philosophy*, New York: Macmillan Publishing, ch. 12, pp. 154–173.

—— [1991]: *The Existence of God*, 3rd ed., New York: Oxford University Press.

—— [1996]: *Is There a God?* New York: Oxford University Press.

—— [1997]: *Simplicity as Evidence of Truth*, Milwaukee, WI: Marquette University Press.

—— [1998]: *Providence and the Problem of Evil*, Oxford: Clarendon Press.

—— [1998]: "Response to Derek Parfit," in P. Van Inwagen & D. W. Zimmerman (eds.), *Metaphysics: The Big Questions*. Oxford: Oxford University Press.

—— [2000]: "Reply to Grünbaum's 'A New Critique of Theological Interpretations of Physical Cosmology,'" *British Journal for the Philosophy of Science* 51, pp. 481–485.

—— [2001]: *Epistemic Justification*, New York: Oxford University Press.

Taylor, Richard [1966]: *Action and Purpose*, Englewood Cliffs, NJ: Prentice-Hall.

Timoshenko, Stephen, and Goodier, James N. [1951]: *Theory of Elasticity*, New York: McGraw-Hill.

Torretti, Roberto [1983]: *Relativity and Geometry*, New York: Pergamon.

Unger, Peter [1984]: "Minimizing Arbitrariness: Toward a Metaphysics of Infinitely Many Isolated Concrete Worlds," *Midwest Studies in Philosophy* 9, pp. 29–51.

van Fraassen, Bas C. [1980]: *The Scientific Image*. Oxford: Clarendon Press.

von Hayek, Friedrich A. [1952]: *The Counter-Revolution of Science: Studies on the Abuse of Reason*, Glencoe, IL: Free Press.

Vilenkin, Alexander [1982]: "Creation of Universes From Nothing," *Physics Letters B* 117, nos. 1–2, pp. 25–28.

—— [1983]: "Birth of Inflationary Universes," *Physical Review* D27, pp. 2848–2855.

Wald, Robert M. [1984]: *General Relativity*. Chicago: University of Chicago Press.

Watkins, John W. [1978]: "Corroboration and the Problem of Content-Comparison," in G. Radnitzky & G. Andersson (eds.), *Progress and Rationality in Science*, Boston Studies in the Philosophy of Science 58, Dordrecht, The Netherlands: D. Reidel, pp. 339–378.

Weigel, George [1992]: "The New Anti-Catholicism," *Commentary*, June, pp. 25–31.

Weinberg, Steven [1993]: *Dreams of a Final Theory*, New York: Vintage Books.

—— [1999]: "A Designer Universe?" *New York Review of Books* 46, no. 16, October 21, pp. 46–48.

Weisskopf, V. [1989]: "The Origin of the Universe," *Bull. Am. Acad. Arts Sci.* 42, p. 22.

Weyl, Hermann [1950]: *Space-Time-Matter*, New York: Dover Publications.

Worrall, John [1997]: "Is the Idea of Scientific Explanation Unduly Anthropocentric? The Lessons of the Anthropic Principle," in *Discussion Paper Series from the London School of Economics*, DP25/96.

Wright, Robert [1988]: "Why Men Are Still Beasts," *New Republic* no. 3834 (July 11), pp. 27–32.

Zamansky, Harold S. [1958]: "An Investigation of the Psychoanalytic Theory of Paranoid Delusions," *Journal of Personality* 26, pp. 410–425.

■ INDEX*

accretion (of matter), 174, 204–207, 225–226
ad hoc assumptions, 10, 54, 208, 244
Adler, Alfred, 11, 45, 49, 55
analytic/synthetic distinction, 63
anthropic principle, 218, 221, 246–250
anthropocentrism, 122–123, 143, 244
anthropomorphism, 237–238, 244
Aquinas, Thomas, 158, 171–173, 201, 205, 209–210, 216, 222, 224, 233
Aristotle, 159, 175, 208–209, 223, 225–226, 238–239, 261
atheism, 117–120, 125, 135, 138–142, 146, 181, 184, 193, 196, 201–202, 208, 223, 234–235, 261
Augustinus, Aurelius (St. Augustine), 81, 95, 110, 111, 201–202, 216
auxiliary assumptions, 2, 10, 18, 39, 48–49, 62–63, 65, 69, 73

Bacon, Francis, 11–12, 13–17, 19–20, 28, 34–35, 38, 53, 56, 192
Barth, Karl, 117
Bartley, William, 48–49, 57, 58
Bayes' Theorem, 10, 15, 22, 25, 30–35, 37, 42n31, 189, 193–194, 228, 239, 245–246, 250–255. *See also* probability
Bergin, Allen. E., 19
Bergson, Henri, 156–157, 159–160, 166, 210, 224
Berkeley, George, 205
Berofsky, Bernard, 108n1
biocritical values, 246, 248–249
Bondi, Hermann, 138, 174, 190, 204, 206, 208, 224
Bonhoeffer, Dietrich 144, 145n
Brandt, Willy, 144

brute facts, 152, 163, 169, 184, 186–187, 190–193, 195, 235–236, 240, 248
Buber, Martin, 3, 117, 121–125, 251
Buckley, Michael, 134, 203, 206, 214, 241, 244

Cain, Seymour, 126–128, 143–144, 146
Calvin, Johannes, 81, 95, 110, 111
Campbell, Charles Arthur, 82, 89, 107
Cantor, Georg, 232
Carlson, Erik, 161, 162
Carnap, Rudolf, 51, 67, 68, 253, 255n11
Carter, Brandon, 218
Carter, Stephen L., 119, 132
Carter, Quentin, 218
Cassirer, Ernst, 108n1
causation
    of belief, 94–95, 102, 104, 106, 111, 112, 250
    causal hypotheses, 13–14, 21, 30, 34, 52–54, 192
    historical, 3, 81, 110–113
    versus meaning connections (psychology), 2, 43–44
    multiple, 41n13
Cioffi, Frank, 44
Clarke, Samuel, 235
Cohen, Hermann, 117
conceptualization
    in scientific theories, 1, 59, 69, 80, 203, 223–224, 229–232, 260
    in theology, 116, 120, 154, 172, 203, 213, 221, 223–224, 226–228, 241. *See also* theism
    in Leibniz, 160–162, 167–168, 227
confirmation (of theories), 10, 64. *See also* Popper, on corroboration
    versus explanation, 193

* This Index was compiled and provided by Leanne Longwill.

prediction
of human behavior, 3, 80–81, 84,
110–113
in the natural sciences/physics, 79
"risky." *See* Popper
probability (calculus), 11, 22–26, 28–34,
42n31, 155, 191, 193, 246–249,
250–255. *See also* Bayes' Theorem
Pruss, Alexander, 228
psychoanalysis, theory of, 1, 9, 10–11, 14,
16–22, 43–57, 97
and empirical validation, 2, 18, 26,
44–45, 47–49, 56
See also statistics/statistical method
psychotherapy, 16–17, 97
causal efficacy of, 18–19, 20
Putnam, Hilary, 65

quantum physics, 89, 155–156, 163, 179, 202
and freedom of will, 98–101
indeterminacy, 98, 100–101, 107–108
quantum cosmologies, 216–218, 242,
260, 262
See also statistics/statistical method
Quine, Willard Van Orman, 1, 2, 5n4,
62–74, 76
Quinn, Philip, 3, 152, 164, 175–176,
186–191, 194–195, 201–207, 213–216,
221–222, 227–228, 232, 236–245, 250,
254

Rachman, Stanley, 18
regress (infinite)
of explanations, 163, 177, 182–183, 236
Rehnquist, William H., 119
Reichenbach, Hans, 30, 32, 67–68, 72
Religion. *See* theism
Rescher, Nicholas, 160, 167, 223
Ricoeur, Paul, 43
Roshwald, Mordecai, 110–113
Rusher, William A., 115
Russell, Bertrand, 119, 122, 139, 238
Ryle, Gilbert, 85

Safire, William, 123, 127–128
Salmon, Wesley, 27–28, 42n31, 51,
192–194, 246, 251–253, 254

Samuelson, Norbert, 158
Santayana, George, 238
Satan, 131, 176, 230, 243, 253
Schick, Theodore, 247
Schlick, Moritz, 85, 88
Schneerson, Menachem M., 125, 127
Schopenhauer, Arthur, 157, 179–180, 182
"scientism," 43, 175–176, 191, 228
Scriven, Michael, 108n4
Secular Humanism. *See* humanism, secular
Sellars, Wilfrid, 80–81, 86, 88, 102
Smith, Quentin, 217–218, 231–232, 235,
245, 249
Smoot, George, 203
Solzhenitsyn, Alexander, 121, 130–131,
136
Specker, Ernst P., 107–108
Spencer, Herbert, 131
Spinoza, Baruch de, 45, 95, 111
Stachel, John, 211
statistics/statistical method
and human behavior, 1, 79, 83, 85,
89–90
in quantum physics, 98–101, 226
quasi-statistical prediction (in
psychoanalysis), 48
statistical hypothesizing, 21
statistical validation (of theories), 47
Swinburne, Richard, 3, 152–157, 160–173,
181–196, 201, 205, 210, 221–232,
236–246, 248–255

Taylor, Richard, 79, 108n14
Teitelbaum, Joel, 124, 125, 127
theism
and agent causation, 133, 152, 183,
186–187, 193, 206–207, 212–213, 236,
241, 243, 262
fatwa (Islamic), 128
moral sterility of, 3, 126, 128–133,
140, 142
psychopathology of, 118, 140
versus rational/scientific explanation,
3, 133, 151–156, 162–164, 167, 169,
170, 175, 177–178, 181–195, 201, 203,
206–209, 218, 221, 225–228, 232,
236–249, 251, 254